The Juan Pardo Expeditions

With Documents Relating to the Pardo Expeditions
Transcribed, Translated, and Annotated by Paul E. Hoffman

The Juan Pardo Expeditions

Explorations of the Carolinas and
Tennessee, 1566–1568

Charles Hudson

Afterword by David G. Moore, Robin A. Beck Jr., and
Christopher B. Rodning

The University of Alabama Press
Tuscaloosa

Copyright © 1990 Charles Hudson
Preface and Afterword copyright © 2005 The University of Alabama Press
Tuscaloosa, Alabama 35487-0380
All rights reserved
Manufactured in the United States of America

Originally published by Smithsonian Institution Press

Cover design by Erin Bradley Dangar
Times New Roman

∞
The paper on which this book is printed meets the minimum requirements of American National Standard for Information Science—Permanence of Paper for Printed Library Materials, ANSI Z39.48-1984.

Library of Congress Cataloging-in-Publication Data

Hudson, Charles M.
  The Juan Pardo expeditions : exploration of the Carolinas and Tennessee, 1566–1568, / Charles Hudson ; with documents relating to the Pardo expeditions transcribed, translated, and annotated by Paul E. Hoffman ; preface to the revised edition by Charles Hudson ; afterword by David G. Moore, Robin A. Beck Jr., and Christopher B. Rodning.— Rev. ed.
    p. cm.
  Includes bibliographical references and index.
  ISBN 0-8173-5190-6 (pbk. : alk. paper)
  1. Catawba Indians—History—16th century. 2. Pardo, Juan, 16th cent. 3. Explorers—Southern States—Biography. 4. America—Discovery and exploration—Spanish. 5. Southern States—History. I. Hoffman, Paul E., 1943– II. Title.
  E99.C24.P374 2005
  975'.01—dc22
                           2004020693

Contents

Illustrations vii
Preface to 2005 Edition ix
Preface to First Edition xi

Part I   The Juan Pardo Expeditions

  1   Early Spanish Exploration   *3*

  2   Juan Pardo's Two Expeditions   *23*
  Pardo's First Expedition: December 1, 1566 to March 7, 1567
  Moyano's Foray: April 1567
  Pardo's Second Expedition: September 1, 1567 to March 2, 1568

  3   The Indians   *51*
  The Mississippian Transformation
  Social Structure of Chiefdoms in the Carolinas and Tennessee
  Polities, Cultures, Languages
    Cofitachequi
    Joara
    Guatari
    The Cherokees
    Coosa
  Economic Patterns

  4   The Foundations of Greater Florida   *125*
  Outfitting the Second Expedition

The Road to Zacatecas
Dugout Canoes
Pacifying the Indians
The Houses the Indians Built
The Forts the Spaniards Built
The Missionaries
Prospecting for Precious Metals and Gems

5 The Failure of Greater Florida   169
Misconceptions about the Land and the Indians
The Failure of the Forts
The Shrinking of Florida
The Decline and Coalescence of the Indians
Los Diamantes and La Gran Copala

Part II The Pardo Documents

The "Long" Bandera Relation: AGI, Santo Domingo 224   205

The "Short" Bandera Relation: AGI, Patronato 19, R. 20   297

The Pardo Relation: AGI, Patronato 19, R. 22 (document 1)   305

The Martinez Relation: AGI, Patronato 19, R. 22 (document 2)   317

Three New Documents from the Pardo Expeditions: AGI, Contratación 2929 No. 2, R. 7   323

Part III Afterword

Pardo, Joara, and Fort San Juan Revisited   343
    David G. Moore, Robin A. Beck Jr., and Christopher B. Rodning

Index   351
Errata   363
About the Authors   365

Illustrations

1. Early European exploration of North America. 7
2. Eastern portion of the route of the Hernando de Soto expedition (1539–40) and the route of the Tristán de Luna expedition (1559–61). 9
3. Juan Pardo's first expedition (December 1566 to March 7, 1567) and Hernando Moyano's foray (Spring 1567). 24
4. Page one of Juan de la Bandera's "long relation." 30
5. The signature page of Juan de la Bandera's "short relation." 31
6. Towns and locations of Juan Pardo's second expedition (September 1, 1567, to March 2, 1568). 33
7. Zimmerman's Island. 37
8. Social honor accorded a sixteenth-century Timucuan chief. 54
9. Formal order of march assumed by the Timucuan chief Holata Outina embarking on a military expedition. 56
10. The mound at the Town Creek site near Mt. Gilead, North Carolina. 57
11. Defensive palisade around the Town Creek site. 57
12. The bride of a Timucuan chief carried on a litter. 64
13. Some late prehistoric/early historic archaeological phases and sites, with selected historic Indian towns. 69
14. Sociogram of Cofitachequi and its hinterland. 75
15. Sociogram of Santa Elena and its hinterland. 80
16. A four-sided Pisgah phase house floor. 85
17. Sociogram of Joara and its hinterland. 89
18. A circular house floor at the Upper Sauratown site (31Skla). 92
19. Timucuan male transvestites carrying packbaskets of food. 99
20. Sociogram of Coosa and its hinterland. 104
21. Timucuan Indians transporting food in a dugout canoe. 133

22. The murder of the Frenchman Pierre Gambié by Timucuan Indians.   136
23. Wedge and chisel from Santa Elena.   137
24. Knives from Santa Elena.   138
25. Spherical turquoise blue glass beads from the Hampton site, Rhea County, Tennessee.   139
26. Gilded metal ball button from Santa Elena.   140
27. Conch shell drinking cup from the Hixon site, Hamilton County, Tennessee.   141
28. Indians dressed in matchcoats.   142
29. Senkaitschi, the Yuchi "king."   143
30. Forts built by Pardo and houses built by the Indians on the "road to Zacatecas."   144
31. Wattle and daub construction in a reconstructed house at the Town Creek site, North Carolina.   145
32. Reconstructed house at the Town Creek site, North Carolina.   145
33. Reconstructed corncrib at the Chucalissa site, Memphis, Tennessee.   146
34. A harquebusier with a lighted matchcord.   147
35. Devils carrying away a naked Indian.   154
36. Freshwater pearls from the Hixon site, Hamilton County, Tennessee.   157
37. How the Indians were reputed to have extracted gold from streams in the "Apalatcy Mountains."   158
38. Quartz crystals from the Toqua site, Monroe County, Tennessee.   161
39. Cut mica discs from Cofitachequi.   162
40. Native copper ax from Long Island, Roane County, Tennessee.   163
41. Sixteenth-century Spanish misconceptions about the geography of North America.   171
42. Timucuan warriors desecrating their enemy dead.   178
43. Locations and movements of certain Southeastern Indians ca. 1700.   186

Preface to 2005 Edition

The main body of the text of this revised edition is unaltered. A list of errata from the hardcover edition is appended. New to this edition is an afterword by David Moore, Robin Beck, and Christopher Rodning briefly summarizing research on the Berry site and its identification as the probable site of Joara. Also new is an index, regrettably omitted in the hardcover edition.

Research on the Berry site has located the site of Joara some 20 miles or so northeast of where my colleagues and I initially located it. As Robin Beck realized in his 1997 article in *Southeastern Archaeology* 6(2):162-69, this new location for Joara makes for alterations in both the Soto and Pardo routes as we initially understood them.

It means that Soto traveled not west from Joara, but northwest, crossing the mountains to reach the upper Nolichucky River. Hence, Guasili and Canasoga on the map at fig. 2, p. 9 of this present book should be moved to the Nolichucky River.

I argue in this book that Moyano traveled northwest from Joara via the Toe River to the upper Nolichucky River to attack the Chiscas, whereas Beck argues that Moyano traveled more to the north and encountered the Chiscas at the salines at or near present Saltville, Virginia. This direction of travel is plausible, though what is proposed is a long distance for Moyano to have traveled. Moreover, the Chiscas are so poorly described in the documents—is Chisca a town or a cluster of towns, or is it a native linguistic, cultural, or social category?— that I think it prudent to not locate them in so precisely delimited an area. Finally, the Chiscas could have been traders in salt and copper without having produced either of these substances.

What is now clear is that Soto did not encounter the Chiscas on the upper Nolichucky, as I previously had thought. What is notable is that later in his expedition Soto sent Juan de Villalobos and Francisco de Silvera back to the north to locate the Chiscas. They did so and returned by dugout canoe to rendezvous with Soto at Bussell Island in the mouth of the Little Tennessee River. This argues for the Chiscas being located on the Holston River or its tributaries.

Distracted and misled by the historical importance of the trail down the French Broad River in the late eighteenth and nineteenth centuries, my colleagues and I placed Cauchi on the French Broad River near present Marshall, North Carolina. I never had great confidence in this location because it did not jibe with archaeological evidence. David Moore could find no supporting archaeological evidence for Cauchi at Marshall, and Richard Polhemus made a determined search for Tanasqui at the junction of the Nolichucky and French Broad Rivers and found insufficient supporting archaeological evidence for its being there.

I once considered locating Cauchi at or near the Garden Creek site near present Canton, North Carolina. Beck places it squarely in this area, and I think that he is correct in doing so. From here I think it likely that Pardo and his men went right down Big Pigeon River to its junction with the French Broad, and this implies that Tanasqui was located somewhere in this general vicinity.

July 2004
C. H.
Danielsville, Georgia

Preface to First Edition

My first attempt to answer some of the questions that have led me to write this book was in 1963–64, when I was a graduate student at the University of North Carolina. At that time I was struggling to develop a theoretical framework I could use to make sense of the Catawba Indians of South Carolina. Trained as a social anthropologist, I found myself doing research on the Catawbas not so much to answer a question I had asked, but more because fieldwork was required for the Ph.D. in social anthropology and, as it so happened, funds had become available for someone to do fieldwork on the Catawbas.

It did not take long for me to see that I would not be able to do the kind of research on the Catawbas that social anthropologists had done on more exotic people in far-flung parts of the world. Having spent my childhood in a very small and very rural town in Kentucky, I found that the Catawbas in many respects resembled people who were already familiar to me. What was interesting about the Catawbas, it seemed to me, was the history that had made them what they were. But this put me into another dilemma. I had not been trained as a historian.

I have always been ambivalent about the book based on this research—*The Catawba Nation* (1970). I knew at the time the book was published that many questions about the Catawbas remained unanswered. But I felt that I had gone some distance in developing an approach to understanding them and other native peoples in the Southeast. I understood, for example, the fallacies in the attempts of James Mooney, Frank Speck, and John R. Swanton to classify the native peoples of the Southeast into quasi-linguistic groups established largely on the assumption that people who were similar in their languages were also similar in their culture and even in their genes. In these schemes, the Catawbas were classified as "Eastern Siouans" along with an extensive list of other peoples on the basis of flimsy and erroneously used evidence.

I could not proceed in my research without some kind of organizing framework, but it was with no great confidence that I used A. L. Kroeber's classification of North American Indian cultures, which was based on natural areas. Like many anthropologists before him, Kroeber realized that the cultures of preindustrial people have been powerfully shaped by their environments, and that this could provide the basis for a classification on surer grounds than those proposed by Mooney, Speck, and Swanton. Along with using Kroeber's classification as a "point of departure," I also advocated quite another way of apprehending the Catawbas and their neighbors. Both in the early historical literature and in the preliminary archaeological information that had been accumulated by the late 1960s, it appeared that societies of two levels of complexity had existed in the area where the Catawbas lived. Using terms coined for quite different purposes by Joffre Coe and Elman R. Service, I referred to these two levels of societies as "hill tribes" and "chiefdoms." The former were the societies of the Carolina Piedmont, and the latter were the Mississippian societies of the Wateree and Pee Dee rivers. I realized that the Catawbas were ultimately to be understood within the context of these chiefdoms and tribes, and within the context of a broadly conceived social history, but in 1970 my thinking had not gone beyond these realizations.

I was also cognizant of the fact that several Spanish explorers had penetrated the interior of the Southeast in the sixteenth century, but I had no idea of where they had gone. And when I tried to make connections between the activities and observations of these sixteenth-century Spaniards and the Catawbas I had come to know, I could make none. I was simply not able to evaluate or use this body of information.

The Juan Pardo Expeditions

Part I
The Juan Pardo Expeditions

I
Early Spanish Exploration

After Pedro Menéndez de Avilés founded St. Augustine in September 1565 and quickly crushed the French Huguenots, who in the previous year had established a fort near the mouth of the St. Johns River, he took a detachment of soldiers north to found yet another town—Santa Elena, on the point of land at the southern end of Parris Island, near present-day Beaufort, South Carolina. Menéndez intended to found a vast colony in the region the Spaniards called *La Florida*, so that he would have dominion over not only the Atlantic coast from Newfoundland to the Florida peninsula, but over the interior of North America all the way to Mexico, and Santa Elena was to be his capital city. Between December 1566 and March 1568, as part of this imperial design, Menéndez sent Captain Juan Pardo from Santa Elena on two expeditions into the interior. In both instances, Pardo departed from Santa Elena in the autumn with over a hundred men and with orders to explore the interior, subdue the Indians, and to establish a road to the Spanish silver mines in Zacatecas, Mexico. His assignment must have seemed even more impossible when he was told to accomplish all of this and return to Santa Elena by the following spring.

In the course of time, Menéndez and the officials of Florida who succeeded him had to settle for a far more modest empire in North America. The practical limits of what Spain could achieve were not known to Pardo and his soldiers, but they were soon to learn just how deadly it could be to exceed these limits.

The Pardo expeditions were failures. They do not compare with Cortés's expedition into the Valley of Mexico; and in terms of the expanse of territory explored, they are slight next to the expedition of Hernando de Soto, which was itself a colossal failure. Pardo's expeditions were failures even when measured against the standard of what Menéndez commanded him to do.

The Pardo expeditions are disappointing to students of Southeastern Indian culture because the documents that were produced by Pardo's activity contain little descriptive information about the culture of the Indians who were encountered. But for those who are interested in the social history of the sixteenth-century American South, the Pardo expeditions have proven to be crucial. From Pardo's written account of his expeditions, particularly from a detailed account by Juan de la Bandera, Pardo's scribe and notary, it has been possible to reconstruct where Pardo went and to locate—sometimes precisely—the Indian towns he visited.[1] An accurate reconstruction of the routes Pardo followed is important, not only because it reveals the locations of many previously unknown sixteenth-century Indian towns, but even more because Pardo visited five of the towns visited by Hernando de Soto in 1540. The most important of these were Cofitachequi, near Camden, South Carolina, and Chiaha, on Zimmerman's Island in the French Broad River near Dandridge, Tennessee. Both of these locations are many miles to the northeast of where most scholars have thought they were located.[2]

With these signposts in the interior precisely located, it has become possible to reconstruct the routes of both the Hernando de Soto expedition of 1539–43 and the Tristán de Luna expedition of 1559–61.[3] The documents of the Soto and Luna expeditions, together with those of the Pardo expeditions, contain most of the information we are ever likely to possess on the history of the sixteenth-century Southeastern Indians of the interior. The only other source of information is archaeological. Accurate details of where Soto, Luna, and Pardo went will make it possible to combine this historical and archaeological information, from which we may expect a kind of multiplier effect: When combined, the two sources of information are *far* richer than when taken separately.

Thus, the history of the Pardo expeditions is the key to a door that once appeared to be forever locked. Behind this door lies an entire world that until now has only been glimpsed. It is an unknown South—a land and a time when the large and powerful Indian chiefdoms that held dominion in the South collided for the first time with people from the other side of the world.[4] For the Southeastern Indians, the Spaniards would hardly have been more alien if they had come from Mars.

Santa Elena, the ill-fated town from which Pardo departed on both of his expeditions, was located on an island whose wretched soil was barely suitable for farming, and it flooded when struck by the storms that plague the Atlantic coast. Apparently, the cape or point of Santa Elena was first named between 1521 and 1526, during the course of the slaving and colonizing activities of Lucas Vásquez de Ayllón, although its location was not accurately mapped. After a series of failures at colonizing the southeastern United States, authorities in Spain decided that Santa Elena was the most favorable location for founding the principal city of yet another colony.[5] But the authorities were wrong. The colonists at Santa Elena seldom did better than subsist, and at times they did not even fare this well. Eventually the town was abandoned for military, and possibly political, reasons, and the site on which the town stood was so forgotten that its precise location could only be discovered through historical and archaeological research.[6]

Ironically, St. Augustine, whose site was not so much selected as happened upon, survived and became the center of Spanish Florida. Although it, too, had a troubled career, St. Augustine was the first Spanish success in North America after a remarkable string of failures. After stunning conquests elsewhere, conquistador after conquistador was frustrated and defeated by the Southeast and its people.

Following Columbus's discovery of the New World, Spain quickly established colonies on the larger islands of the Caribbean: Hispaniola, Puerto Rico, Jamaica, and Cuba. From these ports Spain explored other islands in the Caribbean as well as the coasts of the southeastern United States and Central America. For about a dozen years, however, settlement was limited to the islands, where in due course Spaniards succeeded in building an economy based on sugar production and cattle ranching. These plantations and ranches were worked by native Indian slaves as well as slaves imported from Africa.

The real impetus for Spain to explore and colonize the mainland came in 1519 when Hernán Cortés encountered the wealthy and populous Aztec empire. Not only did Cortés discover gold in the form of jewelry and ritual objects, which he could take by force, but he also found a large population habituated to subordination, who could be put to work on plantations and in mines. Cortés became fabulously wealthy, the most powerful private citizen in the New World, and he was granted the title of marquis.[7]

If such wealth and power could be had in New Spain, then it was possible that as much or even more could be found in the land to the north of the Caribbean Islands. It was a land whose shape, dimensions, and natural features came to light very slowly. The continent was first sighted by Juan Ponce de León, who sailed along the coast of southern Florida in 1513. More was learned about this area in 1519, when Alonso Alvarez de Pineda sailed along the Gulf Coast from the western coast of Florida around to Vera Cruz. Observations made on this voyage were later incorporated into the first map of the Gulf Coast.[8]

Knowledge of the interior was far more difficult to come by. Was this land a continent, or was it a series of large islands? This question could only be answered through exploration and colonization. Juan Ponce de León made the first attempt to found a colony in North America in 1521. Making a landfall at an unknown location on the southwestern coast of the Florida peninsula, he put ashore with two hundred colonists, horses, livestock, and tools. But the local Indians wanted no invaders in their land. A large number of them soon attacked, killing many Spaniards and wounding others. Ponce himself was seriously wounded. Disheartened, the Spaniards abandoned the colony and returned to Cuba, where Ponce died.

Yet another colonial venture was set in motion in 1521 when two ships dropped anchor off the Atlantic coast of the Lower South. One of these ships was owned by Lucas Vásquez de Ayllón, an official in Santo Domingo. The Spaniards went ashore, where they had a friendly encounter with the Indians, some of whom they persuaded to come out and visit their ships. But when the Indians canoed out and climbed aboard, the Spaniards promptly enslaved about 60 of them and sailed for Santo Domingo.[9] One of the ships sank en route, and most of the Indians on the other ship died, but at least one of them survived, to be baptized Francisco de Chicora. He not only survived, but learned Spanish and went with Ayllón to Spain. Partly because of Francisco's extravagant accounts of his native land, Ayllón determined to colonize it.[10]

In 1526 Ayllón sailed from Santo Domingo with a fleet of six ships carrying 500 colonists, as well as horses, farm animals, and slaves. They are thought to have first landed near the place where Francisco de Chicora was enslaved, although they did not succeed in finding the province of Chicora, whence Francisco had come (see figure 1). Their landing was at the mouth of a river they named the Rió de Jordán, said to be located at 33 2/3° N.[11] The Rió de Jordán was possibly the South Santee River. As they entered its mouth, their flagship ran aground and sank. They reconnoitered the surrounding land and found it unsuitable for colonization. At the first opportunity Francisco escaped, and perhaps as a consequence the local Indians refused to have anything to do with Ayllón and his people.

The Spaniards located a site that appeared to be more favorable for a colony about 40 to 45 leagues to the west (actually, to the southwest). It

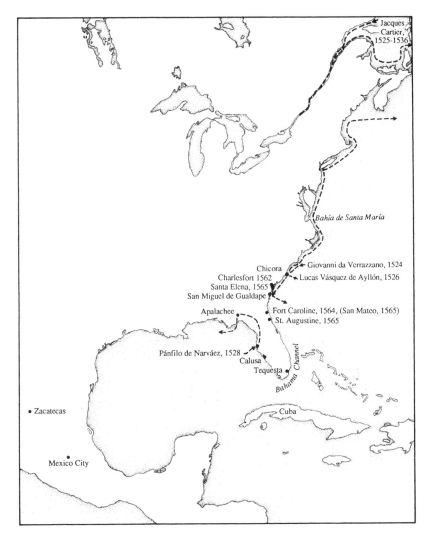

Figure 1. Early European exploration of North America.

was near the mouth of a large river they named the Río de Gualdape, probably the Savannah River. All of the Spaniards moved to this new location, where Allyón founded and built the town of San Miguel de Gualdape. The site of this town has not yet been located, but it was possibly on Sapelo Sound, in the area Spaniards would later call Guale.[12] The colonists soon began to starve, and thus weakened, fell ill. Many died, including Ayllón himself. After a few months they abandoned the colony

Early Spanish Exploration

and returned to Santo Domingo. According to one source, only about 150 of the 500 colonists survived.

Two years later Pánfilo de Narváez attempted to establish a colony on the Gulf Coast. He landed near Tampa Bay in April 1528 with 400 people. After landing, he marched northward to the province of Apalachee, in and around present Tallahassee. But the Indians of Apalachee were so unremittingly bellicose, and the land was so unpromising, Narváez and his followers soon retreated to the Gulf Coast, where they built four crude flatboats, and the 250 or so who were still alive attempted to sail around the Gulf coast to Mexico. But their number soon began to dwindle. Some died of thirst, others of hunger. Some were killed by Indians and others by illness. Some of them drowned. Eventually several boats were blown ashore on the Texas coast, but the only members of the expedition who survived and returned to Mexico were Alvar Núñez Cabeza de Vaca and three others, who miraculously survived nine hard years among the Indians.[13]

This string of failures dampened Spain's interest in North America, but it was revived by yet another spectacular conquest. Francisco Pizarro sailed southward from Panama in December 1530, landing on the coast of what is now Ecuador. He and his soldiers proceeded to fight their way overland to Cuzco, where he discovered the capital of the Inca empire, another native state rich in gold and other treasures. Once they had subdued most of the natives, the conquerors began fighting among themselves, and their conflict escalated into a civil war.[14] But fabulous fortunes had already been made, particularly by one Hernando de Soto, who in 1535 left Peru for Spain a wealthy man.[15]

Soto was not content with the wealth he had amassed in Peru; he wanted a domain of his own. Ponce de León, Lucas Vásquez de Ayllón, and Pánfilo de Narváez had failed in Florida, but none of them had penetrated the interior. Both in Mexico and Peru, Spaniards had found riches only after reaching the interior. On April 20, 1537, Soto received his *asiento* to conquer La Florida. Cabeza de Vaca returned to Spain in the summer of that year, and when Soto offered him a place in his expedition, Cabeza de Vaca declined. Cabeza de Vaca also showed a reticence in talking about some of the things he had seen, a reticence Soto interpreted as concealment, and this fired even more Soto's hope of discovering his own golden empire.

Soto landed at Tampa Bay on March 25, 1539, with a large expedition (see figure 2). Like Narváez before him, he proceeded northward to the chiefdom of Apalachee, where he spent the winter. The following spring

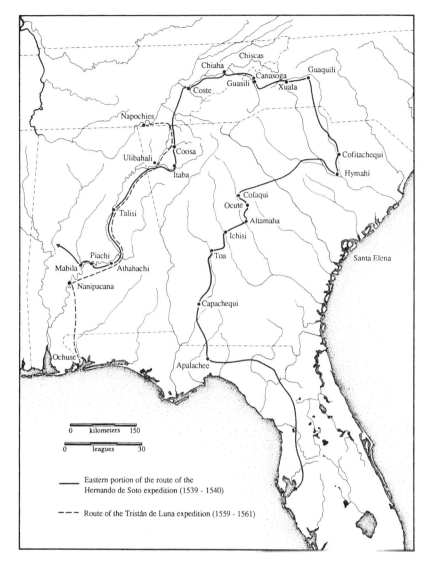

Figure 2. Eastern portion of the route of the Hernando de Soto expedition (1539–40) and the route of the Tristán de Luna expedition (1559–61).

he set out toward the north, where he visited the small chiefdom of Capachequi, to the southwest of Albany, Georgia. From there he went to the chiefdom of Toa, on the upper Flint River. Next he visited the chiefdom of Ichisi on the Ocmulgee River, the chiefdom of Ocute on the Oconee River, and the chiefdom of Cofitachequi on the Catawba-Wateree River in

Early Spanish Exploration

South Carolina. He expected there to be gold and silver at Cofitachequi, but not only did he find no precious metals, he found very little food, because the people of Cofitachequi had been struck by an epidemic disease, probably of European origin, and it had caused many deaths. He did obtain a quantity of freshwater pearls at Cofitachequi.[16]

From Cofitachequi Soto headed northward, following a trail that ran parallel with the Catawba-Wateree River. He crossed the Blue Ridge Mountains and picked up a trail along the French Broad River, which led him to the chiefdom of Chiaha in the Tennessee Valley. From Chiaha he headed southwest to the large and powerful chiefdom of Coosa, whose main town was near present-day Carters, Georgia. The ruler of Coosa was a paramount chief whose power extended north over Chiaha, as well as south over the chiefdom of Talisi, the center of which was near present-day Childersburg, Alabama. From Coosa, Soto proceeded southwest to the chiefdom of Tascaluza on the upper Alabama River. In this chiefdom, at the town of Mabila, in the vicinity of present-day Selma, Alabama, Soto fought the greatest battle of his expedition. It was a costly battle for the Spaniards; for the Indian combatants, it was a horrendous disaster. From Mabila, Soto continued on through Mississippi and Arkansas, encountering small and large chiefdoms all the way. But his expedition steadily wore down and finally failed. Soto himself died of a fever on the banks of the Mississippi River. He lost his fortune, his life, and about half of his army. His only discovery was that there was no populous native state in North America like the one Cortés had conquered in Mexico and the one Pizarro had conquered in Peru.

Soto found no precious metals along his route, although he heard and saw tantalizing hints that such metals might exist there. One of his soldiers was said to have found "a trace of gold" in the Catawba-Wateree River near the central towns of Cofitachequi.[17] Also, some of Soto's men were convinced that pieces of the great quantity of copper jewelry and religious objects they found at Cofitachequi contained traces of gold, but they lacked the means to do a metallurgical test.[18] At Chiaha, Soto was told that the Chiscas who lived to the north produced and traded gold, and he had apparently heard the same story in Cofitachequi. He sent two men north from Chiaha to investigate the Chiscas, but what they found there is not clear. The various chroniclers of the Soto expedition differ in their accounts of what these two men discovered. The Gentleman of Elvas only says that the country between Chiaha and Chisca was so mountainous and sparsely populated that an army could not march through it; he says nothing about discovering any metals.[19] Rondrigo Ranjel merely says that they brought back "good news."[20] Garcilaso de la Vega, always prone to exaggeration, reports that they found three copper mines, and that they felt a further search might reveal gold and silver.[21] Luis de Velasco, Viceroy of New Spain, was probably referring to this same incident when he wrote in July of 1560 that gold could be procured in this general region, and that some of Soto's men said they had bartered for it there.[22] Except for the freshwater pearls found at Cofitachequi and elsewhere, none of the Soto chroniclers mentions the discovery of any precious stones, even though they showed the Indians the stones set in their rings and asked if they knew where any like them could be found in their land.[23]

After the failure of the Soto expedition, the possibility of finding precious metals and gems in the Southeast ceased to be a strong motive for exploration. The Spanish certainly found no gold and silver among the Indians of La Florida, as they had in Mexico and Peru, although there was a chance that it could be mined. But other motives for colonization began to gain strength, particularly the desire to prevent other European powers, especially France, from moving in.

In the first half of the sixteenth century, France acquired considerable knowledge of the geography of the east coast of North America. The first details came from Giovanni da Verrazzano, a Florentine whose voyage was financed by France and by an association of merchants, mostly Italian. In March 1524 Verrazzano sighted the coast of North America, probably just south of Cape Fear, North Carolina (see figure 1). He first sailed south for about 160 miles, and then doubled back and went northward, making observations and compiling a map as he sailed along the coast. Searching for a passage through which ships might voyage to the Orient, he sailed along the Carolina Outer Banks where, from what he could see, he concluded that the continent became a narrow isthmus here, and that an ocean leading to China and India lay on the other side. He made his way north, evidently reaching Newfoundland before recrossing the Atlantic to return to France.[24] Verrazzano raised false hopes about a northern passage to the Orient, but from his voyage the French learned that the North American continent was, in fact, a continent, and not a series of islands.

The French subsequently learned about the northern coast in far more detail from the voyages of Jacques Cartier. In 1534 Cartier explored the western coast of Newfoundland, the northern coast of Prince Edward Island, and the coast of the Gaspé Peninsula. On his next voyage, in 1535–36, Cartier sailed into the mouth of the St. Lawrence River and explored it as far as the site of present-day Montreal (see figure 1), where he learned that the St. Lawrence was not, as he had hoped, a northern passage to the

Orient. He built a fort on the river and spent a hard winter there, with many of his men dying from starvation and scurvy. Cartier returned for a third time in 1541, but this time the commander of the expedition was a Protestant, Jean François de la Roque, seigneur de Roberval. The Spanish, who regarded Protestants as heretics, must have been affronted when they learned that Roverval was in command. Again a party of Frenchmen spent the winter on the St. Lawrence River and returned home the next year. No permanent settlement had been founded, but Cartier and Roberval had successfully challenged Spain's exclusive right to North America.

France had demonstrated to Spain its ability to plant a colony in North America, if only fleetingly, in an extremely inhospitable climate. Even more troubling to Spain, French privateers and pirates had begun landing on the lower Atlantic and Gulf coasts to take on water and fuel, and to trade, at least on a limited basis, with the Indians. French privateers had not only attacked Spanish ships at sea, but they sacked and burned Cartagena, Santiago de Cuba, and Havana.[25] Philip II determined that Spanish colonies had to be founded in North America to preempt French colonization and to prevent French ships from taking on provisions; the colonies were also to offer refuge to shipwrecked Spaniards, to evangelize the Indians, and to protect the Indies fleet as it sailed through the Bahama Channel bound for Spain.[26]

Luis de Velasco, the viceroy of New Spain, selected Tristán de Luna y Arellano to lead yet another colonization effort. Velasco and Luna had the benefit of a map and at least two written accounts of the Soto expedition to help them plan their venture, and several of Luna's men appear to have been members of the Soto expedition. Despite this hard-won intelligence from the Soto expedition, Velasco and his contemporaries had some serious misconceptions about the geography of the Southeast, misconceptions that led them to make costly mistakes. They believed, for example, that from a town established at Ochuse (Pensacola Bay)—a harbor whose existence was known from the exploration of members of the Soto expedition—it would be relatively easy to send a force of men overland to Coosa, where another town would be founded, and from there to the Point of Santa Elena on the Atlantic coast, where a third town was to be founded.[27] They believed that it was only a short distance from Coosa to the Point of Santa Elena, and that in fact the Indians of Coosa were in communication with the Indians near the Point of Santa Elena. After these towns were established, they expected to build a road connecting all three.[28]

Like all previous Spanish attempts at colonizing the Southeast, Luna's colony was in trouble almost from the beginning. He set out with thirteen ships from Mexico on June 11, 1559. But in approaching the land to the north, he missed the entrance to Pensacola Bay and landed at Mobile Bay. Because he thought Pensacola would be a safer port, Luna moved his people there on August 14 (see figure 2). While they were still unloading their supplies from the ships, a hurricane struck and destroyed all but three of Luna's ships. Moreover, it soon became clear that they would not be able to obtain food from the local Indians, who were few in number and not very cooperative.

Because prospects for food looked better further inland, in February 1560 Luna moved his colony north to the Alabama River, where the settlers took up residence in an Indian town called Nanipacana, on the south side of the river, not far from where Soto had fought the battle of Mabila, in which thousands of Indians had been killed. No doubt remembering Soto and his army, the Indians in and around Nanipacana responded to Luna's incursion by simply abandoning their towns and moving elsewhere. They skirmished with the Spaniards from time to time, but Luna's men had no sustained contact with them.

Luna's colonists again faced starvation. He sent a detachment of cavalry and infantry along with two Dominican friars to find Coosa, where Soto had found cooperative Indians and plenty of food. This detachment slowly made its way from Nanipacana to Atache (Soto's Athahachi), and from there they retraced Soto's route northward, paralleling the Coosa River past Talisi and Ulibahali to the main town of Coosa. Here they found food, though not in plenty. They took pains to gain the cooperation of the Indians and to avoid offending them, even to the point of sending a detachment of soldiers to assist Coosa's warriors in chastising the Napochies, a tributary chiefdom on the Tennessee River near present-day Chattanooga, who had rebeled and were not paying their customary tribute to Coosa. While all this was taking place, Luna's hard-pressed colonists were forced to abandon Nanipacana and return to Pensacola Bay, where they began quarreling among themselves. By now it was all too evident that the colony was a failure. The colonists who had gone to Coosa soon rejoined them.

Even though Luna had failed to establish towns at Ochuse and Coosa, Velasco worked hard to found a colony at the Point of Santa Elena, fearing that the French would establish a colony in that area. Perrenot de Chantonne, the Spanish ambassador to France, had warned Philip II that the French were formulating such a plan.[29] Hence, Velasco ordered Angel de Villafañe to take some of the Luna colonists to the Point of Santa Elena by ship. Leaving behind 50 men at Pensacola Bay (who were later removed),

Early Spanish Exploration

Villafañe took the rest to Havana, where most of them found reasons to excuse themselves from the attempt to colonize Santa Elena. Only about seventy went with Villafañe in his fleet of four small ships. Villafañe sailed up the coast, supposedly reaching the Point of Santa Elena at 33° N. But the harbor to the north of the Point, according to Villafañe, was unsuitable for a colony. He sailed further north, but had to turn around when a hurricane destroyed two of his ships. He reached Havana on July 9, 1561.[30]

Philip II appreciated the advantages to be gained from founding a Spanish colony at the Point of Santa Elena, but how was this to be achieved in a land that seemed to chew up conquistadors and spit out the pieces?[31] The very next year his worst fears were realized—a French colony was founded on the south Atlantic coast. The peace of Câteau-Cambrésis brought a pause to the Hapsburg-Valois rivalry, thus freeing France to follow up on the exploration of Verrazzano. Those who were chosen to found the colony were Huguenots, Protestants who had no love for Spanish Catholics. One motive for establishing the French colony was to set up a base of operations from which they could raid convoys of Spanish ships laden with treasure, as they sailed northward along the coast after having passed through the Bahama Channel.[32]

Jean Ribaut was chosen to establish this colony. In February 1562 he departed from France in three ships with about 150 men. In April Ribaut entered the mouth of the St. Johns River, where he erected a stone column claiming the land for France. He then surveyed the coast to the north and selected Port Royal Sound for the site of his colony. There he erected a second stone column, and on a waterway that entered Port Royal Sound he built a small fort—Charlesfort. The exact site of this fort has not yet been discovered.

Ribaut left behind a small force of twenty-eight or so men to garrison the fort while he returned to France for reinforcements.[33] During the winter, the men at Charlesfort ran short of food and were close to starving. They mutinied, built a small boat, and set sail for France. One of them, however, a young boy named Guillaume Rouffi, chose to remain behind with the Indians. On the open sea, suffering terribly from hunger and thirst, some of the men on the boat resorted to cannibalism, if death did not claim them first. Finally the survivors were rescued at sea by an English ship, and they were returned to France in 1563.

In February 1563 Philip II learned of the existence of Charlesfort, which had been built near the Point of Santa Elena.[34] But communication was slow, and more than a year passed before officials in the Caribbean could investigate. At the first opportunity, they sent a ship commanded by Hernando Manrique de Rojas to reconnoiter the coast. Rojas located the site of Charlesfort, and, guided by Orista Indians, he picked up and interrogated Guillaume Rouffi. From Rojas's report, it was clear that although Charlesfort had failed, the French posed a real threat.

Meanwhile, the French had decided to found another colony. The new expedition set sail from France on April 22, 1564, under the command of René de Laudonnière. This colony was to be located on the southern end of the territory they had previously explored. With about three hundred colonists, Laudonnière built Fort Caroline on the south bank of the St. Johns River at some distance above its mouth, in the territory of the Timucuan chief Saturiba. Those stationed at the fort had been ordered to explore more of both the coast and the interior than had been done by the men who had been stationed at Charlesfort.

The French found that the Indians who lived nearby possessed some objects of gold and silver, which had come from Spanish ships that had wrecked along the coast.[35] Beyond this, the French were persuaded that some of this precious metal had been obtained by Indians from sandy streams in the mountains of "Apalatey."[36] But along with these encouraging discoveries, the men at Fort Caroline were soon to experience the same troubles that had befallen all previous Europeans in the Southeast. Winter brought starvation, and when they tried to force the Indians to share their food, relations between the French and the Indians deteriorated.

In the fall and winter of 1564, some of Laudonnière's men mutinied, seized several of his ships, and set sail for the Caribbean to try their hand at piracy. Some of them were captured by Spaniards and taken to Havana for interrogation. Guillaume Rouffi—whose name the Spaniards hispanicized as Guillermo Rufín—was on hand in Havana to serve as translator for the Spaniards. From this interrogation the Spaniards learned that the French had established yet another settlement, this one further south than the first, strategically located to prey upon Spanish ships. The prisoners described the fort and its armaments in detail. This news reached Seville in March of 1565.[37]

Even before he received news of this second French colony, Philip II had already selected a man whom he believed was capable of founding a colony in La Florida, and who, if need be, could root out French heretics. This was Pedro Menéndez de Avilés, who was born and bred on Spain's northern coast, where he had been a privateer preying upon French ships and also hunting down and attacking French privateers. Beginning in 1554, he had served as captain-general of the Indies fleet, and he was responsible

Early Spanish Exploration

for managing and protecting ships as they sailed to Spain with their loads of New World bullion.

Menéndez had some definite ideas about the problems in Florida. Menéndez was of the opinion that the French not only posed a threat to Spanish ships, but might incite slaves in the Caribbean to rebel. He thought that the French heretics and the American Indians held similar devilish beliefs, which might lead them to form an alliance. He also believed that 80 leagues north of Bahía de Santa María (Chesapeake Bay) there was a great inlet and arm of the sea extending inland some 1,200 miles (see figure 41). From it one could portage to another waterway that led to the Spanish silver mines in Zacatecas and perhaps to the Pacific Ocean. By 1565 the rich silver mines in the New World had become far more important to the Spanish economy than gold, and a French colony could pose a threat to these mines.[38] Menéndez also had some strong ideas about how a Spanish colony, with its main town at the Point of Santa Elena, should be established.[39]

Thus, Menéndez was selected to lead yet another attempt at colonization in the Southeast. This was to be a military and colonial effort jointly funded by Menéndez himself and by the crown. Once the colony was established, Menéndez was to be given considerable authority both to make and to enforce laws. One of his duties was to evangelize the Indians. If he could then defend the lands he had conquered, he would be entitled to profit from them. To begin with, Menéndez was to have the title of marquis and to rule and profit from a personal agricultural estate covering 25 square leagues—more than 5,500 square miles.[40]

When Philip II learned that a new French colony in Florida was in place and that it would soon be receiving supplies and reinforcements, preparations for Menéndez's mission took on a new urgency. The core of Menéndez's organization consisted of his kinsmen by blood and marriage and long-time business associates from northern Spain. With startling aggressiveness, Menéndez assembled the necessary money, men, supplies, and ships. On June 27, 1565, his fleet sailed out of the harbor at Cádiz, and at about the same time a second fleet departed from northern Spain under the command of Esteban de las Alas. The two fleets were supposed to rendezvous in the Canary Islands, and from there sail to the Indies.

But Alas was late in sailing, and the two fleets did not meet up. Menéndez therefore set out from the Canary Islands alone, but a storm scattered and damaged his fleet, including his command ship, the *San Pelayo*. He reached Puerto Rico on August 8, 1565, with his fleet seriously crippled. In a desperate attempt to discover the location of the French fort before it could be reinforced, Menéndez put his army in order, repaired his damaged ships as best he could, and on August 15 set sail with eight hundred people aboard five ships.[41]

As it turned out, the French reinforcements under Jean Ribaut arrived at Fort Caroline at about the same time Menéndez was sighting the Florida coast. By the time Menéndez reached the mouth of the St. Johns River, on September 4, 1565, he found that Ribaut had anchored his ships in the mouth of the river and was unloading them. Darkness had fallen and Menéndez sailed right into the midst of the French ships. Members of the two fleets exchanged insults, shots were fired, and the French ships cut their cables and sailed away in order to keep from being entrapped in the mouth of the river. Menéndez pursued, but with his ships damaged he was unable to overtake Ribaut's fleet.[42]

Menéndez then sailed south to a site he had visited earlier. There he established a town that was to become St. Augustine. In a nearby Indian town he hurriedly threw up a fortification. Ribaut now decided to try a naval attack against Menéndez, but a storm drove Ribaut's fleet southward. Menéndez, guessing that the best of the French fighting men would be aboard the ships, immediately moved north with five hundred men to launch a daring overland attack. Guided by Indians, they marched through terrain made difficult by heavy rain. On September 20, his army came upon Fort Caroline, surprised the garrison, and killed almost all of the men who were in the fort. They spared the women and children, and about forty-five Frenchmen escaped by climbing over the stockade and fleeing into the woods and into the river, where some of them found refuge on three small ships anchored there.[43]

Leaving behind some of his men at Fort Caroline—which the Spaniards renamed San Mateo—Menéndez returned to St. Augustine. Here he learned that Ribaut's ships had, one by one, been driven ashore on the Florida coast. The Indians killed some of the survivors. Others were reported to be marching northward, up the coast. Menéndez took his soldiers to intercept them, and when they surrendered he executed most of them, making captives of the rest.[44] It only remained to track down those Frenchmen who had attempted to find refuge among the Indians.

With his enemies defeated, Menéndez could now begin establishing the towns, missions, and outposts of his colony. High on his agenda was Santa Elena, which was to be his capital city. In April 1566, after reinforcing St. Augustine and San Mateo, Menéndez sailed northward with 150 men in two small boats, exploring the inlets among the islands of the Georgia coast. He explored Sapelo Sound, which he thought was a promising location for an outpost, leaving behind a detachment of eight men among the Guale Indians on the southern part of Saint Catherines Island. Menéndez persuaded the Guale Indians to release into his custody two Orista Indians whom they held captive. These Indians came from the area where Menéndez was to build his town. Continuing, they came to the Point of Santa Elena—Tybee Island—and here they entered Port Royal Sound. Menéndez selected an elevated site on the southern end of Parris Island for the location of his city of Santa Elena.

Menéndez and his men soon encountered the nearby Orista Indians, with whom they established good relations by releasing to them their two kinsmen who had been held captive by their enemies, the Guale. Menéndez laid out the city of Santa Elena and picked the site for Fort San Salvador, the first of six forts to be built at this location in due course. Menéndez named Esteban de las Alas the civil and military leader of the city and placed about fifty men under his command. Within fifteen days the fort was completed.[45]

Meanwhile, back in Spain, Philip II ordered Sancho de Archiniega to transport reinforcements to Menéndez. Archiniega sailed on April 19, 1566 with seventeen ships carrying 1,500 men who were to be stationed in various parts of the Indies.[46] When this fleet anchored at St. Augustine on June 29, among those who disembarked was Captain Juan Pardo, with a company of 250 men. Two of Archiniega's ships were detailed to transport Pardo and his company along with supplies to Santa Elena. They arrived at Santa Elena around July 18 to find it in bad shape. On June 4 a group of soldiers had mutinied, seized a ship, and had fled to Cuba, leaving behind Esteban de las Alas with twenty-seven soldiers. They were without arms, and they had very little food. Captain Pardo and his company had arrived none too soon. They set to work immediately and built a second fort, San Felipe, which was presumably more secure than the first.[47]

Notes

1. Chester B. DePratter, Charles M. Hudson, and Marvin T. Smith, "The Route of Juan Pardo's Explorations in the Interior Southeast, 1566–1568," *Florida Historical Quarterly* 62(1983):125–58; Charles Hudson, "Juan Pardo's Excursion beyond Chiaha," *Tennessee Anthropologist* 12(1987):74–87.
2. Until recently, the most authoritative reconstruction of the Soto route was by members of the U.S. De Soto Expedition Commission, whose report was published in 1939. It has recently been reprinted: John R. Swanton, *Final Report of the United States De Soto Expedition Commission* (Washington, D.C.: Smithsonian Institution Press, 1985), pp. 185–86, 207–8, 348b–48c.

3. Charles Hudson and Jerald Milanich, *Hernando de Soto and the Indians of Florida*, Ms.; Charles Hudson, Marvin Smith, and Chester DePratter, "The Hernando de Soto Expedition: From Apalachee to Chiaha," *Southeastern Archaeology* 3(1984):65–77; Chester DePratter, Charles Hudson, and Marvin Smith, "The Hernando de Soto Expedition: From Chiaha to Mabila," in *Alabama and the Borderlands, from Prehistory to Statehood*, ed. Reid R. Badger and Lawrence A. Clayton (University: University of Alabama Press, 1985), pp. 108–26; Charles Hudson, Marvin T. Smith, and Chester B. DePratter, "The Hernando de Soto Expedition: From Mabila to the Mississippi River," in *Towns and Temples along the Mississippi*, ed. David Dye (University: University of Alabama Press), in press; Charles Hudson, "De Soto in Arkansas: A Brief Synopsis," *Field Notes: Newsletter of the Arkansas Archaeological Society*, 205(July/August, 1985):3–112; Charles Hudson, Marvin T. Smith, Chester B. DePratter, and Emilia Kelly, "The Tristán de Luna Expedition, 1559–1561," *Southeastern Archaeology* 8(1989):31–45.
4. Charles Hudson, "An Unknown South: Spanish Explorers and Southeastern Chiefdoms," in *Visions and Revisions: Ethnohistoric Perspectives on Southern Cultures*, ed. George Sabo and William Schneider, Proceedings of the Southern Anthropological Society, no. 20 (Athens: University of Georgia Press, 1987), pp. 6–24.
5. Paul E. Hoffman, "The Chicora Legend and Franco-Spanish Rivalry in La Florida," *Florida Historical Quarterly* 62(1983–84):419–38.
6. Stanley South, "The Discovery of Santa Elena," Research Manuscript Series 165, Institute of Archaeology and Anthropology, University of South Carolina, 1980; "Exploring Santa Elena, 1981," Research Manuscript Series 184, Institute of Archaeology and Anthropology, University of South Carolina, 1982.
7. Charles Gibson, *Spain in America* (New York: Harper, 1966), pp. 24–28.
8. Robert S. Weddle, *Spanish Sea: The Gulf of Mexico in North American Discovery, 1500–1685* (College Station: Texas A&M University Press, 1985), pp. 99–102.
9. Francisco López de Gómara, *Historia general de los Indias*, (Madrid, 1932), vol. 1, pp. 89–90, translated in *New American World: A Documentary History of North America to 1612*, ed. David B. Quinn (New York: Arno Press, 1979), vol. 1, p. 248.
10. Peter Martyr, *De Orbe Novo*, ed. and trans. F. A. MacNutt (New York, 1912), vol. 2, pp. 254–71.
11. Gonzalo Fernandez de Oviedo, *Historia general y natural de las Indias* (Madrid, 1959), vol. 4, pp. 325–30, translated in Quinn, *New American World*, vol. 1, pp. 260–64.
12. Paul E. Hoffman, *A New Andalucia and a Way to the Orient: A History of the Southeast during the Sixteenth Century* (Baton Rouge: Louisiana State University Press), in press. Hoffman's reconstruction of the founding of the

Ayllón colony is quite different from that of Paul Quattlebaum, who identifies the Jordan River as the Cape Fear and the river of Gualdape as Winyaw Bay. See Paul Quattlebaum, *The Land Called Chicora* (Gainesville: University of Florida Press, 1956).

13. Weddle, *Spanish Sea*, pp. 185-207.
14. John Hemming, *The Conquest of the Incas* (New York: Harcourt Brace Jovanovich, 1970).
15. Miguel Albornoz, *Hernando De Soto: El Amadís de la Florida* (Madrid: Ediciones de la Revista de Occidente, 1971), pp. 227-40.
16. For a detailed reconstruction of the route of the Soto expedition, see the references in note 3.
17. Rodrigo Ranjel, "A Narrative of De Soto's Expedition," in *Narratives of the Career of Hernando De Soto*, ed. Edward Gaylord Bourne, trans. Buckingham Smith (New York: Allerton, 1922), vol. 2, p. 102.
18. A Gentleman of Elvas, "Narrative," in *Narratives of the Career of Hernando De Soto*, trans. Buckingham Smith (New York: Bradford Club, 1866), p. 72; Garcilaso de la Vega, *The Florida of the Inca*, trans. John G. Varner and Jeanette J. Varner (Austin: University of Texas Press, 1962), pp. 311-12.
19. Elvas, "Narrative."
20. Ranjel, "Narrative," p. 110.
21. Garcilaso, *Florida*, p. 340.
22. Herbert I. Priestly, ed. *The Luna Papers* (Deland, Fla.: Florida State Historical Society, 1928), vol. 1, pp. 186-87.
23. Garcilaso, *Florida*, p. 311.
24. David B. Quinn, *North America from Earliest Discovery to First Settlements: The Norse Voyages to 1612* (New York: Harper, 1977), pp. 154-58.
25. Woodbury Lowery, *The Spanish Settlements within the Present Limits of the United States, 1562-1574* (New York: G. P. Putnam's Sons, 1911), pp. 22; Paul E. Hoffman, *The Spanish Crown and the Defense of the Caribbean, 1535-1585* (Baton Rouge: Lousiana State University Press, 1980, pp. 64-70).
26. Paul E. Hoffman, "Legend, Religious Idealism, and Colonies: The Point of Santa Elena in History, 1552-1566," *South Carolina Historical Magazine* 84(1983):59-71.
27. The Soto chroniclers generally spelled the name of this place "Achuse." Ranjel, "Narrative," p. 81; Elvas, "Narrative," p. 49; Garcilaso, *Florida*, pp. 247-48.
28. Hudson et al., "Luna Expedition."
29. Lowery, *Spanish Settlements*, p. 25.
30. Quinn, *North America*, pp. 237-38.
31. Hoffman, "Legend," p. 70.
32. Quinn, *North America*, pp. 241-42.
33. Lowery, *Spanish Settlements*, p. 35.

34. Ibid., pp. 44–45.
35. Ibid., p. 42.
36. Ibid., p. 84; René de Laudonnière, *A Notable History, Containing Four Voyages Made by Certain French Captains into Florida* (Paris, 1586), reprinted in Quinn, *New American World*, vol. 2, p. 281.
37. Eugene Lyon, *The Enterprise of Florida* (Gainesville: University Press of Florida, 1976), pp. 38–41.
38. Gibson, *Spain in America*, p. 121.
39. Lyon, *Enterprise*, pp. 41–43.
40. Ibid., p. 51.
41. Lowery, *Spanish Settlements*, pp. 152–53.
42. Ibid., pp. 156–58.
43. Lyon, *Enterprise*, pp. 115–22.
44. Ibid., pp. 125–27; Eugene Lyon, "The Captives of Florida," *Florida Historical Quarterly* 50(1971):1–24.
45. Eugene Lyon, "Santa Elena: A Brief History of the Colony, 1566–1587," Research Manuscript Series 193, Institute of Archaeology and Anthropology, University of South Carolina, 1984, p. 2.
46. Lyon, *Enterprise*, p. 147.
47. Lyon, "Santa Elena," p. 2.

2
Juan Pardo's Two Expeditions

Menéndez went to inspect Santa Elena in August 1566. He found that Esteban de las Alas and Juan Pardo had taken control and had managed to ease the problems caused by the mutiny, although their supply of food was inadequate. On this visit, Menéndez formally designated Santa Elena the principal city of his colony, and he appointed Alas to serve as his chief lieutenant.[1] It was partly to achieve his colonial ambitions, but also to relieve the food shortage in Santa Elena that Menéndez ordered Pardo to take half of his company of soldiers and go into the interior. He instructed Pardo to explore the country, pacify the Indians, evangelize them, and bring them under the dominion of Spain. He also told Pardo to find a road to the Spanish silver mines in Zacatecas, Mexico (see figure 1). Pardo was to provision his men with food paid as tribute by the Indians.

As it turned out, this was only the first of two expeditions Pardo would lead into the interior. The second one took place the following year and covered more territory than the first. And because the documentary record of the second expedition is more detailed, the route Pardo followed on that expedition can be reconstructed more precisely.

Pardo's First Expedition: December 1, 1566, to March 7, 1567

Pardo departed from Santa Elena on Saint Andrew's day, December 1, 1566, with 125 soldiers.[2] The expedition proceeded north by northwest, following a trail between the Coosawhatchie and Salkehatchie rivers (see figure 3).[3] Even though Menéndez had ordered Pardo to find a road to

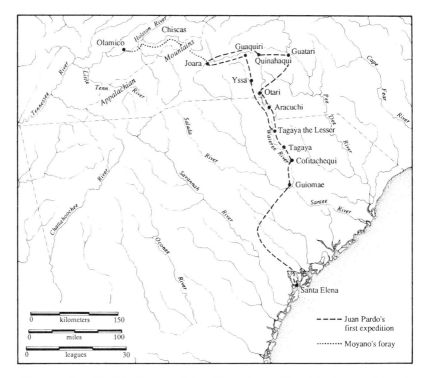

Figure 3. Juan Pardo's first expedition (December 1566 to March 7, 1567) and Hernando Moyano's foray (Spring 1567).

Zacatecas, which lay to the west, Pardo traveled north. The reason he did so was that sixteenth-century expeditions, being unable to carry along a sufficient supply of food, had to obtain it, by persuasion or force, from the people they encountered. Even though Pardo may have wished to go west from Santa Elena, he was constrained to go northward, where Indian guides must have told him that he would find sizable Indian towns with supplies of food.

After several days of travel, nearing the headwaters of the coastal plain rivers, the Spaniards turned toward the northeast and proceeded to cross several swamps before arriving at a large river, the Congaree, probably at the Indian town of Guiomae.[4] For this entire distance, they encountered few Indians, and many of those they did encounter had already sent representatives to Santa Elena, where they had been informed that they were now subjects of Spain.[5] At Guiomae and elsewhere, Pardo gave a prepared speech to the Indians, translated from Spanish to an Indian language by his interpreter, in which he told them that they were subjects of Spain and of the pope. On most of these stops, judging from what occurred on the second expedition, he must also have instructed the Indians to build houses for the Spaniards to use when they visited the towns and to lay up stores of corn exclusively for Spanish use.

From Guiomae they proceeded northward, following the western bank of the Catawba-Wateree River until they crossed the river and came to Cofitachequi (also called Canos) near present-day Camden, South Carolina. From here they continued up the Catawba-Wateree River. Now marching along the eastern bank, they came to Tagaya. They continued on to Tagaya the Lesser, probably located near the Wateree River in the vicinity of present-day Fishing Creek Dam. Here they apparently forded the river, and then continued northward, now along a trail to the west of the river.

They next came to Yssa, on the South Fork of the Catawba River near Lincolnton, North Carolina, where a great chief lived. Many Indians assembled at this place to hear Pardo's speech.[6] The direction of their travel beyond Yssa is unknown, although it was probably to the west or northwest, where they came next to a tributary town of Yssa.[7] They then continued on through an uninhabited area where they had to bivouac for the night.

The next place they came to was Joara, a very important town near present-day Marion, North Carolina, at the foot of the Blue Ridge Mountains. As will be explained later, this was the same town the Soto chroniclers called "Xuala." The land around Joara impressed the Spaniards as being very good.[8] Again, many Indians assembled and listened to Pardo's speech. Because snow had already fallen in the mountains, Pardo could not continue his expedition. He remained in Joara for two weeks, during which time one or more of his men attempted to teach the Indians about Christianity.[9] He also built a small fort at Joara, which he named Fort San Juan. When he departed, he garrisoned the fort with thirty men under the command of Sergeant Hernando Moyano de Morales. He provisioned them with a supply of powder, matchcord, and lead balls for their matchlock guns.[10]

When Pardo departed from Joara, he followed the Catawba River in a "northerly" direction (actually, he traveled east by northeast). He came to two towns, one of which was Quinahaqui, on the Catawba River near the present-day Catawba, North Carolina. He spent four days in one of these towns and two days in the other. He describes the land in this area as being very good, with excellent fields of bottomland (*muy jentiles begas*).[11]

From Quinahaqui, Pardo set out toward the east, traveling through uninhabited land for two days. He arrived at Guatari, near present-day Salisbury, North Carolina, where he found more than thirty *caciques* and many Indians assembled, and he gave them his usual speech. His company, now numbering about ninety-five men, remained here for about fifteen or sixteen days. While he was there a runner brought a letter from Esteban de las Alas in Santa Elena calling him back to be on hand in case of a French reprisal against Menéndez for his massacre of their countrymen.[12]

Pardo departed from Guatari the next day, leaving behind his chaplain, Sebastian Montero, and four soldiers to evangelize the Indians.[13] Traveling southwestwardly, Pardo says that they came to Guatariyati[qui], although he probably meant to write "Otariyatiqui." The town in question was probably in the vicinity of present-day Charlotte, North Carolina.[14] From here they continued on to Aracuchi, in the vicinity of present-day Van Wyck, South Carolina, where the land seemed very good.[15] At this town, Pardo again gave his speech to a large number of caciques and Indians. Now traveling southward, Pardo again came to Tagaya the Lesser, and from here he traveled the same trail that he had followed while coming into the interior. With his force now down to about ninety soldiers, Pardo arrived in Santa Elena on March 7, 1567.[16]

Moyano's Foray: April 1567

Sergeant Moyano did not see Pardo again for about nine months, although the two of them exchanged letters carried by messengers. Moyano did not remain idle during this time, but it is difficult to reconstruct his activities precisely because he apparently left no written account of his actions. Perhaps more than any of Pardo's men, Moyano was interested in discovering precious metals and gems. Moyano was in Joara for about two months—during part of January, February, and part of March—and there are indications that he engaged in prospecting, perhaps making one or more forays to mineral sites reported to him by Indians.

Moyano also participated in the political rivalries among the Indians, a strategy frequently used by Spaniards in their conquest of parts of the New World. Moyano's military exploits are known mainly from an account written by the soldier Francisco Martinez. Some of Martinez's information is inaccurate, and some of it appears to be exaggerated. It appears to be secondhand information that he obtained from at least two letters from Moyano that he read or heard being discussed.

About thirty days after Pardo had returned to Santa Elena, presumably in early April 1567, he received a letter from Moyano, who reported having fought a battle with some Chisca Indians. These were the same Chiscas whose land was reconnoitered by two of Soto's soldiers. He sent them there when he was told that they mined gold.[17] Moyano took fifteen of his soldiers and attacked a Chisca town, burning the houses and killing many of the inhabitants, while only two of his own men were slightly wounded.[18] The territory of the Chiscas appears to have been in and to the other side of the mountains north of Joara, including the area along the upper course of the Nolichucky River. However, there is no description of the route they traveled to this first Chisca town, so that its location is unknown.

In this action, as in another to be discussed presently, Moyano's soldiers were accompanied by a force of Indian warriors. When Juan de Ribas, one of Pardo's soldiers, was questioned in 1602, he said that Moyano had helped an Indian chief defeat a rival.[19] Ribas does not say who this chief was nor how many Indian warriors were involved. But since Moyano had been in residence at Joara, it seems reasonable to suppose that he allied himself with the chief of Joara; and because this man was politically powerful, it is probable that a large force of Indian warriors assisted Moyano in his exploits. In his letter describing this first action, Moyano is reported to have told Pardo that if ordered to do so, he would push on from Joara and make further explorations.

Moyano's involvement in aiding one interior chief against another is confirmed by Jaime Martinez, who served as an accountant in Florida from 1571 until about 1579. During this time Moyano told Martinez about his exploits, although Martinez was under the mistaken impression that Moyano had aided the chief of "Latana" (i.e., "Latama," derived from Altamaha) against the chief of "Cossa" (i.e., Coosa). In reference to this, Martinez noted that the Indians were as cruel to other Indians they killed or captured in war as they were to Spaniards. Specifically, Moyano told Jaime Martinez about an incident in which his Indian allies overtook some of their traditional enemies. When they succeeded in felling one of these with arrows, darts, and warclubs, made of very hard wood, they leaped upon him and with a

sharp flint fastened to a wooden handle cut a circle around his scalp and then astride the wretched Indian pulled out his hair by the roots, leaving the Indian dead. The one that does this is said to be *tasigaya*—valiant. I have seen many of these scalps which the Indian wear fastened about their legs, though they do not wear stockings.

This is a very early account of the practice of scalping.[20]

Pardo sent Moyano a reply saying that he should station ten of his men and a squad leader in the fort at Joara and take the rest to make further explorations. But before Pardo's letter arrived at Joara, a "mountain cacique," presumably also a Chisca, threatened to kill and eat all of the Spaniards and Moyano's dog as well.[21] It may be significant that when Soto traveled through the area, his men, hungry for meat, would sometimes eat the Indians' dogs. Hence, this threat may have been an allusion, perhaps humorous, to the Spaniards' habit of eating dogs. But it is also possible that Moyano's dog was one of the large war dogs the Spaniards used against the Indians, in which case this threat may have been the mountain cacique's way of indicating just how ferocious he was.[22]

Moyano did not wait for Pardo's reply. He took twenty of his men and went on the offensive, and the subjects and allies of Joara must have gone with him (see figure 3). He spent four days crossing the mountains, evidently following a trail northward from Joara to the North Toe River. At Yellow Mountain this trail forked, and Moyano probably took the left trail, which followed the Toe River to where it empties into the Nolichucky River. On modern maps, the Clinchfield Railroad follows on or near this trail.

When the Spaniards reached the enemy town, they were suprised at how strongly it was fortified. It was surrounded by a high wooden palisade having only one small door, which was protected by barricades.[23] In order to approach this door, Moyano built a movable shield to protect himself and some of his men from the Indians' arrows. They succeeded in gaining entry, although Moyano sustained a wound in the mouth, and nine other soldiers were slightly wounded. Once the Spaniards were inside the palisade, the Indians retreated into their earth-covered houses, from which they would dash out and skirmish with the Spaniards. Eventually the Spaniards reached the entrances of the houses and set fire to them, killing many who were inside. One estimate—probably exaggerated—placed the number of Indian dead at 1,500.[24]

Judging from Moyano's travel time from Joara to the town he destroyed, and his travel time from this town to Chiaha, which was the next place he visited, the most likely location for the destroyed town was on the upper Nolichucky River. A possible location is the Plum Grove site near present-day Embreeville, Tennessee. Little excavation has been done at this site, but it does appear to date to the sixteenth century, and there is a three-acre area there that was densely occupied and probably palisaded.[25] This may also be the site of the Chisca village that was reconnoitered by two members of the Soto expedition in 1540.

This was the only aggressive action of any magnitude taken by the members of Pardo's two expeditions. There are several possible reasons why this action occurred. It may be that Moyano received a serious threat and concluded that a military offensive was his best course of action. Or, in view of his interest in prospecting, it may be that he was aware of the rumors of gold circulating among the members of the Soto expedition which pointed to the Chiscas and was therefore strongly motivated to go and take a look for himself. This possibility is strengthened by the testimony of Juan de Ribas, who was questioned some thirty years after the event, and who said that for assisting the cacique against his rival, Moyano received a payment in gold.[26] Ribas may have been mistaken in saying that Moyano received a payment, but the possibility of finding gold may well have been a factor in Moyano's decision to launch the attack.

While Moyano was still at the scene of this battle, he evidently received Pardo's letter giving him permission to explore further. He set out in a west-by-southwest direction, traveling for four days, apparently through uninhabited or sparsely inhabited country, before coming to the chiefdom of Chiaha. The main town of Chiaha was on Zimmerman's Island, in the French Broad River, near present-day Dandridge, Tennessee.[27] Moyano found the town defended by a strong palisade and square towers. Many warriors were in the town, but no women or children were present. Their absence may indicate that the Indians were contemplating a suprise attack, such as the one waged against Hernando de Soto at Mabila, but no such attack occurred.[28] Initially, Moyano's men were well fed and well treated. What Moyano did next is not altogether clear, although he appears to have spent several days exploring the territory of Chiaha. Whatever the nature of his activities, he returned to the main town of Chiaha, where he built a small fort in which he awaited Pardo's arrival.

Pardo's Second Expedition: September 1, 1567, to March 2, 1568

In early May 1567, Menéndez again visited Santa Elena. On this occasion Pardo no doubt gave him an account of his first expedition, and he must have told him about Moyano's activities as well. Perhaps encouraged by what Pardo said about the quality of the land in the interior, Menéndez at this point began to make plans for the founding of his agricultural estate there.[29]

Menéndez ordered Pardo to lead a second expedition into the interior. On May 25 he commanded Pardo to depart at the beginning of September

Figure 4. Page one of Juan de la Bandera's "long relation" (Published with the authorization of the Archivo General de Indias, which reserves all rights to successive reproductions and publications).

with as many as 120 soldiers, harquebusiers, and crossbowmen. He was to find the most direct road to the Spanish silver mines in Zacatecas, Mexico. He was to establish friendly relations with the caciques of the country, impress upon them that they were subjects of Spain, and ask them whether they wanted a monk to be sent to them to teach them Christian doctrine. He was to take pains to see that his men got sufficient rest and recreation

Figure 5. The signature page of Juan de la Bandera's "short relation" (Published with the authorization of the Archivo General de Indias, which reserves all rights to successive reproductions and publications).

en route and make sure that they behaved in a Christian manner. In every place he visited he was to take possession of the land in the name of the king of Spain. Menéndez ordered Juan de la Bandera to make a formal record (see figures 4 and 5) of all these transactions. Upon arriving at the first Christian village in Zacatecas, Pardo was to send Bandera to the viceroy of New Spain and the *audiencia* of New Galicia to inform them of the journey and of its purpose. Pardo might also entreat the viceroy for the services of a monk who could accompany his men on their return and then be left in the interior to evangelize the Indians. As in the first expedition, Pardo and his men were to do all this and return to Santa Elena by March, to be on hand in case of a French attack.[30]

To serve as translator, Pardo took along Guillermo Rufín, the French boy from Jean Ribaut's failed colony. Rufín was said to be able to understand the languages of all of the Indians, although in fact he could not, because in several places they had to use Indian interpreters as intermediaries in a chain of translation.

On September 1, 1567, Pardo again departed from Santa Elena with a company of men.[31] Whether it numbered more or fewer than the 120 he was authorized to take is not known. The first leg of their journey was by boat, and at the end of the first day they reached Uscamacu, a town on an island that was surrounded by rivers (see figure 6). This town was probably on the northern end of Port Royal Island. It was 5 leagues from Santa Elena.[32] The soil where Uscamacu stood was sandy but good for growing corn, and grapes thrived there. Bandera says that good clay for making pots and roof-tiles could be obtained nearby.[33]

On September 2 Pardo proceeded to Ahoya, a village said to be on an island. Some places in the vicinity of Ahoya were surrounded by rivers, but the rest of the area, like the mainland, was fairly good for corn, and had many grapevines with abundant vine-shoots. However, Ahoya may have been on a neck of land surrounded by tidal rivers, so that it only appeared to be an island. Ahoya was possibly located on Scott's Neck, in the vicinity of present-day Sheldon.

On September 3 Pardo and his men reached Ahoyabe, a small village that was subject to Ahoya. Ahoyabe was probably on the Coosawhatchie River or the Salkehatchie River, in the general vicinity of the present town of Cummings. Here the quality of the soil was similar to that at Ahoya.

On September 4 they went to Cozao, a town under the control of a rather important cacique. The area around Cozao contained much good land like that seen before, but the soil was stony in many places. At Cozao, one could cultivate corn, wheat, barley, and all kinds of fruits and vegetables because it was "a land of sweet rivers and brooks" and the soil was fairly good for everything. Cozao was possibly located near Coosawhatchie River or Jackson Branch in the general vicinity of present-day Fairfax. The stony soil indicated that they had reached the edge of the Aiken Plateau, where small pebbles first occur after the sandy expanse of the Coastal Plain. The "sweet rivers and brooks" presumably means they had reached an area where the streams flowed more swiftly.[34]

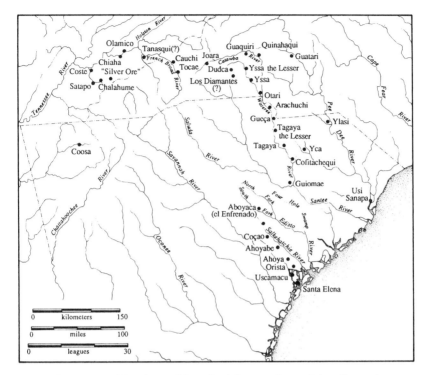

Figure 6. Towns and locations of Juan Pardo's second expedition (September 1, 1567, to March 2, 1568).

The next day, September 5, they entered a small town that was affiliated with or was tributary to Cozao. Here the arable land was good, but there was little of it. This town, whose name was not recorded, was probably located on the Salkehatchie River near present-day Ulmer.

On September 6, they arrived at a place Bandera calls *el Enfrenado*, literally "reined in." Much of the soil there was poor, although there were many tracts of good soil, like the ones already encountered. It was probably located near the South Fork of the Edisto River. Why the Spaniards referred to the town by this name rather than an Indian name is not clear. It is possible that this was the place they called Aboyaca on their return trip. The next day they traveled further, and, coming to no town, they slept in the open, somewhere to the northeast of present-day Orangeburg, near the headwaters of Four Hole Swamp.

On September 8 they reached Guiomae, which, as previously mentioned, was located just north of the junction of the Congaree and Wateree rivers, near present-day Wateree, South Carolina. They evidently reached

Pardo's Two Expeditions

this town by crossing the Congaree River in Indian dugout canoes. Guiomae is the town called Aymay or Hymahi in the Soto narratives. It was the place Soto came to after spending many days crossing an uninhabited area and after fording the Congaree River near present-day Columbia.[35]

The land at Guiomae, like that at Cozao, could be cultivated, but it was more fertile. Nearby there were several large and deep swamps where the land was flat and poorly drained. These were the enormous swamps that lay along the Congaree, Wateree, and Santee rivers.

After resting in Guiomae, they departed on September 9, now traveling toward the north.[36] Their first night on the trail, September 10, they slept in the open, and the next day they arrived at Cofitachequi, also called Canos and Canosi. This was the same Cofitachequi that Soto had visited in 1540, and, just like Pardo, Soto had taken two days to get there from Guiomae. The site of Cofitachequi was at or near the McDowell or Mulberry site, at the mouth of Pine Tree Creek, near Camden, South Carolina. Like Soto before him, Pardo and his men must have crossed the Wateree River in Indian canoes to reach Cofitachequi.[37]

Between Guiomae and Cofitachequi they encountered the transition from the flat coastal plain to the hilly country near the Fall Line. The country is described as having some small swamps that could be crossed on foot, even by a boy. The country had high cliffs, stones, and boulders, but also some low spots. The soil was reddish and of better quality than they had seen before. The land around Cofitachequi was good for corn. Several varieties of grapes grew in profusion there, though they varied in quality. It was an area in which a major town could be founded.[38]

On September 13, Pardo departed from Cofitachequi and went to Tagaya. Now traveling up the east side of the Wateree River, the Spaniards crossed the Fall Line, which lies just south of Wateree Dam, and from here on they encountered no more swamps. The land was relatively flat, with black and red soil and an abundance of good water in springs and streams. The trees here seemed less dense than those in the terrain previously seen. Tagaya was probably located on or near lower Beaver Creek.[39]

On September 14 they traveled to Tagaya the Lesser *(Tagaya Chico)* and to Gueça, where the land was again abundant and good.[40] Gueça was probably on the Wateree River north of Lancaster, South Carolina, and perhaps on Waxhaw Creek. From Gueça they traveled to Aracuchi, which was probably located on the Catawba River in the vicinity of present-day Rock Hill, South Carolina.

They spent one or two days traveling from Aracuchi to Otari, which was perhaps on the Catawba River or one of its tributaries, in the general vicinity of Charlotte, North Carolina. Otari was located where at least one important trail intersected the trail Pardo was following. This intersecting trail led northeast to Guatari, some 15 or 16 leagues away.[41]

While they were in Otari, Pardo and his men were visited by two female cacicas from Guatari, who were named Guatari Mico and Orata Chiquini. On his previous expedition, Pardo had left behind in Guatari his chaplain Sebastian Montero and several soldiers. Two soldiers—Anton Muñoz and Francisco de Apalategui—accompanied the two cacicas.

Exactly when the expedition departed from Otari is not known, but it was probably on September 18 or 19. In any case, they traveled two or three days to their next stop, Quinahaqui, which they reached on September 20. Quinahaqui was located on the Catawba River, possibly near the present town of Catawba.

The next day Pardo went from Quinahaqui to Guaquiri, a town located in the vicinity of present-day Hickory, perhaps on the Catawba River or else at the head of the South Fork of the Catawba River. In all probability, this was the town of Guaquili visited by Soto in 1540.

After leaving Guaquiri, Pardo spent three days (September 22 to 24) traveling to Joara, located near Marion, North Carolina, where he had built Fort San Juan on his first expedition. This was the same town the Soto chroniclers called "Xuala."[42] Joara was situated at the foot of a range of mountains—the Blue Ridge Mountains—and was surrounded by rivers. To Pardo and his men it was very good country, as good as the best in Spain. Upon arriving, Pardo learned that Sergeant Moyano and his men were in uneasy circumstances in Chiaha—they were surrounded by Indians.[43] Pardo put Corporal Lucas de Caniçares in command of Fort San Juan and ordered him to remain there. He left some corn and ammunition for the men in the fort. It is not clear whether, upon leaving, Pardo reinforced the contingent of ten men who had previously been stationed at Joara by Moyano.

While Pardo was in Joara, a detachment of five men from Santa Elena caught up with him. They were carrying iron tools and other trade goods from San Mateo that Pedro Menéndez Marques and Estebano de las Alas were to have delivered to him at Santa Elena. But when Menéndez Marques and Alas arrived, they found that Pardo had left thirteen days earlier. Hence they sent these five men in pursuit of Pardo.[44]

With Moyano's peril in mind, Pardo departed in some haste from Joara, probably on September 29. The next town, Tocae, was three days away, implying that they arrived there on October 1. In getting there, Pardo followed the Catawba River to its headwaters and crossed the Blue Ridge Mountains through Swannanoa Gap. Here he picked up the headwaters of the Swannanoa River and followed it to its junction with the French Broad River. Tocae was located on the French Broad River in the vicinity of present-day Asheville.

The land at Tocae is described as being very good, better in fact than the Sierra Morena, the mountains to the north of the Guadalquivir River in Andalucia. In the surrounding country Pardo and his men found grapes, chestnuts, walnuts, and many other fruits. Nearby were many fields of bottomland, which was not craggy or rough. The soil was very good at Tocae, and one could grow anything on it.[45] Because of his concern for Moyano's safety, Pardo remained in Tocae only four hours before continuing his journey, and he and his men probably camped in the open that night.

The next day, October 2, Pardo and his men followed a trail that ran along the French Broad River to Cauchi, a town probably near Marshall, North Carolina, perhaps on Ivy Creek.[46] Cauchi seems to have been the town that the Soto chroniclers called Guasili.[47] Why the name was different, or changed, is not clear. Even though the alluvial lands alongside the French Broad River in this area are quite narrow, they impressed the Spaniards as being very rich.

Departing Cauchi on October 3, they continued following a trail along the French Broad River, which grew ever larger as it descended the mountains. They traveled through uninhabited country for three days and then came out of the mountains to the rich alluvial soils of the Tennessee Valley, reaching the town of Tanasqui on October 6. The Spaniards compared this land to that of Andalucia.[48]

The town of Tanasqui was situated in the V formed by the junction of two sizable rivers. One was the French Broad; the other was either the Pigeon River or the Nolichucky. The open side of the town was fortified by a log palisade with three defensive towers. The Spaniards waded across one of these rivers, waist deep in some places, with some difficulty. After traveling a good distance further, they entered the town.[49] Presumably because of something he saw at Tanasqui, Pardo believed that silver and gold could be had in this general area.[50]

On October 7 they reached the chiefdom of Chiaha, whose principal town was called Olamico.[51] Here Pardo found Moyano and his men. They were not in immediate danger, although the Indians had apparently isolated them in their fort.[52]

Olamico was situated on Zimmerman's Island, now under Douglas

Figure 7. Zimmerman's Island. Note the mound toward the left side of the upper end of the island. Aerial photo taken in 1925. (Photograph courtesy of Tennessee Valley Authority.)

Lake, in the French Broad River near present-day Dandridge, Tennessee. This same town was visited by Soto in 1540. One of the chroniclers of the Soto expedition, the Gentleman of Elvas, wrote a detailed description of the island: "The town was isolated between the two arms of a river, and seated near one of them. Above it, at a distance of two crossbow shots, the water divided, and united a league below. The vale between, from side to side, was the width of a crossbow shot and in others two. The branches were very wide, and both were fordable: along their shores were very rich meadow-lands having many maize-fields."[53] On Zimmerman's Island (see figure 7) a 30-foot-high mound, around which the town must have been located, was situated some 550 to 660 yards from the upstream end of the island, and since a crossbow shot was a distance of about 300 yards, the location of the town is consistent with Elvas's description. The island was likewise about 550 to 600 yards across at its widest, and it was about $2\frac{1}{2}$ miles long—somewhat less than a league.[54]

Olamico was well defended. Not only was it on an island, but it was surrounded by a strong palisade. Presumably the fort that Moyano built was on the island, and perhaps it was near the town of Olamico. The Spaniards praised the land near Olamico extravagantly, saying it was a rich and broad land, surrounded by beautiful rivers, with many small towns 1, 2, or 3 leagues apart. It was a place where many grapevines and persimmon trees grew. It was a *tierra de ángeles*—a land of angels.

After having rested in Olamico for five days, Pardo and his men set out again, intent on going to Coosa and to the silver mines of San Martín in Zacatecas.[55] On October 13 they struck out directly west from Olamico, but later turned to the southwest, traveling 5 leagues and bivouacking at the end of the day in level country. Their camp was probably at or near the junction of Walden Creek and Cove Creek. Three chiefs bearing gifts of food met them there.

The next day, October 14, they traveled 5 leagues further over very rough country. They probably followed Walden Creek to its head, crossed a watershed, and on the other side picked up the headwaters of Reed Creek, which they followed to its junction with Little River. Then they picked up Hesse Creek and followed it until they ascended it, or its tributary Flat Creek, climbing to the top of a spur of Chilhowee Mountain. That night they camped in the area known as The Flats and Lake in the Sky.

Earlier that day they had caught sight of the highest mountains yet seen on their travels—the Great Smoky Mountains. Perhaps they were looking at Clingman's Dome, Thunderhead Mountain, and Gregory Bald. On top of the "very high mountain" that they reached at the end of the day, Pardo picked up a small reddish stone that he gave to Andres Suarez, who was a "melter of gold and silver."[56] Although there is no evidence that Suarez performed a test on this stone, he said that it "might be silver."

On October 15, Pardo and his men continued their journey, probably along a trail through Happy Valley. This led them to the town of Chalahume at the Chilhowee archaeological site near the junction of Abrams Creek and the Little Tennessee River. To Bandera, this town had as beautiful a location as Córdoba, with large plains and good grapes growing nearby.[57] The surrounding area does in fact resemble the land around Córdoba, which is situated in a narrow valley of the upper Guadalquivir River and lies close to a range of mountains.

The next day, on October 16, they traveled 2 leagues, reaching the town of Satapo, situated at the Citico archaeological site near the junction of Citico Creek and the Little Tennessee River. Later in the day, as it began to grow dark, sentinels who had been posted at two high parts of the village (presumably on bastions or watchtowers) heard a great noise coming from some Indians who had gathered outside the village. The sentinels reported to Pardo what they had heard, and he put his company on alert.

Around midnight, an Indian entered the house in which Guillermo Rufín was sleeping. Rufín knew this Indian because he had been traveling with the company for two or three days. Either his native language was intelligible to Rufín or else this man was a bilingual who spoke a language that was intelligible to Rufín. The man told Rufín that if he could arrange for Pardo to give him an axe, he would tell him about a plot by the Indians of Satapo, Coosa, Uchi, Casque, and Olamico.[58] The Indian told Pardo that the great chief of Coosa and his allies were planning to launch three ambushes on Pardo's company as they traveled southward from Satapo to Coosa. Moreover, all the chiefs had agreed not to give Pardo any food or anything else unless they were compensated for it.

Pardo also learned from this Indian informer that the Spaniards were not following the easiest trail to Coosa. The easiest road ran along the river that Olamico was situated on. Presumably this was the trail Soto had followed in going from Chiaha (Olamico) to Coste. This suggestion is consistent with the fact that none of the Soto chroniclers complained that the trail they followed from Chiaha to Coste was as difficult or rough as the one Pardo's men had followed.

That same night, the soldier Alonso Velas asked an Indian interpreter who had been traveling with them to tell him what the Indians of Satapo were planning. This Indian, a native of Oluga (location unknown), told Velas that the great chief of Coosa, who lived on the route Pardo was planning to take, had decided that he would not give the Spaniards any food unless they paid for it, and that he and his allies would wage war against the Spaniards. Moreover, this Indian said that it would trouble him greatly if he had to go to Coosa, because five of his "brothers" had gone in the company of Soto, and the people of Coosa and neighboring towns had attacked and killed Soto and his men and had enslaved his "brothers."

Early the next day, October 17, Pardo called for a meeting with the chief of Satapo. He asked this chief to provide him with burden bearers. The chief dissembled. He pretended to go off to fetch burden bearers, but he came back with none, giving transparent excuses that he had hastily made up. Pardo also noted that the Indian men who had accompanied him from

Chiaha and also the men of the town of Satapo were nowhere to be seen—only the women and children of Satapo were in sight. With this, Pardo knew that a plot was indeed afoot.

Pardo assembled his officers. He told them what he had learned of the plot and asked them whether they thought the expedition should continue to Coosa or turn back. The officers expressed various opinions, but ultimately they decided that it would be best to return to Chiaha, a three-day's journey away, and that they should take a different route from the one by which they had come.

They departed from Satapo on that same day and traveling over very rough trails arrived at the town of Chiaha on October 19. On this return journey, they must have followed a trail that lay to the northwest of Chilhowee Mountain, probably the trail known in the eighteenth century as "The Great Indian Warpath."[59] This led them to the town of Chiaha.

They departed from the town of Chiaha on October 20 and arrived at Olamico later that same day. Immediately upon arriving, Pardo laid out a fort "in Olamico." Bandera says that they began building it, "and after four days it was finished." However, Pardo spent only one full day at Olamico—October 21. Hence, Bandera's meaning is not clear. Perhaps Pardo began building the fort with his full company and assumed that the small detachment he left behind could finish the construction in another three days. Or it may be that they merely strengthened the fort Moyano had built.

Pardo and his men departed Olamico on October 22 and arrived in Cauchi on October 27, after spending six days going up the mountains—one day longer than it had taken them to come down. As soon as Pardo arrived at Cauchi, he laid out Fort San Pablo. The work was finished on October 30. After this, Pardo and his men rested for a few days.

Leaving a small detachment at Fort San Pablo, Pardo and his men traveled from Cauchi to Tocae on November 1, and they apparently rested there the following day. On November 3 they again began to travel, going 5 leagues before camping for the night in level country. They probably camped on the Swannanoa River near the present town of Azalea. On November 4 they went 5 leagues further, where they spent the night near a stream in a ravine. They were still on the Swannanoa River, probably near present-day Ridgecrest, North Carolina. On November 5 they went 4 leagues further and bivouacked. Clearly, Pardo and his men were exhausted. Had they not been, they could easily have reached Joara before this day ended. As it was, some Indians from Joara came to their camp, bringing them food. That night they probably camped east of present-day Old Fort, North Carolina. On November 6 they arrived at Joara.

Because they had traveled for so long and had been so poorly provisioned, they rested at Joara for a considerable period. During this time, Pardo met with a large number of caciques who came to Joara. However, there was to be no rest for Moyano and Andres Suarez, whom Pardo sent out from Joara on November 7 to investigate a crystal mine. They did find the mine, and they broke off a sample of the crystal with the point of a mattock and carried it back to Joara on November 8.

Before leaving Joara and Fort San Juan, Pardo garrisoned the fort with his ensign, Albert Escudero de Villamar, and thirty of his men.[60] Pardo and his company departed from Joara on November 24 and traveled 5 leagues before bivouacking. They probably traveled southeast from Joara. There are three routes they could have followed: a trail on or near the Clinchfield Railroad; a trail on or near State Highway 226; or a trail that paralleled the Southern Railroad to about present-day Morganton and then turned southeast along State Highway 1924.

The next day, November 25, they continued their journey, traveling another 5 leagues before reaching Dudca, a small village (*lugarejo*) subject to the chief of Yssa. If they followed a trail on or near the Clinchfield Railroad, Dudca would have been in the vicinity of present-day Ellenboro, in eastern Rutherford County. If they followed a trail on or near Highway 226, Dudca would have been in the vicinity of present-day Polkville, in northwestern Cleveland County. If their trail followed near the Southern Railroad and State Highway 1924, Dudca would have been near the town of Camp Creek in the southeastern corner of Burke County.

On the morning of November 26, as they departed from Dudca going toward Yssa, they traveled for a quarter of a league, and after making a left-hand turn on the trail and then a right-hand turn, guided by Moyano and Suarez, they came to the "crystal mine" that the two had previously visited. If Dudca was in the vicinity of Polkville, the crystals could have been garnet, a gem that is known to occur in that area. If Dudca was in the vicinity of Ellenboro, the crystals were probably beryl or quartz. If Dudca was near Camp Creek, the crystals were probably corundum, which is known to occur near Carpenter's Knob in northern Cleveland County.[61]

They marked the mine without specifying how they did so. Later in the same day, after traveling a quarter of a league beyond this crystal mine, Pardo found other crystals on the right-hand side of the trail that seemed to him to be the beginning of another mine. It is not clear when they took leave of these crystal mines, but they arrived in Yssa on November 28. Yssa was probably located on the South Fork of the Catawba River, in the vicinity of present-day Lincolnton. Here they rested and dealt with several Indians who came in to meet with Pardo. Two of these Indians were again from Guatari. Pardo gave the two socketed axes with orders to take them back to Guatari and to cut down some trees with which to build a fort when he arrived there.

On December 10 Pardo was informed that a league downstream from Yssa, and on the opposite side of the river, there was yet another source of crystals. He sent Moyano and Bandera to investigate, and they did in fact find some crystals with sharp and faceted points, which appeared to be of a different kind from the ones they had previously found. These were perhaps crystals in a deposit of hornblende schist known to occur in this area.

On December 11 they departed from Yssa and traveled 3 leagues to Yssa the Lesser. This town was probably in the vicinity of Maiden, North Carolina. On December 12 they traveled 5 leagues, reaching Quinahaqui.

On December 14 they traveled 5 leagues further and bivouacked in level country. This would have been near the Iredell-Rowan county line. The next day they traveled 6 leagues, arriving at Guatari Mico.[62] This was a country of both mountains and flat, arable land. The houses in Guatari were large, earth-covered structures, with circular floor plans. A large river, the Yadkin, ran by Guatari and was said to empty into the sea near Sanapa[63] and Usi, where salt was made. It was said that a ship could sail up this river for 20 leagues.[64] Pardo again was greeted by the female chiefs of Guatari. He told them to command their people to come and help him build a fort.

On December 16, several caciques arrived, although they did not show up until late in the morning. Pardo laid out the fort and they began to hurriedly build two bastions, so that they would at least have something completed in case they were suddenly summoned back to help defend Santa Elena. Construction was completed on December 20, after five days of work. When no summons from Santa Elena came, Pardo ordered the Indians to appear three days later, and they began to work on the larger fort. They built four tall cavaliers of thick wood and dirt, with walls of tall poles and dirt. Construction on Fort Santiago was finished on January 6, 1568.

Leaving a small detachment at Fort Santiago, Pardo and the rest of his men departed from Guatari on January 8. They traveled 5 leagues, and then bivouacked near present-day China Grove. They continued for three more days through level country, traveling 5 leagues per day, first reaching the vicinity of present-day Harrisburg, then the vicinity of present-day Charlotte, and finally the town of Aracuchi.

Upon arriving at Aracuchi, Pardo split his company into two groups. He wanted to go to the east to Ylasi to meet with some caciques he had not met before. But in order to spare some of his men from the extra travel this would entail, he sent them directly south to Cofitachequi. These men were Fernan Gomez, Pedro Alonso, Luís Jimenez, and Pedro de Barcena. With the help of Indian burden bearers, they carried some munitions with them.

On January 12 Pardo and the remainder of his company set out for Ylasi. They traveled 4 leagues per day for a total of five days before reaching Ylasi on January 16. Their journey was to the southeast through uninhabited country, through present-day Lancaster and Chesterfield counties, following a trail that probably ran near State Highway 9 for most of its distance. Ylasi was in the vicinity of Cheraw, South Carolina. Because it rained very hard, Pardo and his men remained in Ylasi for four days. During this time they met with several Indians who came in to hear Pardo's usual speech.

On January 21 Pardo and his company departed from Ylasi and headed toward Cofitachequi. As they set out, they had to wade across a swamp about a league wide (3.46 miles), with water almost to their knees, and sometimes higher. This was probably the lower course of Thompson Creek, backed up by the rain-swollen Pee Dee River. Because the water was covered with ice, many of the soldiers suffered injuries while wading through it. Even though they had to cross this and other swamps, they were still able to travel 5 leagues on this day. They bivouacked at an uninhabited place, probably between the present towns of Patrick and Middendorf. The next day they traveled 6 leagues, reaching the village of Yca, which was probably located on Lynches River or Little Lynches River, near the present town of Bethune.

The following morning, presumably after they had traveled for 3 leagues or so, Pardo again split up his company. He sent a corporal with twenty men and twenty Indians southward to Guiomae, probably following a trail that ran near the present Southern Railroad line. When this contingent reached the vicinity of Guiomae, they must have crossed the Wateree River in canoes. In Guiomae, they were to sack up a quantity of corn. Pardo commanded Moyano to go along with them and to take this corn to Coçao and guard it until the others arrived. Pardo and his men then continued along the remaining 2 leagues to Cofitachequi, arriving there on January 23.

The cacique of Cofitachequi was absent when Pardo arrived. When the cacique showed up on January 26, Pardo ordered him to fetch several canoes in which to load corn from the storehouse at Cofitachequi so that it could be taken downstream to Guiomae. Pardo commanded Sergeant Pedro de Hermosa to go with three soldiers in the canoes, presumably along with the Indians, and to store the corn in the Spaniard's storehouse in Guiomae. Here they would meet with some of the soldiers who had already been sent ahead.

Precisely what Pardo did after January 26 is not clear, but he remained about two weeks in Cofitachequi. Probably he spent this time building yet another fort—Fort Santo Tomás. On February 11, Pardo, in the company of his men and some Indians, arrived in Guiomae with several sacks of corn from Cofitachequi. On February 12, Pardo called together his company along with some of the Indians who had been traveling with them, and he dismissed almost all of his Indian interpreters.

They probably departed from Guiomae on February 12 and arrived at Aboyaca on February 14. The chief came out to meet them, giving them some meat and a quantity of wild roots, which the Spaniards called "sweet potatoes" (*batatas*). Clearly, they wanted to conserve the corn they were carrying for the people at Santa Elena.

On February 15 they had a particularly hard day because they had to carry their burdens across many swamps, including three that were very deep and wide: the South Fork of the Edisto River, Little Salkehatchie River, and Salkehatchie River. Miraculously, none of the men stumbled and fell while crossing these swamps, nor did any of them drop a load they were carrying into the water. Even with all of these difficult crossings, they claimed to have traveled 7 leagues on this day, bivouacking at the end of the day in an uninhabited place. They were at this time probably near present-day Ulmer, South Carolina.

On February 16 they reached Coçao, where again the Indians fed them wild roots together with meat and acorns. Here Pardo was met not only by Moyano, whom he had sent ahead, but also by the chief of Orista and thirty of his Indians, and by the chief of Uscamacu and some of his Indians. Pardo was relieved to find them waiting there, because he had heard rumors that Coçao, Orista, and Uscamacu had rebelled against Spanish rule. The Indians brought with them baskets and deerskin sacks, and loaded up 60 bushels (*fanegas*) of corn that had been stored at Coçao. The Spaniards departed immediately afterward and traveled 2 leagues before stopping for the night. During the night, Pardo took the precaution of posting a heavier guard than usual to protect the corn.

On February 17 they traveled 5 leagues further, camping for the night in an uninhabited place. They were at this time in the vicinity of present-day Early Branch, South Carolina. The next day they arrived at Ahoya, which was 7 leagues from Santa Elena. Beyond Ahoya, one had to travel by water.

Pardo intended to build a house and a fort at Ahoya, but when he arrived there he learned that the people of Ahoya had killed a Spanish corporal. Evidently the Indians had rebelled when commanded to ferry people and supplies back and forth to Santa Elena in their canoes. The village of Ahoya had been burned, the Indians had fled, and the cacique had been imprisoned in Santa Elena. The cacique of Orista then asked or consented to have the fort built in his village, where he promised to keep canoes in readiness for the Spaniards.

Eventually some of the *mandadores* and Indians of Ahoya came to meet with Pardo. He ordered them to go and tell their people that they should return to their village and live in it. If they did not do so, they would be considered traitors, and the Spaniards would make war on them. Pardo gave them gifts, and they promised they would go and tell their people what Pardo had said.

On February 19, Pardo and his company departed from Ahoya and carried the bags of corn on their shoulders for about a league to a point where they could be loaded into canoes that had come from Orista. They probably met these canoes somewhere on the Coosaw River. Enough canoes were available to carry all of the corn and all of Pardo's company. At about two o'clock in the afternoon they landed within a quarter of a league of Orista, which was located about 5 leagues from Santa Elena.[65]

When they arrived at Orista, they put the corn they had been carrying into a large building that "belonged" to the chief.[66] The next day, on February 20, Pardo ordered his men to build a stronghouse, and they began to cut and saw the wood that would be needed. Construction was completed on March 2. Pardo named the fort Nuestra Señora, and he named the town Buena Esperança. He commanded Corporal Gaspar Rodriguez to serve as warden of the fort, with twelve soldiers in his company.

On February 28 Pardo had to take measures to stretch the food supply because conditions were becoming desperate at Santa Elena. Pardo may have exaggerated somewhat when he claimed that there was not enough food in Santa Elena to support fifty men for two months, and he alone had a hundred men in his company at Santa Elena. He commanded Sergeant Pedro de Hermosa to take thirty soldiers to Fort Santo Tomás at Cofitachequi and to remain there as warden of the fort.

He sent three others to Guando Orata, who had a house full of corn for the Spaniards, and asked them to try, with the help of the Guando Indians, to bring as much of this corn as they could to Buena Esperança. Moreover, they were to try to persuade these Indians to sow a large quantity of corn the next year. Guando must have been located east of Orista on the coast.

On March 1 several chiefs came to Buena Esperança bringing food—Yetencunbe Orata, Ahoyabe Orata, Coçapoye Orata, and Uscamacu Orata, and some of their leading men. Pardo gave them gifts, and he assembled the chiefs and some of their leading men in a large house, where they customarily met. He asked them to treat well the soldiers he was leaving in the stronghouse and to sow for them a large field of corn the next year. He instructed them all to obey Orista Orata, whom they should regard as their chief. He again stressed that the people of Ahoya who had fled their village should now return. Pardo said that he would return to Orista in twenty days, and the Indians present should see to it that all the people be present on that occasion, including those of Ahoya. Pardo would then punish those who were at fault in the rebellion, whereas he would consider all the others to be his brothers. The next day, on March 2, in canoes that the Indians provided, Pardo and his men made their way from Orista to Santa Elena. They arrived at Santa Elena at about three o'clock in the afternoon.

On this second expedition, Pardo had spent six months exploring the interior. He had, of course, failed to reach the silver mines in Zacatecas, but on the other hand he had some notable achievements to his credit. He had led a detachment of over a hundred soldiers on foot all the way from Santa Elena to the upper Tennessee Valley. He had met with Indian chiefs all along the way, getting from them promises of obedience (never mind how sincere) and giving them presents. He discovered a location that seemed to contain silver ore (though in fact it did not), and he located three deposits of crystals. In addition to the fort he had built at Joara on his first expedition, he built four small forts and a larger one at Guatari, stationing detachments of soldiers at all of them. He managed to do all this without losing a single man, although he would undoubtedly have suffered casualties if he had not turned back at Satapo. The only Indian casualties were those who were killed and injured when Moyano attacked the Chiscas.

Pardo did his part to implement Menéndez's plan for occupying the interior of *La Florida*. The plan itself, however, was foolhardy. The forts were too small and too far from the coast. The Indians had no reason to submit to the demands of the soldiers who manned the forts once the main body of the Spaniards had left. Few of these unfortunate men would ever see Santa Elena again.

Notes

1. Eugene Lyon, *The Enterprise of Florida: Pedro Menéndez de Avilés and the Spanish Conquest of 1565–1568* (Gainesville: University Presses of Florida, 1976), p. 166.

2. There are three short accounts of the Pardo expeditions and one long account. The short accounts are (1) Pardo's narrative (apparently from memory) of both of his expeditions, published in Eugenio Ruidiaz y Caravia's *La Florida, su Conquista y Colonización par Pedro Menéndez de Avilés* (Madrid, 1894); (2) a highly condensed précis of what appears to be a secondhand account of the first expedition by a soldier, Francisco Martinez, also published in Ruidiaz's *La Florida*; and (3) a brief account of the second expedition by Juan de la Bandera, the official notary and scribe for that expedition. This short account by Bandera was published by Buckingham Smith in his *Colleción de varios Documentos para la Historia de la Florida y Tierras Adyacentes* (London, 1857), and it was later included in Ruidiaz's *La Florida*.

These three acounts were translated into English and edited by Herbert E. Ketcham in "Three Sixteenth-Century Spanish Chronicles Relating to Georgia," *Georgia Historical Quarterly* 38(1954):66–82. They were also translated by Gerald W. Wade and edited by Stanley J. Folmsbee and Madeline Kneberg Lewis in "Journals of the Juan Pardo Expeditions, 1566–1567," *East Tennessee Historical Society's Publications* 37(1965):106–121. All three of these are included in Part II of this book in new transcriptions and translations by Paul E. Hoffman. Hereafter Hoffman's translations will be referred to as *Pardo*, *Martinez*, and *Bandera I*. Because all three of these documents are quite short, page numbers will be omitted from citations in the footnotes that follow.

The fourth document, Juan de la Bandera's long and detailed account of the second expedition, is also included in Part II of this book. It has been transcribed and translated by Paul Hoffman and it is published here for the first time. Hereafter it will be referred to as *Bandera II*. All references to this document will cite the pagination of the original.

The events recounted in this chapter generally follow the routes that were reconstructed in Chester B. DePratter, Charles M. Hudson, and Marvin T. Smith, "The Route of Juan Pardo's Explorations in the Interior Southeast, 1566–1568," *Florida Historical Quarterly* 62(1983):125–58; and in Charles Hudson, "Juan Pardo's Excursion beyond Chiaha," *Tennessee Anthropologist*, 12(1987):74–87. In the present work the probable locations of some Indian towns on the routes of exploration have been slightly changed from the locations given in the article by DePratter, Hudson, and Smith.

3. This direction and route is based on the assumption that Pardo followed the same trail on both the first and second expeditions.
4. Pardo.
5. That Pardo considered the Indians in the area from Santa Elena to Guiomae to be of little interest is also evident in the account of the second expedition in the Bandera II document, folio 2.
6. Pardo spells it *Ysa*.
7. The location of this town is unknown.

8. Here Pardo spells this *Juada*.
9. Sebastian Montero was probably one of those who attempted to teach Christianity to the people of Joara. See infra.
10. Pardo; Martinez; Bandera II, folio 2.
11. Pardo evidently had a lapse of memory when he said that he went from Quinahaqui to another town, also on the river, where he stayed for two days. His first stop was probably at Guaquiri, which may not have been on the Catawba River, and his second stop was probably at Quinahaqui, which was definitely on the Catawba River (see infra).
12. Pardo. The Bandera II document incorrectly says that this letter arrived when Pardo was at Joara (see folio 2). He *was* in the general vicinity of Joara, and perhaps that is what was meant.
13. Michael V. Gannon, "Sebastian Montero, Pioneer American Missionary, 1566–1572," *Catholic Historical Review* 51 (1965–66), pp. 335–53.
14. The root *yatika* meant "speaker" or "interpreter" in several Southeastern languages. Hence, *Otariyatiqui* may have been the Otari interpreter, and the town in question may have been the town of Otari (see infra), or else a subsidiary town of Otari.
15. Pardo spells it *Aracu[c]hilli*.
16. Bandera II, folio 2.
17. Chester B. DePratter, Charles M. Hudson, and Marvin Smith, "The Hernando de Soto Expedition: From Chiaha to Mabila," in *Alabama and the Borderlands*, ed. R. Reid Badger and Lawrence A. Clayton, (University, Alabama: University of Alabama Press, 1985), p. 111.
18. Francisco Martinez says fifty houses were burned and a thousand Indians were killed. Surely this is an exaggeration.
19. Charles W. Arnade, *Florida on Trial, 1593–1602* (Coral Gables, Fla.: University of Miami Press 1959), p. 40.
20. Jaime Martinez, "Brief Narrative of the Martyrdom of the Fathers and Brothers of the Society of Jesus, slain by the Jacan Indians of Florida," comp. Ruben Vargas, trans. Aloysius J. Owen, *The First Jesuit Mission in Florida*, U.S. Catholic Historical Society Records and Studies, vol. 25 (New York, 1935), p. 143.
21. Francisco Martinez, p. 117. Francisco Martinez's account of this incident implies that a second letter from Moyano reached Santa Elena. This second letter had to have arrived at Santa Elena on or before July 11, 1567, the date of Martinez's account.
22. John G. Varner and Jeannette J. Varner, *Dogs of the Conquest* (Norman: University of Oklahoma Press, 1983).
23. Martinez. The Spanish reads: *"una pequena puerta con sus traveses."*
24. Ibid.
25. Roy S. Dickens, Jr., "Preliminary Report on Archaeological Investigations at the Plum Grove Site (40Wg17), Washington County, Tennessee," Georgia State University, 1980.

26. Arnade, *Florida on Trial*, p. 40.
27. Martinez mistakenly says that the island lay between two very large rivers.
28. Martinez may exaggerate in saying that 3,000 warriors were inside the town. Still, this is close to the number of warriors who assembled at Mabila to mount a surprise attack against Soto.
29. Lyon, *Enterprise*, p. 182.
30. Bandera II, folio 1 verso.
31. Unless otherwise noted, this account of Pardo's second expedition is based on information in the long Bandera document (Bandera II).
32. Felix Zubillaga, *Monumenta Antiquae Floridae (1566–1572)* (Rome: Monumenta Historica Soc. Iesu, 1946), p. 326. Here spelled *Escamacu*.
33. Bandera I. This deposit of clay could be an aid in locating precisely the site of Uscamacu.
34. Ibid.
35. Hudson et al., "Apalachee to Chiaha," p. 72.
36. Pardo says they rested two days.
37. Hudson et al., "Apalachee to Chiaha," pp. 72–73. Also see Steven G. Baker, "Cofitachique: Fair Province of Carolina," M.A. thesis, University of South Carolina, Columbia, 1974, *passim*.
38. Bandera I.
39. Ibid.
40. Pardo.
41. Bandera I.
42. Hudson et al., "Apalachee to Chiaha," p. 73.
43. Pardo.
44. Pedro Menéndez Marques to Pedro Menéndez de Avilés, Havana, March 28, 1568, AGI, SD115, fol. 206. I am grateful to Paul Hoffman for this reference. The letters say the man caught up with Pardo at Chisca, but it had to have been at Joara.
45. Bandera I.
46. An archaeological survey in and around the town of Marshall, North Carolina, failed to turn up any evidence of a late prehistoric occupation. David Moore, personal communication.
47. Hudson et al., "Apalachee to Chiaha," p. 74.
48. Bandera I.
49. Richard Polhemus has conducted excavations at a site near the confluence of the Pigeon and French Broad rivers that could have been Tanasqui. Thus far, however, the archaeological evidence is inconclusive.
50. Pardo.
51. Also spelled *Olameco*, *Lameco* (Pardo), and *Solameco* (Bandera I).
52. Pardo.
53. Gentleman of Elvas, "Narrative," in *Narratives of De Soto in the Conquest of Florida*, trans. Buckingham Smith (New York: Bradford Club, 1866), p. 70.
54. DePratter et al., "Pardo's Explorations," pp. 145–46.

55. When my colleagues and I wrote our initial paper on the Juan Pardo expeditions (1983), we were not confident that our reconstruction of Pardo's activities beyond Chiaha was correct. While our paper was in press I worked out the reconstruction that is summarized in this chapter. See Charles Hudson, "Juan Pardo's Excursion beyond Chiaha."
56. Ibid. In his shorter account Bandera calls this stone *humo de metal*. It may be relevant that in the town of Zacatecas, where the richest silver strike in the New World had been made, many of the houses were built out of a reddish stone. P. J. Bakewell, *Silver Mining and Society in Colonial Mexico. Zacatecas 1546–1700* (Cambridge: Cambridge University Press, 1971), p. 49.
57. Bandera I.
58. Pardo, who learned of the conspiracy when he first reached Chiaha, remembered the conspirators as *Chisca*, *Carrosa*, and *Costeheycoza*. The last name is obviously a mistranscription of *Costehe y Coza* (Pardo). Bandera, who wrote down statements and depositions at the time the events occurred, is more to be trusted.
59. William E. Myer, "Indian Trails of the Southeast," *42nd Annual Report of the Bureau of American Ethnology* (Washington, D.C.: Government Printing Office, 1928), pp. 749–51.
60. Bandera II, folio 18 and folio 18 verso.
61. W. F. Wilson and B. J. McKenzie, *Mineral Collecting Sites in North Carolina*, North Carolina Department of Natural Resources and Community Development, Information circular 24 (Raleigh, 1978). pp. 29–30, 42–43, 66.
62. It is perhaps noteworthy that in his account of his travels Pardo omits all mention of the discovery of these crystal deposits (Pardo). In fact, he says that he spent four days in traveling from Joara to Guatari. Counting the detour and stayover at Yssa, he actually took 22 days to travel from Joara to Guatari.
63. In the text this is spelled "Sauapa," almost certainly a mistranscription of "Sanapa."
64. Bandera I.
65. Zubillaga, *Floridae*, p. 399.
66. An incident that occurred in July of 1568, when Pardo took Father Juan Rogel to visit the village of Escamacu, may help pinpoint the location of Orista. When they arrived at Escamacu they found that the soldiers who were stationed at the stronghouse at Orista were mistreating the Indians. During the night, which they spent at Escamacu, some women at Orista began yelling to the people of Escamacu to come and help them because they were being abused by the Spanish soldiers who were stationed there. This means that Orista and Escamacu were within shouting distance of each other, or nearly so, on either side of a body of water, probably Whale Branch. If either of these sites can be located, it should be an easy matter to locate the other. See Zubillaga, *Floridae*, p. 326.

3
The Indians

Juan Pardo's second expedition opens a window through which one can glimpse the Indians of the Carolinas and Tennessee in the sixteenth century. Hernando de Soto visited many of these same Indians in 1540. But Soto was intent on the discovery of riches, and if riches could be discovered, he was intent on conquest. When persuasion failed, he was quick to use force, and if force did not suffice, then he used brutality and terrorism. In general, the chroniclers of the Soto expedition only mention individual Indians if they were notable for their political power or valor. None of the Soto chroniclers had any reason to collect information systematically. But Pardo's mission into the interior of the Southeast was on quite a different footing. He was under orders to cultivate the friendship of the Indians and to evangelize them. Moreover, because Pardo was to take possession of the land, Menéndez ordered the notary, Juan de la Bandera, to make a formal record of all that he witnessed. This Bandera did, writing down a series of records in repetitive legalistic language, with each record signed by witnesses who were under oath.

As a piece of historical literature, these records—which constitute the long Bandera document—cannot stand comparison with the chronicles of the Soto expedition, neither with the florid prose of Garcilaso de la Vega nor even with the sober prose of the Gentleman of Elvas or Rodrigo Ranjel. But the Bandera document contains information on the Indians that was to some degree systematically collected. In many instances Bandera tells how many leagues Pardo and his men traveled on a particular day; he reports on the character of the landscape; and occasionally he tells in what direction they traveled. When Indians came to have an audience with Pardo, Bandera almost always lists the names of the most important ones. And when Pardo gave gifts to the Indians, Bandera tells what they were—each axe, wedge, chisel, bead, button, each piece of red or green taffeta. From this information one can sketch the outlines of a number of native polities, some of which were quite large and complex.[1] These were people who, after their population plummeted in the late sixteenth century and throughout the seventeenth century, coalesced to form the far simpler and less numerous Creeks, Cherokees, Yuchis, and Catawbas of the eighteenth and nineteenth centuries.

Most of Pardo's dealings with the Indians were highly formalized. In each of the towns Pardo visited, he would assemble the headman or chief of the town along with other tributary or neighboring chiefs, and he would then deliver a standard address to them. He asked them whether they would like to become Christians and give their obedience to the pope and the king of Spain, and whether they would like to have a Christian sent to instruct them. Guillermo Rufín translated what Pardo said into the language he had learned from the Indians of Orista. This was either the language spoken by the Orista Indians or else a Creek lingua franca.[2] In some cases, Rufín's translation was wholly or partly intelligible to the Indians with whom Pardo dealt, but in other cases Indian translators who understood the Indian language that Rufín spoke would then translate it into quite different languages. In every case, according to Bandera, the assembled Indians after hearing Pardo's address would indicate their obeisance by saying *yaa*, and when they agreed enthusiastically, they would say *yaa* repeatedly. According to Bandera, *yaa* meant "I am content to do what you command me to do."[3] His speech finished, Pardo would then formally take possession of the land. In addition, he generally told the Indians that they were to lay up a supply of corn for the Spaniards' exclusive use, and that they were to remove none of this corn except by order of the king of Spain or one of his deputies. To seal these agreements and alliances, Pardo gave presents to the chiefs and to other important Indians, and in most instances, Bandera made a record of what gifts were given to whom.

The Mississippian Transformation

The sixteenth-century Spanish explorers encountered the Indians of the Southeast at a most interesting juncture in their social evolution. Only a few centuries before the Spaniards arrived, the Indians of the Southeast were caught up in a series of changes that fundamentally altered their way of life. Archaeologists, who refer to the new way of life as the Mississippian culture or tradition, at first defined it in terms of certain material traits, such as the manufacture of pottery tempered with ground-up mussel shell and the construction of truncated pyramidal mounds and large towns. Other traits, including symbolic motifs and presumably ideas that were associated with them, were at times widespread, if not universal among Mississippian peoples.

In time, however, such a definition of the Mississippian phenomenon was seen to be faulty. Archaeologists learned that in some places the Indians had begun manufacturing shell-tempered pottery several centuries before they began building truncated pyramidal mounds or large towns. In other places, they never did temper their pottery with shell, but rather with sand, grit, or soapstone. Even though the initial definition of the Mississippian phenomenon began to wear thin, and then to fall apart, it remained clear that after about A.D. 800 a big change occurred that caused the native people of the late prehistoric Southeast to become more similar to each other.

The Mississippian phenomenon—marked by a pervasive and progressive sameness among Indians over a large area of the Southeast—was not so much a "culture" as a fundamental economic and social transformation. A necessary but not sufficient condition for this transformation is that the people of the Southeast had started to cultivate corn for a substantial part of their diet. This was accompanied by an apparent increase in population. Once they became dependent upon corn, they had to cultivate it whether they wanted to or not. By A.D. 800, native people of the eastern United States had already been practicing horticulture for many centuries, cultivating such native plants as marsh elder, chenopodium, sumpweed, maygrass, knotweed, and sunflower. They had domesticated several of these plants, and in some places their societies had attained considerable complexity. But none of these cultigens possessed the transforming potential of corn. Corn is a demanding plant. It not only draws heavily on soil nutrients, but it requires a heavy investment of labor if one is to enjoy a high yield. This meant that the Indians had either to utilize extremely rich and scarce soils, such as alluvial bottomlands (particularly levee ridges), which could withstand repeated cultivation, or else they had to invest labor in constantly opening up new fields by means of slash-and-burn cultivation. No matter what soils they used, they had to labor long and hard in preparing the fields, hoeing the weeds down, keeping birds and vermin away until the corn was harvested, and protecting it after it was harvested.

With dependency on corn horticulture established, the most crucial as-

Figure 8. Social honor accorded a sixteenth-century Timucuan chief. Engraving by Theodore de Bry, after a painting by Jacques Le Moyne de Morgues. In this and other scenes painted by Le Moyne (and engraved by de Bry) it should be noted that he probably painted from memory and that he makes use of stylized European artistic conventions. But he was an eyewitness observer, and his pictures contain factual information. (Photograph courtesy of the Smithsonian Institution.)

pect of the Mississippian transformation was that it entailed a fundamental change in social structure. Before the transformation, most of the people of the Southeast were organized into segmentary or "tribal" societies, in which one's fundamental allegiance was to a group of kinsmen. But with the transformation they became increasingly organized into chiefdoms, in which one's allegiance was to a chief, whose position was recognized through deference, payments in the form of tribute or labor, and/or military service. A segmentary or "tribal" society consisted of a number of kin groups, all of equal status. The members of a chiefdom also belonged to kin groups, but were subordinate to the chief, together with his kinsmen and retainers, in terms of power and prestige (see figure 8). Moreover, a distinct chain of command existed between the common people and their chief. The actual power a chief possessed varied, so that a chiefdom could be weakly or strongly stratified, and the power of a particular chief could increase or decrease during his or her lifetime.

What caused the first Mississippian chiefdoms to emerge in the Southeast is still a topic of debate, but as already stated, the cultivation of corn was a necessary but not sufficient condition. Mississippian chiefdoms did not develop in the Southeast where corn was not cultivated. The non-agricultural Calusa of southern Florida were organized into a chiefdom, but none would say that theirs resembled the Mississippian chiefdoms to the north. From all appearances, the intensive cultivation of corn (and probably dependency on corn) and organization into Mississippian chiefdoms occurred simultaneously in some places, such as southeastern Missouri and northeastern Arkansas.[4] In other places the cultivation of corn did not automatically or quickly lead to organization into chiefdoms. The Iroquoian- and Algonkian-speaking people of the northeastern United States were cultivating corn at the time of earliest European contact, but they do not appear to have ever been organized into chiefdoms like those in the Southeast. (Only the Powhatans had what could be called a chiefdom, and it was short-lived.) Indeed, there is evidence that the Southeastern Indians cultivated corn on a small scale for several centuries before the Mississippian transformation.[5]

Whatever precipitated this change to a more centralized social structure, chiefdoms are militarily more potent than less centralized societies (see figure 9). Consequently when a society in the Southeast organized itself into a chiefdom and began flexing its military muscles, its various neighbors would have had two choices. They could succumb and become tributary to it, in which case they would soon become familiar with the workings and potentialities of the social order of the chiefdom, or else they could form their own chiefdoms and launch a military resistance. Either way, there was an adaptive advantage in organizing society into a chiefdom, which would explain why this form of social structure spread rapidly throughout the Southeast to usher in the Mississippian era.[6]

Once a society organized itself into a chiefdom, its social dynamics were fundamentally changed both within and without. No longer was the society an acephalous entity composed of kin groups of equal status. It took on a stratified structure in which the chief and his kinsmen stood above all other kin groups. Since succession to the office of chief appears to have been based on kinship, it would have been in the interest of a chief and his kinsmen to heighten and enhance the difference between themselves and commoners. Historical evidence from the sixteenth- and seventeenth-century Southeast shows that one device used to reinforce this difference was a mythological charter, which held that the chief and his blood kin were direct descendants of the sun, a principal upper-world deity in the belief system of many—perhaps all—Southeastern Indians.

Figure 9. Formal order of march assumed by the Timucuan chief Holata Outina embarking on a military expedition. Engraving by Theodore de Bry after a painting by Jacques Le Moyne. (Photograph courtesy of the Smithsonian Institution.)

Thus substructure mounds may have been built so that the chief could live and conduct his affairs above the mundane level of the everyday and the ordinary, as befitted a person of the upper world. Both public buildings and the residences of chiefs were built on top of truncated pyramidal mounds (see figure 10). Periodically these mounds were enlarged, so that their size alone was an indication of a chief's claim to power.

The Mississippian transformation also affected the social dynamics between societies. Chiefdoms possessed a heightened capacity for conducting warfare, and the evidence from works of art and the defensive palisades built around towns suggests that one of the concomitants of the Mississippian transformation was intensified warfare (see figure 11). In a chiefdom, war can be waged not merely to avenge the killing of a blood kinsman, as is the case in kin-based societies, but on behalf of one's chiefdom, and this made it possibile for one chiefdom to subjugate its defeated enemies to tributary status. A chief who succeeded in doing this could then become a paramount chief, and his defeated rivals would then become subsidiary chiefs. Regardless of the way in which a chief might fall under the authority of another, nascent paramount chiefdoms were likely to be conflict-

Figure 10. The mound at the Town Creek site near Mt. Gilead, North Carolina. The building on top of the mound is a reconstruction. (Author's photograph.)

Figure 11. Defensive palisade around the Town Creek site. (Author's photograph.)

The Indians

ridden and prone to disintegration.[7] Their means of political and economic subjugation were simply too rudimentary to achieve long-term stability. Hence, one of the great organizational challenges to Mississippian people must have been to develop ways of overcoming or reducing this intrinsic instability.

If this reasoning is sound, then it follows that within the geographical area where corn horticulture came to be practiced, certain traits should have appeared seriatim: (1) first corn horticulture, then (2) dependency on corn, then (3) organization into chiefdoms, and finally (4) the elaboration of symbols of hierarchical status, including the construction of substructure mounds. (Flat-topped mounds built in the Woodland period appear to have been intended for purposes other than enhancing the power of a chief.) Moreover, these features should be most pronounced where the Mississippian transformation was oldest, and they should be rudimentary where the transformation was recent.

In the archaeological period that preceded the Mississippian transformation, there is no evidence of abrupt or forceful change. This was the Woodland period, which dates from about 500 B.C. to about A.D. 800. It is clear, however, that in the Woodland period—and even toward the end of the earlier Archaic period—the winds of change were beginning to blow. Woodland people primarily hunted and gathered wild food, but in some places they began practicing a kind of easy horticulture to supplement their wild foods. In their gardens they cultivated squash, a cultigen that had been domesticated in Mesoamerica, as well as several of their own cultigens, such as sunflower, marsh elder, chenopodium, knotweed, and maygrass. Toward the end of the Woodland period they began cultivating a little corn. Moreover, there are indications that their societies were becoming more complex, with greater differentiation of social status, and with influential leaders—"big men"—arising through personal ambition and achievement. Some of these Woodland societies may have just reached the chiefdom level of organization. They may have resembled the Calusa, a hunting and gathering society of south Florida, which was dominated by either a big man or a chief in the later half of the sixteenth century.[8]

Social differentiation or favoritism is most clearly to be seen in Woodland mortuary practices. Some individuals were buried with far more elaborate grave goods than others were. Also, some few individuals were buried in or near earthen mounds which appear to have been built as monuments for the dead. In some places Woodland people began heaping up earth to build other kinds of mounds, some in the shapes of animals and others in various geometric shapes. And some Woodland people became unusually interested in exotic materials from faraway places, such as obsidian, mica, grizzly bear teeth, marine shells, sharks teeth, and so on.

Whatever the meaning or importance of these changes during the Woodland period, they do not appear to have entailed a fundamental economic and social transformation. Indeed, many Woodland changes were only elaborations of elements that were already present in the Late Archaic. Moreover, the centers of these Woodland cultures were in the Mississippi and Ohio valleys. For whatever reason, few of these more elaborate Woodland developments reached the areas which Pardo later explored.

In the area Pardo explored, life appears to have gone on with no major changes or dislocations for many millenia. But after A.D. 800, with the first occurrence of the Mississippian transformation in the Mississippi Valley, momentous forces for change were unleashed. It is probably safe to say that the Mississippian transformation was the most abrupt and difficult change the Southeastern Indians experienced until the coming of the Europeans.

The earliest signs of the Mississippian way of life appeared between A.D. 700 and 800 in the Cairo Lowlands of southeastern Missouri.[9] From here the impetus for change spread to other parts of the Southeast. Although it is difficult for archaeologists to reconstruct how the Mississippian way of life spread to other areas, it is reasonable to think that it involved both movements of population as well as the stimulation of indigenous populations to adopt new cultural and social forms.[10] The earliest appearance of the Mississippian way of life near the areas explored by Pardo was about 300 years after it first appeared in the central Mississippi Valley: By about A.D. 1000, Mississippian societies were present on the Upper Tennessee River and in northwestern and central Georgia.

But, according to archaeological evidence, it was not until a century or century and a half later (circa A.D. 1100) that Mississippian societies existed in the Appalachian highlands and on the Savannah River.[11] And it would be another 50 to 100 years (circa A.D. 1200) before the Mississippian way of life spread eastward to the Wateree and Pee Dee rivers.[12] Thus by A.D. 1300 there were a series of Mississippian societies whose principal towns were in the area of rich soils around the Fall Line zones of the Flint, Ocmulgee, Oconee, Savannah, Catawba-Wateree, and Pee Dee rivers. The Pee Dee River marked the easternmost penetration of the Mississippian way of life.

What the people in all these areas had in common about A.D. 1300 was a new mode of economic production and social organization. But these societies differed one from another in many cultural details, as attested by archaeological evidence. Thus, by A.D. 1300 the people who lived on the upper Tennessee and lower French Broad rivers possessed a culture that archaeologists term "Dallas" (see figure 13). Those in the Blue Ridge Mountains and adjacent Piedmont and Ridge and Valley areas possessed a culture termed "Pisgah." Yet another culture, "Lamar," existed after A.D. 1400 in several variants or "phases" from northern Alabama across Georgia to the Atlantic Coast and to the Pee Dee River in South Carolina. For the most part, these archaeological cultures and phases are defined by the kind of pottery the people made, but other aspects of material culture are also used to distinguish them.[13]

As we have seen, chiefdoms do have centralized leadership in the person of a chief, and the chief and his or her relatives may be set apart and above other people in the chiefdom. But chiefs possessed no more than rudimentary means of sanctioning or repressing their people, so that typically they reigned more than they ruled.[14] This would seem to be the reason why chiefdoms are politically unstable, and archaeological evidence indicates that instability was a fact of life in Mississippian chiefdoms. A persistent Mississippian pattern is that towns that were the centers of chiefdoms rose in power, then declined, and in some instances were abandoned altogether, while new centers of power arose elsewhere.[15]

Shortly before the arrival of the Europeans, a social collapse had occurred to the west of the area explored by Pardo, and it clearly had a profound long-term effect on the Indians on all sides. After about A.D. 1400 the Fall Line zone of the Oconee, Savannah, and Wateree rivers was inhabited by Mississippian people whose cultures were Lamar variants. The people on the Oconee River were Duvall phase, those on the Savannah were Rembert, Hollywood, and Irene phases, and those on the Wateree and Pee Dee rivers were Pee Dee phase.[16] Then, around A.D. 1450, most of the territory along the Savannah River was abandoned. Except for the headwaters of the Savannah, the country on both sides of the entire length of the river became a wilderness.[17]

Subsequently, the cultures of the people on the Oconee and Wateree rivers gradually changed. On the Oconee River, Duvall phase changed into Iron Horse phase, and this in turn changed into Dyar phase.[18] On the Wateree River, Pee Dee phase changed into Mulberry phase.[19] On the upper Savannah, Rembert changed into Tugalo phase. Hernando de Soto encountered Indians who were representatives of Dyar and Mulberry phases during the course of his exploration in 1540. But the central and lower Savannah River was devoid of people. The Savannah lay in the middle of what the Soto chroniclers called the "wilderness of Ocute," which measured over 130 miles from side to side.[20]

The causes of the demise of the chiefdom on the Savannah River are not known at present, and perhaps they can never be known with certainty. But some of the historical consequences are known. Namely, when Hernando de Soto reached Ocute (Dyar phase) he was told that the people of Ocute had "forever" been at war with Cofitachequi (Mulberry phase) and went about constantly armed in case they should be attacked. Such sustained and bitter enmity must have had a powerful effect on the social destiny of Cofitachequi, the paramount chiefdom Pardo became most familiar with.

Social Structure of Chiefdoms in the Carolinas and Tennessee

A detailed study of the documents of the Pardo expeditions along with those of the Soto and Luna expeditions makes it possible to reconstruct the broad outlines of the social structure of several sixteenth-century Southeastern chiefdoms. Previous research on Southeastern chiefdoms has generally been based on archaeological evidence alone, or else on archaeological evidence amplified by ethnographic parallels.[21] Archaeological information on chiefdoms necessarily covers a vastly greater span of time than historical information can cover, and it can in principle be obtained from the entire geographical region in which Southeastern chiefdoms existed, whereas early historical coverage is temporally thin and geographically uneven. Nonetheless, certain aspects of social phenomena are poorly "fossilized" in the kinds of information archaeologists can recover, and some aspects are not "fossilized" at all.

At the same time, the documentary evidence used by historians is adventitious. The fullness of a historian's account of the past depends on what gets committed to writing, and after that it depends on what written material survives. None of the sixteenth-century explorers was conducting sociological research on Indian societies. Moreover, by the time of the Pardo expeditions, the chiefdoms were already in decline, and some appear to have been in steep decline. Hence, this historical documentation does not lend itself to "scientific" inquiry. But by reading the documents closely, wringing out all the information they contain, one can gain valuable insights into the internal structure and operation of a number of chiefdoms as exemplified in actual instances of human behavior.

Consistent with the hiearchial organization that is typical of chiefdoms, three levels of political authority are discernible among the people with whom Pardo had dealings. From lowest to highest, the three levels were *orata*, *mico*, and grand chief (*cacique grande*), the last one being a position for which no Indian word was recorded by Bandera. The Spanish regarded both oratas and micos to be caciques—chiefs.

The orata appears to have been a village headman, or if not this, then the headman of the smallest social unit, however constituted. Even the tiny, ramshackle community of Aboyaca had a resident orata. In some cases the orata appears to have held sway over a single village, but in other cases he held sway over several villages. As we shall see, Yssa Orata had power over several subsidiary villages. These oratas almost always took the names of the dominant villages of their territory. Thus, the headman of Aboyaca was called Aboyaca Orata. The only exception was that the headman of Guiomae was called Emae Orata, but this may be no more than an instance of inconsistent perception or spelling of the same word.[22]

The next position in the hierarchy was known as mico, a title that continued to be used by Muskogean-speaking Indians into the nineteenth century, although its connotative meaning had probably changed. Perhaps the best English translation of mico would be "chief."[23] Pardo dealt with about eighty oratas whom Bandera names and as many as thirty-nine others whom he does not name. But Pardo met only three micos: Guatari Mico, Joara Mico, and Olamico. In a marginal note, Bandera defines a mico as a great lord (*un gran señor*), whereas an orata was a minor lord (*un menor señor*). In another place, when speaking of the mico of Guatari, Bandera says that she was above many lesser chiefs who were her vassals (*bassalos*).

Like oratas, micos generally took their name from the societies they governed. Hence, the mico of Guatari was Guatari Mico, and the mico of Joara was Joara Mico. An exception to be discussed later was Olamico, who was mico of the chiefdom of Chiaha.

Above the mico was a chief whom Pardo heard about but did not meet. When he was at Satapo, he learned that further along in the direction he was traveling there was a great chief who was called Coosa (*un cacique grande que se llamaba cosa*), and from all appearances this man was the moving force in the hostilities that were being organized there against Pardo and his men.[24] A similar phrase, *el señor el unico*, was used by Tristán de Luna's men to describe the chief of Coosa.[25] The best English term for this political position would seem to be "paramount chief." At the time of the Soto expedition the paramount chief of Coosa, whose main town was near present-day Carters, Georgia, clearly had power or influence over people to the northeast as far as about Newport, Tennessee, and to the southwest to about Childersburg, Alabama.[26]

The Spaniards' attempts to explain the meaning of these indigenous political positions was colored by their own conception of political power. *Señor* connoted the power to make judgments in legal proceedings and the power to make laws. *Señor* applied to anyone who possessed any degree of such power, from a person who would be addressed as "sir" to the king (*el rey*) himself. A Spaniard might have used the terms *un gran señor* and *un menor señor* to distinguish between a greater and lesser prince in Spain. Both of these princes would have submitted their power to Philip II, who would then have been *señor* to them. But Philip II, *el rey*, submitted to none.

It is notable that Bandera and others do not refer to Coosa as a king. Perhaps to them there was no other king than Philip II, or perhaps it was clear to them that even the most powerful Indian was less than the least European king. But it is possible that in characterizing Coosa as a *cacique grande* they were implying he submitted to no one, whereas oratas and micos submitted to him. None of the sixteenth-century Spanish sources gives an Indian name for the position of paramount chief. It is possible that the Indians had no word for this position. That is, the paramount chief of Coosa was called "Coosa," and it was simply understood that he was peerless.[27]

At the time of the Soto expedition, another paramount chief must have been residing at Cofitachequi. This is the impression one gets from the awed terms used by the young guide, Perico, whom Soto captured in northern Florida. Perico told Soto that Cofitachequi was governed by a female chief who lived in an astonishingly large town. He claimed that she had many tributaries and that her subjects paid tribute to her in gold and clothing.[28] This supposed grandeur was later confirmed by the people of Ocute, yet another paramount chiefdom centered on the Oconee River in central Georgia, who told Soto of their perpetual warfare with Cofitachequi.[29]

When Soto finally arrived at Cofitachequi, the paramount chief fled and took refuge in another town. When Soto sent a detachment to capture her, the Indian who was to guide them to her committed suicide by thrusting an arrow into his heart, presumably so that he would not be the one who would have to betray her.[30] This act suggests that a strong sense of loyalty and honor bound this man to his chief.

Soto hoped to discover another Inca or Aztec civilization at Cofitachequi, but he was soon disabused of this dream; moreover it soon became clear that the population of Cofitachequi had already been decimated by an Old World disease.[31] As Soto approached the first town they came to, the chief—the Lady of Cofitachequi—sent out her niece to meet Soto and his men. She was seated upon a litter being carried on the shoulders of her subjects (see figure 12).[32] But the houses in Talimeco, the ceremonial

Figure 12. The bride of a Timucuan chief carried on a litter. This scene resembles the treatment of chiefs witnessed by the Soto chroniclers at Cofitachequi and at Coosa. Engraving by Theodore de Bry after a painting by Jacques Le Moyne. (Photograph courtesy of the Smithsonian Institution.)

center of the chiefdom, were piled high with corpses, and grass was growing in the streets of some of the towns. The main temple, described in extravagant terms by Garcilaso de la Vega, was situated on top of one of the mounds in Talimeco. Within this temple, and nearby, the people of Cofitachequi kept the remains of their revered dead, as well as certain kinds of wealth, such as pearls and deerskins, and weapons of war. It was a place where ancient lineage, wealth, military might, and probably religious authority converged, and it was set apart on an earthen mound high above the level of common people and mundane events.[33]

If the decline of Cofitachequi began just before the Soto expedition, it had progressed quite far by Pardo's day. When Pardo arrived at Cofitachequi, or Canos, as it was also called, he found not even a mico, let alone a paramount chief. The highest authority appears to have been Canos Orata. Even so, Cofitachequi was still located in a central, easily accessible location so that Indians from far and wide congregated there to meet with Pardo and to hear his address, and to receive gifts from him.

The degree to which chiefs and paramount chiefs were set apart socially from their subjects is not clear from the Pardo documents. But social distance is implied in their custom of carrying prestigious people on litters. In some places the Indians would come out 4 to 6 leagues from their town to meet Pardo. As a mark of honor, they would carry him seated on a litter (*silla corriendo*) to their town. And as they drew near the town, Indians painted in many colors would come out dancing to meet the Spaniards.[34] Archaeological remains of litters have been found in high-status burials both at the Mississippian sites of Spiro in Oklahoma and at Cahokia in Illinois; hence, as a Mississippian ritual practice, being carried on a litter is known to be old in the southeast.[35]

Regrettably, the Pardo documents do not contain enough information to determine the nature of the relationships between a paramount chief and his tributaries, subjects, and allies. The documents do indicate that the largest political entities in the sixteenth-century Southeast consisted of at least two structural levels. For convenience, we can refer to the smaller of these as a "simple chiefdom" and to the larger as a "paramount chiefdom." A paramount chiefdom, such as Cofitachequi and Coosa, comprised several simple chiefdoms that were governed by micos and oratas. But because of the problems with the evidence discussed above, the term "paramount chiefdom" must be understood as a unit in which some of the constituent chiefdoms may have been more allied with a paramount chief than subject to him, more under his influence than under his power. Hence, "paramount chiefdom" is here used to mean a spatially large political entity that could be tightly or loosely integrated, and the degree of integration could vary with time.

Several other positions in the native social structure are either mentioned or implied in the Bandera document. When Pardo revisited Tocae on his return from the Tennessee Valley, he met with several headmen from mountain towns who had come to ask for details of where they were to take their tribute. Present at this meeting were two *ynahaes*, whom Bandera defines as being like "magistrates or jurymen, as it were, who are in charge of villages."[36] Also, when Pardo was handing out supplies at Olameco, he gave shoes and sandals to the more important Indians, including a pair of shoes to the *yniha* of Olameco, whom he says was like a "sheriff who commands the town."[37] It is probable that these men were similar in more than name to the *henihas* of the eighteenth- and nineteenth-century Creeks.[38] If so, then these *ynahaes* (or *yninas*) encountered by Pardo probably mediated more then they commanded.

None of the Soto chroniclers specifically mention *ynahaes*. However, when Soto reached Cofitachequi, he was met by six principal Indians, who were like "town magistrates." All were between forty and fifty years of age.[39]

The men whom Bandera identifies as *mandadores*, literally "commanders" or "drivers," were probably head warriors or war chiefs. It is probably significant that the first time that Bandera used this term was at Olameco and Satapo, the first places where the Indians began to show signs of resistance. But after this, Bandera notes them frequently, and the implication is that they stood next to the oratas and micos in social rank. On the return journey, Bandera mentions mandadores at Cauchi, Tocae, and at other places. Only one was specifically mentioned by name, the mandador of Yssa, who was called *Mati Yssa*.[40]

While Pardo was at Quinahaqui on his return journey, he was visited by an Indian—one Otariyatiqui Orata—who had come from Otari to see him. This man was clearly a *yatika*, a speaker or interpreter, a functionary who in the nineteenth century was still important among Creeks and other Southeastern Indians.[41]

In addition, Pardo gave gifts to certain individuals who were called *Indios principales*, "principal men." At least some of them held their positions by virtue of their descent from the chief's lineage. It is not possible to say anything more about them.

The Bandera document confirms the probable veracity of what is perhaps the most romanticized episode in the Soto expedition—namely, the discovery that the chief of Cofitachequi was a woman. In fact, Indians at some considerable distance from Cofitachequi knew that this was the case. Soto first learned that Cofitachequi was governed by a woman while he was still in Apalachee, although he heard it from the Indian boy Perico, whose birthplace is unknown. Soto heard it again when he was in Ocute.[42] There is no evidence that Pardo encountered a female chief of Cofitachequi, but it is clear that the chief of Guatari was a woman. In fact, Guatari had at least two principal cacicas, one a mico and the other an orata, and they possessed power over male caciques, who are specifically said to have been their vassals. Pardo gave gifts to two of the sons of these female chiefs. Moreover, although Joara had a male mico, Bandera mentions an old female chief (*la cacica bieja*) at Joara with whom Pardo had dealings. On his return journey, Pardo gave a gift to this woman's son, who was a principal man.

This new evidence from the Pardo expeditions does not prove beyond a reasonable doubt that being female was a prerequisite to holding one of these positions. One cannot rule out the possibility that these females held their positions because no qualified male was available. In fact, Garcilaso describes the chief of Cofitachequi as being a "widow," although if succession was matrilineal, as seems likely, the truth may be that she lacked a brother rather than a husband.[43] In any case, there is no doubt that women could be politically powerful in these chiefdoms on the Catawba-Wateree and Yadkin-Pee Dee rivers. It is also clear that in this area social status could be inherited matrilineally. In fact, the "Lady" of Cofitachequi sent out her niece (presumably her sister's daughter) to meet Soto when he arrived at the main town.[44] From the Soto expedition there is limited evidence that descent was also matrilineal in the Tennessee Valley. Namely, when Soto was at Chiaha, a chief of Chiaha informed him that his uncle was the actual chief, and that the uncle would govern until he, the nephew, came of age.[45] Also, in the chiefdom of Ocute, Soto encountered an old chief named Cofaqui, whose nephew governed for him.[46] Aside from the chiefs of Cofitachequi and Guatari, there is no evidence of female chiefs anywhere else in the Southeast. One must ask whether there is any significance in the fact that both of these female chiefs were to be found in relatively young chiefdoms on the eastern margin of the vast area in which Mississippian chiefdoms occurred.

Polities, Cultures, Languages

The paramount chiefdom Pardo became most familiar with was far from intact at the time of his visit. In 1540, when the Soto expedition first encountered Cofitachequi, it was a *nouveau arriviste* paramount chiefdom on the eastern margin of the vast area that was dominated by Mississippian chiefdoms. Just a few years before Soto arrived, Cofitachequi had been weakened by its first documented experience with Old World diseases. It was in steep decline in 1566–68 when Pardo and his men journeyed through the territory of this once potent chiefdom. It is probable that many, perhaps most, of the seemingly independent towns and polities Pardo encountered in the Coastal Plain and Piedmont had once been tributaries of Cofitachequi.

Pardo also encountered two small chiefdoms in the upper Piedmont: Joara, which had probably once been a tributary of Cofitachequi, and Guatari, which may have been an independent chiefdom. While crossing the mountains he encountered a number of Cherokee-speaking headmen who appear not to have been organized into a chiefdom, although it is possible that some of them paid tribute to Joara.

On the western side of the Blue Ridge Mountains Pardo encountered a paramount chiefdom, Coosa, that was still a going concern at the time of his visit. It may have been in decline, but it was still capable of organizing such a formidable military threat that Pardo found it expedient to cease his explorations and turn back.

Cofitachequi The documents of the Soto expedition give some idea of the northern extent of the power and influence of Cofitachequi, but indicate virtually nothing about its southern extent.[47] Soto took the niece of the chief of Cofitachequi hostage and forced her to go with him as he traveled northward. Indians all the way to the mountains paid deference to her. Even if they were not under her power, they were at least respectful of her. It is probably significant that she made her escape when Soto was traveling along the upper French Broad River, a few miles downstream from present-day Asheville, North Carolina. She may have felt at this point that she was in alien or enemy territory, or if not, she surely would have been if she had continued with the Spaniards as they crossed the mountains.

The principal towns of Cofitachequi lay on the eastern side of the Wateree River. West of the Wateree River, an uninhabited wilderness or buffer zone stretched from Cofitachequi (Mulberry phase) all the way to Ocute, a paramount chiefdom on the Oconee River (see figure 13). When Soto was preparing to cross this buffer zone, the Indians of Ocute warned him that there was no clear trail across it, and that his men would have to carry with them all the food that they would need. They themselves only crossed this area to carry out raids against Cofitachequi, when they traveled little-known, out of the way trails, carrying parched corn as their only food. When Soto finally crossed the buffer zone and arrived at the Broad-Congaree River, he found only some abandoned huts of fishermen or hunters.[48] If Soto had traveled from Ocute northeastward to the headwaters of the Savannah River, he might have encountered people living there. If he had traveled southeastward from Ocute he definitely would have encountered people living on the coast both west and east of the Savannah River. And if he had gone eastward, he would have encountered small populations living in the Coastal Plain east of the Savannah River, in such towns as Coçao and Aboyaca. Similar populations may also have lived in the Coastal Plain west of the Savannah River.[49] This means that the wilderness or buffer zone was widest in the Fall Line area.

Bandera's account of Pardo's second expedition indicates that Eastern Muskogean languages were spoken in the principal towns of the chiefdom of Cofitachequi, but these towns exercised power and influence over towns in which other languages, such as Catawban, were spoken. The etymology of *Cofitachequi* is uncertain. It is probable that *chequi* is Hitchiti *ciki*,

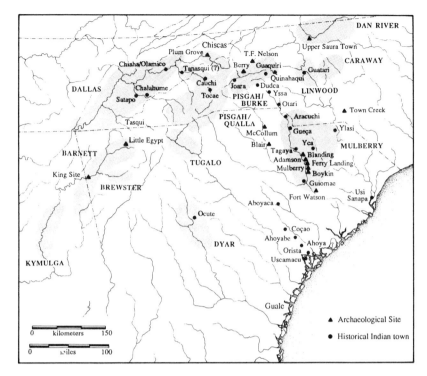

Figure 13. Some late prehistoric/early historic archaeological phases and sites, with selected historic Indian towns. The boundaries of the phases here depicted are both approximate and subject to revision.

"house, town," but the meaning of the first part of this word is unknown.[50] That the principal language of Cofitachequi was Eastern Muskogean is confirmed by the fact that at the time of the Soto expedition the main town, the ceremonial center, was *Talimeco*, which would seem to be Creek *talwa*, "town," compounded with *im-*, "its," and *mí·kko* "chief." This word, meaning "town chieftain" or "chief's town," designated the principal town of the chiefdom.[51]

The young Indian guide whom Soto captured in Apalachee called Cofitachequi *Yupaha*.[52] Presumably, this was the name used to refer to it by some other linguistic group, perhaps the Apalachees or a Timucuan-speaking group.

This Eastern Muskogean identification is strengthened by the fact that Cofitachequi was also called *Canos* in Pardo's day, and this name was probably a variant of *caney*, which Ranjel says is the word that referred to

The Indians

the residence of the chief of Cofitachequi at Talimeco.[53] This is probably the Creek word *kan-osi,* "little earth." More specifically, it was almost certainly related to the Muskogee word for mound, *ekun-hu'lwuce,* "the earth raised up in the air."[54] By the time of Pardo's expeditions, the chief's residence had probably deteriorated, but the mounds still existed.

The central towns of the chiefdom of Cofitachequi lay along the eastern bank of the Wateree River above and below present-day Camden, where there is a series of four documented mound sites, and probably others that are undocumented (see figure 13). The documented mounds are the Blanding site (38KE17) with four mounds situated about 4 miles northwest of Camden, the Adamson site (38KE11) with two mounds about 2 miles west of the center of Camden, the McDowell or Mulberry site (38KE12) with two large mounds and eight small mounds some $2\frac{1}{2}$ miles south of Camden, and the Boykin site (38KE8) with one mound about 3 miles south of the Mulberry site.[55] In addition, there are several large areas of village habitation without mounds in this area, particularly the Ferry Landing site, just downstream from the Adamson site, as well as a village area shown on William Blanding's map of 1845 at the mouth of Town Creek, between the Mulberry and Boykin sites.[56] None of these sites has been extensively investigated by archaeologists. Some excavation was done at the Mulberry site in 1952 by the University of Georgia and the Charleston Museum. The University of South Carolina is now conducting a long-term research program at the Mulberry site and at other sites in the Wateree Valley. With the exception of the Mulberry site, all the others are known only through surface collections.

The archaeological evidence from the Carolinas, while still incomplete, is consistent with the theory that the area occupied by the paramount chiefdom of Cofitachequi was located where a relatively late expansion of the Mississippian way of life occurred. It is quite possible that in places this development entailed an actual movement of people from west to east. In the Piedmont and Fall Line zone, starting at the central Savannah River, there is a tendency for Mississippian sites to be younger as they occur eastward to the Yadkin-Pee Dee River.[57] And judging from the most thoroughly excavated of these sites, Town Creek, these Mississippian occupations do not show continuity with earlier occupations, which implies that these people moved to Town Creek from elsewhere.

The indigenous people on whom these Town Creek pioneers impinged represented a very different cultural tradition. About the beginning of the Christian era, the Piedmont area of upper North and South Carolina was occupied by people who subsisted by hunting and gathering. They fashioned relatively large projectile points that were probably hafted onto darts and thrown by means of spear-throwers. They made a simple clay- and grit-tempered pottery that was characteristically decorated by impressions from paddles that had been wrapped with cord and coarse fabric. As time went on, the cultures of these people gradually changed. They made their pottery and tools in slightly different ways, but their basic mode of subsistence remained the same.[58]

Around A.D. 1000, a more fundamental change occurred. These people began to rely in part on corn agriculture for their subsistence. In the central Piedmont the new culture that accompanied this change has been termed the Uwharrie complex. The people of the Uwharrie complex began building villages—small clusters of houses along the margins of the larger rivers. The houses had circular floor plans, and they were probably made by inserting poles in the ground, which were then bent over to form a dome-shaped structure that was covered with bark and animal skins. Even though they cultivated corn, hunting—by this time with bow and arrow—remained an important source of food. They tempered the clay used in their pottery with crushed quartz, some of which consisted of quite large pieces. It was a simple ware—spherical bowls and jars with conoidal bases—that was still decorated by cord and fabric marking. In addition, a new technique of surface treatment came into vogue. The surfaces of some of the vessels were scraped smooth, and they were sometimes decorated by incising, usually with four to six lines beneath the rims of the vessels. This Uwharrie complex was quite homogeneous, and it was widespread, reaching northeastward into Virginia and southeastward to the upper Catawba and Broad rivers in South Carolina. Archaeological evidence does not indicate that these people were organized into chiefdoms.

The central and southern parts of South Carolina are less well known. But at a comparable time period, people in central South Carolina were making incised, check-stamped, and simple-stamped pottery. And at the Mattassee Lake sites along the lower Santee River people were making simple-stamped pottery between A.D. 810 and 1340. It is not known whether these people were cultivating corn, although by the end of this period they probably were.[59] Again, there is no evidence that these people were organized into chiefdoms.

By about A.D. 1100 the Indians along the Savannah River were cultivating corn. They had begun constructing mounds, and they are thought to have been organized into chiefdoms. They made a variety of complicated stamped pottery that archaeologists term "Savannah." People with a similar culture were probably living on the Broad River in South Carolina at about the same time. Construction could have begun on the McCollum Mound in Chester County and on the Blair Mound in Fairfield County at about this time. Construction of the Fort Watson mound on the lower Santee River, as well as some of the mounds in the Wateree Valley, probably began about A.D. 1200.

Then, about A.D. 1300, two broad pottery horizon styles began emerging on opposite sides of the Savannah River. These are thought to represent some kind of emerging social or cultural difference. Both developed out of Savannah pottery types. On the western side of the Savannah River this is called the Lamar horizon style, and on the eastern side of the Savannah it is called the Chicora horizon style. Both of these pottery traditions made use of complicated stamping as well as other similar features. They differ only in terms of small details, such as the type of decoration on the rims of jars and bowls. The most characteristic development in the Chicora horizon style is the Pee Dee phase. Pee Dee phase pottery is present in the earliest stages of the Mulberry mounds and the Fort Watson mound.

By A.D. 1300, people making Pee Dee pottery reached the upper Pee Dee River. At the Town Creek site, near Mt. Gilead, North Carolina, they built a small substructure mound and a ceremonial center (see figure 10). As mentioned earlier, the archaeological evidence clearly indicates that this was an intrusive occupation. That is, it is an instance of people moving to this place from elsewhere. The ceremonial center had a defensive palisade (see figure 11) built around it, no doubt to protect the people from the indigenous people they had displaced. Pardo led a detachment of men to the town of Ilasi, in the vicinity of present-day Cheraw, and one would expect that a similar mound should be located there. Unless the site of Ilasi has been completely destroyed, an archaeological survey of this area should confirm or deny the existence of such a site. Further south, Pee Dee pottery has been found all the way to the Atlantic coast.[60]

The people who built the mound at Town Creek never pushed further north than the area where the Uwharrie River flows into the Yadkin. The descendants of the people of the Uwharrie complex continued to live north of this area, diversifying somewhat into more localized phases or cultures such as Caraway, Linwood, and Dan River. Hence, at the time of the Pardo expeditions, the descendants of the people of the Uwharrie complex lived on the upper reaches of the Yadkin, Uwharrie, Eno, and Dan rivers.[61] Similar cultures may have occupied the upper Broad River and the south fork of the Catawba River, but the archaeology of these two rivers is at present poorly known.

The Mississippian transformation penetrated no further east than the environs of the Pee Dee River. Why this should have been so is not known. Perhaps by the time of the sixteenth-century Spanish explorers the people along the Pee Dee River had simply not had enough time to press beyond this frontier. Another possibility is that the hostile political situation to the west may have become too debilitating to allow them to extend further to the east. By about 1500, all of the mound centers on the western and southern periphery of the Wateree Valley had been abandoned—McCollum, Blair, and Fort Watson.[62] The culture of the people in this more restricted area is known as Mulberry phase; it was directly derived from Pee Dee phase. It is possible that by about A.D. 1500 Cofitachequi found itself facing hostile people on more than one front—to the west certainly, and probably to the east, and it is possible that the people living to the north and south of the Wateree Valley were not fond of the "Lady" of Cofitachequi.

This, then, is the archaeological background of the paramount chiefdom of Cofitachequi and its neighbors. Like many Mississippian chiefdoms in the late prehistoric Southeast, the center of the chiefdom of Cofitachequi was situated near the rich, alluvial soils which occur just below the Fall Line, the geological boundary separating the Piedmont from the Coastal Plain. On the Wateree River, the Fall Line is located at a series of shoals that occur just downstream from where Wateree Dam is today. The shoals would have been an excellent place for catching fish. Then, further downstream, the river begins its meander zone, which near Camden widens to about 4 miles, and it is about that wide all the way south to its junction with the Congeree. The rich bottomland soil in this meander zone was one of the soil types preferred by late prehistoric Southeastern Indians.

In Pardo's day Cofitachequi or Canos was probably located at the Ferry Landing site or at the Mulberry site, more probably the latter, since the name *Canos* refers to mounds. The Blanding and Adamson sites can be ruled out because both appear to be too early. The Boykin site is a possibility, but it would seem to have been rather too far to the south to have been the site Pardo visited.

Soto examined the contents of a temple in what was probably the first town of Cofitachequi he visited. Here he found axes and rosaries that the people of Cofitachequi had obtained from the Ayllón colony on the Atlantic coast. If this town of Cofitachequi was the same one first discovered by a scouting party that Soto sent out the night before he himself arrived, then the town must have been visible from the opposite bank of the river, because from the opposite side of the river Soto's men could hear the barking of dogs and the crying of children.[63] Perhaps this town was at the

Ferry Landing site. After crossing to this village, Soto went to Talimeco, the village of the female chief, where he found an elaborate temple upon a high mound that was much revered, and that was the center of power. Soto went to this town at the behest of the chief to find pearls and other finery. The home of the chief was very large, and was covered with finely woven cane mats.[64] It is likely that Talimeco was at the Mulberry site. This is consistent with Garcilaso's statement that Talimeco was about a league from the first town they came to.[65] Moreover, this is consistent with Garcilaso's description of Talimeco being "situated on an eminence overlooking the gorge of a river."[66] Of all the mound sites near Camden, the Mulberry site has the steepest, most gorge-like bank.

When Pardo reached Cofitachequi on his second expedition, he met with an extraordinary assemblage of chiefs: Canos Orata, Ylasi Orata, Sanapa Orata, Unuguqua Orata, Vora Orata, Yssa Orata, Catapa Orata, Vehidi Orata, Otari Orata, Uraca Orata, Achini Orata, Ayo Orata, and Canosaca Orata (see figure 14).[67] In addition to these principal chiefs, many others were present who were subject to them. The locations of the towns of some of these chiefs is known, at least generally, while the locations of others can be little better than guesswork. Similarly, the linguistic identity of some of these chiefs can be determined with some confidence, while others cannot confidently be assigned to any linguistic stock.

Canos Orata was, of course, the chief of Cofitachequi. As we have seen, the etymology of *Canos* is probably the Creek word *kan-osi*, "little earth" or perhaps "mound." The etymology of Canosaca is similar. It is probably *kan-os-ak-a*, "ground" + dimunitive + plural + place, or "the mounds place."[68] Where this Canosaca Orata's town was located is unknown, but it is reasonable to suppose that he was from one of the mound sites near Camden.

Ylasi Orata was from the vicinity of Cheraw, South Carolina, some 55 miles northeast of Cofitachequi. This is probably the same as the Ilapi visited by members of the Soto expedition, at which time it was a secondary center of the chiefdom of Cofitachequi, as evidenced by the fact that seven cribs of corn belonging to the chief of Cofitachequi were stored there.[69] In fact, De Soto split his force, sending one contingent of men to Ilapi to seize corn that was stored there.

When Pardo departed from Cofitachequi and traveled northward, coming to Gueça, he was visited by a Herape Orata (cf. Ilapi), who may have been from Ylasi or a town subject to Ylasi. The etymology of these forms is unknown, but it may be related to a later Creek town name that has been variously spelled *Hilibi* or *Hilabi*. It is possible that *Herape* was pronounced by a Catawban-speaker who substituted *r* for *l*.[70]

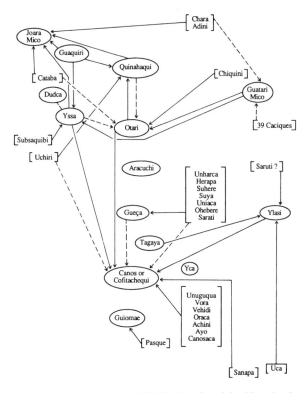

Figure 14. Sociogram of Cofitachequi and its hinterland.

Sanapa Orata was from a town on the coast, and perhaps from the Georgetown region, some 98 miles southeast of Cofitachequi. Bandera says that salt was made at this place.[71] The Sampit River, which empties into Winyah Bay at Georgetown, possibly derived its name from *Sanapa*. The etymology of this place name is uncertain. It could be an earlier form of Eastern Muskogean *sampa*, "basket."[72] The form of the word adopted for the Sampit River may retain -*t*, the Eastern Muskogean subject suffix.[73]

Yssa Orata's main town was near Lincolnton, North Carolina, some 100 miles north of Cofitachequi.[74] Cataba Orata, who appeared in the company of Yssa Orata on more than one occasion, must have had his town near that of Yssa Orata. Both Yssa and Cataba definitely spoke Catawban languages. "Yssa" is derived from Catawban *i·swa*? "river." Some of the descendants of these people were called *Esaws* in early South Carolina. There is no known etymology for *Cataba*, but it is clearly the same name as that used by contemporary Catawba Indians in South Carolina.[75]

Otari Orata's town was probably located west of present-day Charlotte, North Carolina. For reasons that will be discussed later, it is probable that

The Indians 75

the people of Otari spoke a Catawban language, but the word *Otari* has no known etymology.

The locations of the towns of the other six chiefs who met with Pardo at Cofitachequi are unknown, and their linguistic identities are unknown or dubious. *Achini* may be Eastern Muskogean *acina*, "cedar," and *Ayo* may be Eastern Muskogean *a:yo*, "hawk."[76] The linguistic identity of *Vehidi* is unknown, but it is probably the namesake of the Pee Dee River and of the Indians who bore that name in the eighteenth century, and Vehidi Orata's town may have been located on the lower Pee Dee River.[77] The linguistic identities of Unuguqua Orata, Uraca Orata, and Vora Orata are unknown.[78] It is, however, reasonable to suppose that some or all of these had their towns to the south and east of Cofitachequi, in the territory between the Wateree-Santee and Pee Dee rivers.

It is striking that none of the chiefs of the four towns on the Catawba-Wateree River north of Cofitachequi were among those who met with Pardo when he arrived at Cofitachequi. These were Tagaya (near the mouth of Beaver Creek), Tagaya the Lesser (near Fishing Creek Dam), Gueça (near Irwin or Waxhaw Creek), and Aracuchi (near Rock Hill). Their absence may be yet another indication of how far the power of Cofitachequi had declined. It is possible that Eastern Muskogean languages were spoken in all of these towns, but no etymologies are known for their names.

When Pardo reached Gueça, he learned that Gueça Orata had been among the chiefs whom he had previously ordered to build the house for the Spaniards at Cofitachequi. "Gueça" is probably derived from Creek *wi·sa* "huckleberries," and it is almost certainly the name that after the late seventeenth century, the English colonists spelled as "Waxhaw".[79] The town may have been located at a poorly known archaeological site on Waxhaw Creek.[80] Along with Gueça Orata, Pardo gave his formal address to Unharca Orata, Herape Orata, Suhere Orata, Suya Orata, Uniaca Orata, Sarati Orata, and Ohebere Orata. Like the contingent of chiefs Pardo addressed at Cofitachequi, this contingent appears to have included both Eastern Muskogean-speakers and Catawban-speakers. Many of these chiefs probably had their towns in the territory adjacent to the middle courses of the Catawba-Wateree and Yadkin-Pee Dee rivers.

Pardo ordered all of these chiefs to take their tribute in corn to Cofitachequi, even though this requirement must have forced some of them to carry their corn for a considerable distance. But he made an interesting exception. If any of them did not have a supply of corn, they could bring deerskins, and if they did not have any deerskins, they could bring a quantity of salt. This implies that some of these chiefs may not have been able to produce a surplus of corn, and even more interestingly, that some of them had access to salt, which they either produced themselves or received in trade.

Beyond this, very little can be said about the chiefs whom Pardo met at Gueça. But one suspects that *Suhere* may have been the namesake of the Sugeree Indians and Sugar Creek, and if so, Suhere Orata's town may have been on that creek. Sugar Creek runs from Charlotte, North Carolina, southward to join the Catawba River east of Rock Hill, South Carolina. It may have been the Catawba word *suxere*, "to run alongside," perhaps referring to a stream.[81] *Suya* may contain Catawban *su*, "tree," but no etymology can be had for the rest of the word. The *-ere* ending of *Ohebere* suggests that it is Catawban, but no etymology can be had for the first part of the word. The linguistic identity of Unharca and Uniaca is unknown.[82] It is possible that Suhere Orata and perhaps some of the others who came to Gueça were subject to the chief of Guatari, to be discussed later.

Aracuchi, which appears to have been located in the vicinity of Rock Hill, South Carolina, was probably the northernmost town on the Catawba-Wateree River where Eastern Muskogean was spoken. The etymology may be *ara-uchi*, "place where the *ara* tree grows." Creek *ala* or *ala:* means "buckeye."[83] It may be no coincidence that Rock Hill, South Carolina, lies on the southern margin of the habitat of the yellow buckeye (*Aesculus octandra* Marsh).[84] This tree was significant to the Indians because its fruit could be crushed and thrown into water as a fish poison.

When Pardo was returning from the interior, he went out of his way to visit Ylasi, where he expected to meet some chiefs he had not met before. He may also have wanted to reconnoiter Ylasi, because after Pardo had given his talk to the chiefs at Cofitachequi, Ylasi Orata came to him and told Pardo that even though he had helped build the house for the Spaniards to use at Cofitachequi, he preferred to store his corn in his own town, and Pardo granted him permission to do so. Only three chiefs came to see Pardo while he was at Ylasi: Uca Orata, Tagaya Orata, and Saruti Orata. Uca Orata was from the coast, presumably near the mouth of the Pee Dee River.[85] Tagaya Orata came to Ylasi to meet with Pardo for a second time. At this meeting, Pardo recognized him as a cacique as a way of repaying him for services that he had performed for the Spaniards.

"Saruti Orata" is phonologically similar to "Sarati Orata," who met with Pardo at Gueça. However, Bandera was of the opinion that Saruti Orata had not met with Pardo before. Hence these would seem to have been two different places. It is possible that the town of Saruti (or Sarati) was on the

Pee Dee River above Cheraw.[86] It is tempting to place Saruti (or Sarati) at the Town Creek site, although archaeologists have evidence that this site had been abandoned before the time of the Soto expedition.

Clearly, Ylasi was a town of some importance within the chiefdom of Cofitachequi. As we have already seen, while Soto was at Cofitachequi, he divided his forces, sending one contingent to Ylasi (Ilapi) where there was an ample supply of corn. Hence, it is reasonable to conclude that Ylasi was a tributary chiefdom, and that tribute was being collected there both in Soto's and in Pardo's day, although by Pardo's day Cofitachequi no longer had the power to claim any of it. It is probable that Ylasi controlled the Pee Dee River from the vicinity of the Town Creek site southward to some point in the Coastal Plain.

When Pardo was returning from Ylasi to Cofitachequi, he encountered the town of Yca, probably near the forks of Lynch's Creek. *Yca* may be Eastern Muskogean *ika*, "head."[87] The political affiliation of this town is unknown, although its location strongly suggests that it was affiliated with Cofitachequi and with Ylasi.

Little can be said about the relationship of Cofitachequi to the chiefs of the Coastal Plain whose towns were located between Santa Elena and Guiomae, except that some of them spoke mutually intelligible or at least related dialects and languages of Eastern Muskogean. In the chiefdom of Guale, on the coast southwest of Orista, thirty or so chiefs spoke mutually intelligible dialects or languages. Moreover, this language (or dialects of a language) was said to be understood on at least a limited basis for more than 200 leagues into the interior—that is, to the furthest extent of Pardo's penetration.[88] The Guale language was said to be intelligible to the people of Orista, whose language Rufín had learned. Apparently Rufín could converse with many Indians in the interior, and perhaps with those at Cofitachequi, and even with those at Aracuchi. Unfortunately, as judged from several known omissions, Bandera does not always mention when Rufín used Indian translators as intermediaries.

It is likely that along with the people of Guale and Orista, the people of Coçapoye, Ahoya, Ahoyabe, Coçao, Aboyaca, and Guiomae were speakers of Eastern Muskogean languages. *Ahoya* may have been an Eastern Muskogean word meaning "two going."[89] But *Ahoyabe* may contain the Eastern Muskogean *api*, "stalk, stem," which suggests that *Ahoya* might have referred to a fruit or a nut.[90] *Coçao* is reminiscent of *Coça* (Coosa), the large chiefdom whose domain lay in northern Alabama, northwestern Georgia, and the Tennessee Valley, although it is not reasonable to suppose that any political or historical connection existed between them.

But *Coçao* could well have been Muskogean.[91] And the same is true of *Coçapoye* Orata, a chief who met with Pardo at Orista when he was returning to Santa Elena. The final part of this word may be Creek *apo·itá*: "to accumulate plural squat things in a place," that is, a group of dwellings.[92] *Aboyaca* may contain the same element.[93]

Pasque Orata, who visited Pardo in Guiomae, may have been the chief of a town whose name was derived from Eastern Muskogean *apaski*, "parched corn."[94] Since Pasque Orata was required to bring a canoe load of corn to Guiomae, his town was either upstream or downstream from Guiomae. It is tempting to place Pasque in the vicinity of the Fort Watson mound site. It may be a coincidence that Soto found a quantity of cornmeal, possibly *apaski*, when he reached Guiomae (spelled "Aymay" and "Hymahi" by the Soto chroniclers).[95]

Guiomae may have been derived from Creek *ki·o·mays*, "a place where there are mulberries," a wild fruit that grew in profusion along the middle Wateree River.[96] In fact, when Soto arrived in this town he found many mulberry trees loaded with fruit.[97] Unfortunately, the Bandera document does not indicate whether Guiomae was affiliated with Cofitachequi at the time of the Pardo expedition. Emae Orata, who was the chief of Guiomae, did not appear with the chiefs who met with Pardo at Cofitachequi. Also the people of Guiomae did not take corn to Cofitachequi. They had their own storehouse.

At the time of the Soto expedition, Guiomae was clearly subsidiary to Cofitachequi. When Soto got there, he seized four or five Indians of Guiomae and tried to force one of them to tell him where a center of population (i.e., Cofitachequi) could be located. Soto's torture was to no avail, as the man died refusing to give him any information.[98] This would seem to mean that at least some of the people of Guiomae were profoundly protective of the ruling chiefs of Cofitachequi. Another of these Indians, no doubt fearing the same fate, told Soto that Cofitachequi was two days' journey from Guiomae.[99]

Both the land of the Coastal Plain and the people who inhabited it were less impressive to Pardo and his men than were the land and people north of Guiomae. Between Santa Elena and Guiomae there was "rough land and swampy plains."[100] They encountered no micos in the Coastal Plain, which suggests that for whatever reason the Indians there were less centralized than those to the north. Also, when they were returning through the Coastal Plain with their sacks of corn, they first mention the Indians using acorns (*villota*) as food and also a wild root that the Spaniards called *batatas*, "sweet potatoes." This implies that the Indians in the South Carolina

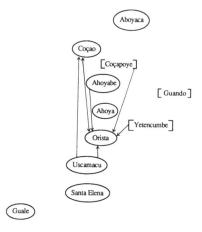

Figure 15. Sociogram of Santa Elena and its hinterland.

Coastal Plain were more dependent upon wild vegetable foods than were the Indians north of Guiomae. But it must be noted that the Indians of the Coastal Plain had been in more sustained contact with more Spaniards than the Indians of the interior, and therefore may have already suffered more depopulation from Old World diseases. Consequently their agricultural production may have been disrupted and they may have had to rely more heavily on wild food.[101]

Another reason why the Coastal Plain Indians did not impress the Spaniards is that there is no clear indication of centralization in the Port Royal area (see figure 15). Orista Orata appears to have stood somewhat above other caciques in power. This is the town to which Guillermo Rufín went after his comrades sailed away, and it is the place where he learned the language he used as a translator. Unfortunately, no etymology can be had for *Orista*.[102] While he was at Orista, Pardo met with Uscamacu Orata, whose name cannot be related to any language family. And he was visited by Yetencumbe Orata. *Yetencumbe* was probably the namesake of the nearby Combahee River, but the word cannot be identified as to language family. Guando, a place mentioned but not visited by Pardo, is also unidentifiable.[103]

Going northward from Santa Elena, Orista was the first town on the mainland, and this was the place where Pardo built a stronghouse. It is also the place where Father Juan Rogel, along with three Spanish boys, built a church and a house and attempted to missionize the Indians. But Rogel had a frustrating time of it. The town of Orista contained only twenty households of Indians. The inhabitants did cultivate corn, but when the acorns became ripe in autumn all but two families moved away from the town and into the woods, splitting up into twelve or thirteen groups. They established residence at 4, 6, 10, and even 20 leagues from Orista. They remained in the woods for nine months of the year, coming together every two months or so to perform rituals. Rogel tried to persuade them to cultivate more corn so they could live in one location for the entire year, but they would not do it. The reason they gave was that the soil was poor and wore out quickly.[104] Rogel says that they were monogamous and that the elders who governed the village customarily met in a large building. There appears to be no evidence that a mico resided at Orista.

In contrast, the chiefdom of Guale, in the St. Catherines Island and Sapelo Sound area, did have a mico, with about 30 caciques who were subject to him.[105] The mission at Guale was under Father Antonio Sedeño and Brother Váez. Only three or four villages of Guale had as many as 190 people, and some did not even have 40 people.[106] In contrast to Orista, the caciques of Guale had three or four or more wives. This arrangement, the chiefs claimed, was for better service in making cornmeal and cooking food. In Guale, one had to go from village to village in dugout canoes, which was very difficult at low tide because of poor landing places. The mud was knee-deep in places, and the Spaniards, who were unaccustomed to traveling in the dugout canoes, had trouble with them and frequently capsized.[107] After Sedeño had been in Guale only a few months, he described it as the worst land that had ever been discovered. For 30 leagues inland, he said, there was little but sand and rivers and swamps. Moreover, he had traveled almost 300 leagues up and down the coast, and the land was all the same. The soil was so miserably poor that the Indians had to move their fields frequently. Sedeño estimated that there were no more than 2,000 people in all of Guale.[108]

The testimony of Francisco of Chicora ("El Chicorano"), the Indian who was captured and enslaved by Spaniards on the South Carolina coast indicates that he was familiar with the chiefdom type of social structure. Francisco was captured in 1521 by slavers employed by Lucas Vásquez de Ayllón who apparently anchored their ships at the mouth of the South Santee River.[109] Francisco's testimony was recorded in Spain by the historian Peter Martyr, who on several occasions invited Francisco and Ayllón to dine with him.[110] Even though Francisco spoke Spanish quite well, according to Martyr, in some instances Francisco was not able to successfully explain himself to Martyr. Another difficulty in interpreting Francisco's testimony is that he appears to have spoken a Catawban language, so that all place names used by him differ from those recorded by Bandera, who depended upon his Muskogean-speaking translator Guillermo Rufín.

Although Francisco was himself from the "district" of Chicora, he spoke at most length about Duahe, a "district" ruled by Datha, a chief of gigantic size who had an equally gigantic wife.[111] Francisco does not specifically say that Chicora was tributary to Duahe, but he does say that several of the towns and societies he named were Duahe's subjects, and that they paid them tribute in kind. Francisco told Martyr that the chief's house was inside the temple; that there was a class of priests who wore a distinctive tonsure and costume; that on ritual occasions these priests stood atop a mound and addressed the people who stood below; and that when the chief wished to visit various parts of his realm he was carried on the shoulders of young men, who ran from place to place; and that common people were required to greet the chief in an elaborately ceremonious way.

It is clear enough that Duahe was a Mississippian chiefdom. The question is, where was the center of this chiefdom located? Most of the place names mentioned by Francisco that can be equated with Bandera place names were located between Guiomae and the coast. These are Sona (cf. Sonapa), Pasqui (cf. Pasque), Huaque (cf. Uca), and Yamiscaron (cf. Guiomae, Aymay, Hymahi). Other possible equivalences are Orixa insyguanin (cf. Orista) and Aranbe (cf. Herape). It would seem that Francisco was more familiar with Indian groups who lived between Guiomae and the coast than he was with people who lived between Guiomae and the mountains.

But this does not help locate the main town of Duahe. If we are to take seriously Martyr's statement that all of the provinces that Francisco mentioned paid tribute to Chief Datha, he could well have been the chief of Cofitachequi. It is also possible that Datha was a secondary chief whose main town was at some unknown location in the coastal plain.

Unfortunately, none of the Pardo documents describes the nature of any relationships that might have existed between polities in the Coastal Plain and the interior. Bishop Calderon's description of the Southern Indians in 1675 specifies that the "province of Escamacu" was subject to the "Mico of Cofatache" (i.e., Cofitachequi).[112] The bishop got part of his information from firsthand observation, but part of it, including the above, was hearsay. However, for what it is worth, this suggests that the power of Cofitachequi extended to the Port Royal area. If this was the case, then the unoccupied area along the mouth of the Savannah River between the Port Royal Indians and Guale was a continuation of the wilderness or buffer zone that existed upriver.

In summary, at the time of the Pardo expedition, Cofitachequi was a paramount chiefdom that had declined, and one might even say disintegrated. In its heyday, it had consisted of a central cluster of towns in the Camden area surrounded in three directions by Muskogean-speaking secondary centers: Guiomae and Pasque to the south, Ylasi to the east, and Gueça to the north. Ylasi controlled the Pee Dee River from the Narrows of the Yadkin to an area far down the Pee Dee River, and perhaps to the mouth of the river.

As we have seen, there is at least some evidence that Cofitachequi was paid tribute from towns in the Coastal Plain all the way from the Santee River southwest to Port Royal Sound. Some but not all of these people were Muskogean-speaking. Some of them appear to have spoken Catawban languages, and some of them may have spoken languages from undocumented language families.[113]

Finally, Cofitachequi had at least attenuated power over towns to the north. The headmen of Otari, Yssa, and Cataba, all of whom were Catawban- speaking, traveled to Cofitachequi to meet with Pardo. As we shall presently see, the power of Cofitachequi may have extended even further north, to the edge of the mountains or even into the mountains, to towns where Cherokee and perhaps Yuchi were spoken.

Joara When Pardo departed from Aracuchi, the northernmost town of Muskogean-speakers on the Catawba River, on his way into the interior during his second expedition, he entered an area that was under the uncertain control of Joara Mico, a chief who appears to have been attempting to expand his sphere of power (see figures 13 and 14). It was a linguistically diverse area, with Iroquoian, Catawban, and possibly Yuchian languages being spoken, and it may have included languages that belonged to families that have never been documented by linguists.

Pardo came to the first of a series of Catawban-speaking towns at Otari, a town west of Charlotte, North Carolina, perhaps on the Catawba River near its junction with the South Fork. Here Bandera says that *Otariyatika* was the name of the chief and also the name of the language of that land from there on.[114] But this was more likely a compound word formed with *Otari*, which was probably a Catawban word,[115] and *yatika*, a word that meant "interpreter" in several Southeastern Indian languages. This suggests that a linguistic boundary lay between Aracuchi and Otari. It is significant that Guatari Mico, whose main town was near Salisbury, North Carolina, and who probably spoke a Catawban language, journeyed south to meet with Pardo when he reached Otari.

Pardo met with several Catawban-speaking chiefs at his next stop, Quinahaqui (see figure 14). Here he met with Yssa Orata and Cataba Orata, whom Bandera says were very important chiefs, as well as with Uchiri Orata. Yssa Orata, as we have already seen, was undoubtedly a Catawban-speaker. His main town was near Lincolnton, North Carolina, and several towns were subject to him, including Dudca, a small town to the west, probably in southern Burke County; Yssa the Lesser, to the north near Maiden, North Carolina; and Subsaquibi, whose location is unknown.

Cataba Orata, another Catawban-speaker, on several occasions appeared before Pardo in the company of Yssa Orata. The word *Cataba*, however, has no known etymology. Pardo never visited the main village of Cataba, but it was probably located near Yssa.

Because *Uchiri* so closely resembles "Yuchi," it is tempting to think that Uchiri Orata may have been a Yuchian-speaking chief. The *-re* ending is Catawban for "be" and the *-ri* ending is a frequent Catawban locative.[116] Because Pardo did not visit Uchiri Orata's town, it is impossible to know where it was located. However, one possible clue to its general location comes from orders which Pardo gave to these chiefs when he was in Quinahaqui. When Pardo had been at Cofitachequi, he had ordered the chiefs of Yssa, Cataba, and Otari to take their corn to Cofitachequi to be stored. But apparently they did not want to comply, and this appears to be what motivated them to meet with Pardo in Quinahaqui. In any case, Pardo rescinded his previous command, ordering Yssa Orata, Cataba Orata, and Quinahaqui Orata to take their corn to Otari to be stored there. But he ordered Uchiri Orata to take his corn to Cofitachequi, implying that Uchiri was located further to the south than Yssa or Cataba, or else on a different river.

Unfortunately, when Soto traveled through this area he did not mention any towns by name until he reached Guaquili (i.e., Guaquiri). However, the Soto chroniclers refer to the people in the general vicinity of Otari and Yssa as the *Chalaque*.[117] This was probably the Muskogean word *čhilokkita*, "people of a different language." Again, this is an indication that a major linguistic boundary lay in this area. Interestingly, to Soto's men the country between Cofitachequi and Xuala (i.e., Joara) seemed very poor, sparsely populated, and with little corn.[118] Elvas says, wrongly no doubt, that here the Indians subsisted solely by hunting and gathering.[119] Soto's men looked for a principal chief in this area, but they found none.[120] The population was so sparse that some of Soto's men feared that they were entering a wilderness like the one they had crossed between Ocute and Cofitachequi.[121]

Figure 16. A four-sided Pisgah phase house floor. Excavated at the Warren Wilson site in Buncombe County, North Carolina. Note the entryway trenches. (Photograph courtesy Research Laboratories of Anthropology, University of North Carolina.)

The archaeology of the Yssa and their neighbors is not well known. However, it is clear enough that the protohistoric culture of the Indians here was a late local variant of the Uwharrie culture.

When Pardo entered the upper Piedmont and the Blue Ridge Mountains, he encountered the bearers of yet another major archaeological tradition— the Pisgah/Qualla phases (see figure 13).[122] The Pisgah phase or culture began forming some time after A.D. 1000, at about the same time that other Southeastern Indian societies in the Carolinas began undergoing the Mississippian transformation. The Pisgah phase had many traits in common with Mississippian cultures elsewhere, including increased dependence on corn agriculture, a nucleated village settlement pattern, substructure mounds, and a more stratified social structure (see figure 16).

Pisgah sites are numerous on the headwaters of rivers in the southern Appalachian mountains. They are most numerous on the headwaters of the Pigeon, the French Broad, and the Little Tennessee rivers. But they also occur in peripheral areas on the headwaters of the Powell, Clinch, Holston, Nolichucky, Catawba, Saluda, Keowee, and Hiwassee rivers. Many of these sites in the northeastern part of this peripheral area appear to date to the earlier part of the Pisgah period.[123]

The Pisgah phase in time gave way to the Qualla phase, although many of the older Pisgah practices continued on in Qualla culture. The transition occurred, according to Roy Dickens, around A.D. 1500,[124] although it is possible that it occurred a half century or so later.[125] The ceramics changed mainly through Lamar and Chicora influences from the south. Large bowls and jars with bold complicated stamping and incising came into vogue during the Qualla phase. Seventeenth-century European trade goods have been found in Qualla sites.[126] Several known eighteenth-century Cherokee sites have been found to contain Qualla materials. Hence, at least some of the people of the Pisgah phase were directly ancestral to the historic Cherokee people, and it is possible that most of the Pisgah people were Iroquoian-speakers.

The time and manner of transition from Pisgah to Qualla is important to know because during the Qualla phase almost all of the peripheral sites in the northern part of the area occupied by Pisgah people were abandoned, and the people moved toward the southwest. Hence, there are no known Qualla sites on the Powell, Clinch, Holston, or French Broad rivers. Only one early Qualla site—Plum Grove—is known to exist on the Nolichucky River, and only one early Qualla site is known to exist on the upper Catawba River.[127] At about the same time that these sites were abandoned, new sites producing similar cultural materials were occupied on the Keowee, Chattooga, Chattahoochee, Little Tennessee, and Hiwassee rivers.[128]

The time of this transition of Pisgah to Qualla is of the utmost importance because Pardo and Moyano visited several of these marginal areas—namely, the upper Catawba, the upper Nolichucky, and the upper French Broad rivers—and they encountered inhabited towns on all of them. The point is that the transition of Pisgah to Qualla may have occurred just after the Soto and Pardo expeditions, and in part as a consequence of them.

Unfortunately, the linguistic evidence contained in the Bandera document is less than one would wish it to be. Joara, the dominant town at the edge of the mountains, was located in the vicinity of Marion, North Carolina, on the upper Catawba River. At present, the most likely location for Joara is the McDowell site (31 Mc 41), where preliminary excavations have revealed Pisgah-like houses with entryway trenches,[129] and pottery that is identical to Pisgah pottery in every respect except tempering—it is tempered with soapstone.[130] In part, Joara must have owed its existence to the fact that two important trails crossed at this location. One trail—Rutherford's War Trace—ran from east to west to the upper French Broad River, where it intersected the Catawba Trail, which led to the territory of the chiefdom of Chiaha. The other trail—the Old Cherokee Path to Virginia—led from Joara to the Chiscas on the upper Nolichucky River, and from there it led to the saline at present-day Saltville, Virginia.[131]

It is possible that the people of Joara may have been salt traders. In 1600, when Governor Don Gonzalo Méndez Canço interviewed several individuals about conditions in the interior, two of those interviewed were Indian women—Teresa Martín and Luisa Méndez—who had come out with Juan Pardo, and who may have been natives of Joara. They both said that at the foot of the mountains there were three to five saltwater springs. Using fire, the Indians extracted great quantities of salt from the water of these springs. Moreover, Luisa Méndez said that these were the only such springs in all that country.[132]

It is clear that both Catawban- and Cherokee-speakers visited Pardo at Joara, but what language was spoken by the people of Joara? Soto's chroniclers recorded *Joara* as *Xuala*, and they recorded *Guaquiri* as *Guaquili*.[133] Hence, they recorded *ž* as *š*, and they recorded *l* as *r*. But it is not clear whether these phonological differences arise because Soto had a Muskogean-speaker as interpreter (Eastern Muskogean lacks *r* but has *l*), or whether it was a sound alternation in the language of Joara because it lacked both *r* and *š*.[134] Catawban can also be ruled out because it lacks both *š* and *ž*. This leaves Cherokee and Yuchi as possibilities. The language of Joara could have been Cherokee, which had both *l* and *r* dialects, so that *l* and *r* could vary in the same word. Cherokee has *j*, but apparently no *š* or *ž*.[135] As well, the people of Joara could have spoken a Yuchian language, which commonly had *š*. Modern Yuchi has *l* but no *r*, but in the sixteenth century it could have had an *r* dialect.[136]

It is possible that "Joara" is the namesake of the eighteenth-century Cherokee town "Jore."[137] It is tempting to think that Guaquiri, only two travel days downstream from Joara, and Quinahaqui, only three travel days downstream from Joara, spoke languages related to the language spoken at Joara. The people of both Joara and Guaquiri could have spoken an *r* dialect of Cherokee, and Cherokee is a language that frequently has *gw* clusters. But *Guaquiri* could also have been a Catawba word beginning with *w*. For example, it could have been Catawba *wakhere*, "blaze." Quinahaqui could have been Iroquoian, since Iroquoian *-aki* means "people." But *-ki* is the definite article "the" in Catawba, so it could have been Catawban as well. Hence, on phonological grounds either Catawban or

Cherokeean languages could have been spoken at Guaquiri and at Quinahaqui.[138]

Joara was a border town in every sense of the word. It was on the border between the Piedmont and the Blue Ridge geographical provinces; it was on the border between Iroquoian, Catawban, and possibly Yuchian languages; and it was on the border of the Pisgah/Qualla, Caraway, and Dan River archaeological phases. The cultural affiliations of Joara can only be settled through archaeological research. The most promising location for Joara is the McDowell site, near Marion, North Carolina. Preliminary excavation at this site has revealed the presence of "Pisgah-like" pottery with soapstone tempering, as well as a strong component of Burke or "Catawba" ware, and even more intriguingly, a small mound. Although much additional archaeological research must be done at the McDowell site and neighboring sites, thus far the archaeological information is generally consistent with the historical evidence.

Pardo visited Joara while on his second expedition into the interior, but on this visit no chiefs other than Joara Mico met with him, perhaps because he was in a hurry to cross the mountains to relieve Moyano. On his return, however, he met with many chiefs. On one occasion at Joara he met with Quinahaqui Orata, two Catapa [i.e., Cataba] Oratas, Guaquiri Orata, and Yssa Chiquito Orata, along with some of their principal men (see figure 14). The fact that these chiefs appeared in each other's company may suggest that they were closely allied with each other politically. Perhaps they were also close linguistically, although in such a diverse language area, one cannot be sure.

On another occasion on his return, Pardo met at Joara with a larger and more heterogeneous group of eighteen chiefs (see figure 17): Atuqui Orata, Osuguen Orata and his mandadores, Aubesan Orata, Guenpuret Orata, Ustehuque (cf. Vastique, *infra*) Orata, three Pundahaques Oratas, Tocae Orata, Guanbuca Orata, Ansuhet Orata, Guaruruquet Orata, Enxuete Orata, Utahaque Orata, Anduque Orata, Jueca Orata, Qunaha Orata, Vastu Orata, and Joara Mico. Bandera says that Joara Mico was the most important of these chiefs, and the inference might be drawn that some of the others were in some way subsidiary to him.

As we shall see, Pardo had already met some of these chiefs when he was returning through the mountains. These were Tocae Orata, Atuqui Orata, Enxuete Orata, Anduque (or Enuque) Orata, Guaruruquet(e) Orata, and Guanbuca Orata (probably the same as Guanuguaca). All of these were probably Cherokee-speakers.[139] On phonological grounds, several others *may* have been Cherokee-speakers, namely Osuguen Orata, Ustehuque

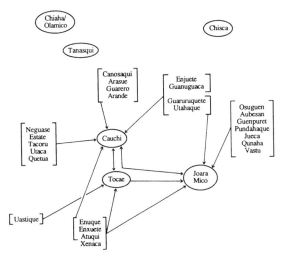

Figure 17. Sociogram of Joara and its hinterland.

Orata, and Ansuhet Orata. The word *Utahaque* may contain Cherokee *uta-*, "mountain." Nothing can be said about the locations of these towns except that most of them were probably in the mountains.

Of the remaining towns, Aubesan, Guenpuret, Pundahaque, and Vastu cannot have been Cherokee because all have labial sounds—*b*, *p*, *v*— which Cherokee lacks. *Guenpuret* may contain Catawban *we*, "town," but *-puret* has no known meaning. *Qunaha* could be Catawban *ku*, "corn," plus *naha*, "eat, ate," with perhaps the meaning "corn-eaters." Jueca may contain a form of Catawba *suk*, "house, mountains."[140] The locations of all these towns are unknown. However, since they arrived in the company of Atuqui Orata, who was probably the chief of a Lower Cherokee town, as well as Enxuete (same as Enjuete and Ansohet) Orata and Anduque (or Enuque) Orata, who were possibly from the same area, it is possible that Guenpuret, Aubesan Pundahaque, Vastu, Qunaha, and Jueca were located on the tributaries of the upper Broad, Saluda, or Savannah rivers.

There are several indications that Joara Mico was extending or widening his power at the time of the Pardo expeditions. The most marked instance came on Pardo's return journey, when he departed from Yssa and went to Quinahaqui. Yssa Orata followed in pursuit of him along with Guaquiri Orata, and they caught up with Pardo at Quinahaqui. Together with Quinahaqui Orata they told Pardo that they were unhappy that he had put them under the domination of Joara Mico (*debaxo del dominio*). Tactfully,

Pardo explained that they were to obey the officer he had left behind at the fort at Joara, and not Joara Mico. To this they said *yaa* many times as a way of expressing their gratitude.

It is possible that Joara Mico was able to ascend in power as Cofitachequi declined. Joara was the most prosperous town Soto had found in the Piedmont, but one does not get the impression that it was a great center of power.[141] In fact, in Soto's day Joara may have been a tributary of Cofitachequi. That is, as Soto moved northward from Cofitachequi, he forced the niece of the cacica to go along. Everywhere they went, the Indians showed deference to and obeyed this woman. However, when they reached Guaxule (evidently the same place as Pardo's Cauchi), they came to the furthest limit of her territory, and using a clever ruse, she escaped and fled to Joara.[142] Garcilaso, for what it is worth, states that Joara was a separate province, but that it was subject to Cofitachequi.[143]

Moreover, there is evidence that Joara Mico was in competition with Guatari Mico. It is striking that when Pardo was entering the country on his second expedition, the two female chiefs of Guatari met Pardo at Otari, and later the interpreters of Guatari came to Yssa to see Pardo, but, as far as the Bandera document indicates, no Indian from Guatari traveled to Joara. As well, no Indian from Joara traveled to Guatari.

An even stronger indication of rivalry between Joara Mico and Guatari Mico came while Pardo was at Joara. Here he was visited by two chiefs, Chara Orata and Adini Orata (see figure 14), both of whom were subjects of Guatari Mico, but who now wished to switch allegiance to Joara Mico. Pardo ordered them to remain subject to Guatari Mico, but he gave them gifts to soothe their feelings. It is impossible to locate these two towns with confidence, but one suspects that they were on the upper Yadkin or South Yadkin rivers, a location that was as close to Joara, or even closer, than to Guatari. Another possibility is that one or both may have been located on the upper Dan River. This incident implies that changing tributary relationship from one chief to another may not have been unusual for these people.

Finally, it would seem that the people of Joara were not only rivals, but enemies of the Chiscas who lived in and to the other side of the mountains, on the upper Nolichucky River and probably on the Holston River. How else could one explain why a contingent of warriors from Joara went with Moyano on his foray against the Chiscas in April 1567?

Evidently, *Chisca* was the name that Muskogean-speakers (and perhaps speakers of other languages) used to refer in general to the people who lived in the mountains and hills of the upper Tennessee Valley. This is what the people of Chiaha called them when Soto was there, and this is the term Sergeant Moyano used.

Archaeologically, the Chiscas appear to have been bearers of the Pisgah culture. A site that may have been occupied by Chiscas is the Plum Grove site (40Wg17) in Washington County, Tennessee. The ceramics have been identified as late Pisgah, but lesser quantities of other ceramics are present. These are said to resemble early Qualla or the black burnished pottery from the upper Catawba River.[144] Similar archaeological materials occur on sites along the Nolichucky River downstream to about present Chucky, beyond which the area along the river was evidently uninhabited up to its junction with the French Broad.[145]

It is clear that the Chiscas traded copper both to the people of Chiaha and to the people of Joara. Soto learned, when he reached Chiaha, that the Chiscas mined copper or "some other metal of that color," and he sent two men to investigate. Both Teresa Martín and Luisa Méndez remembered the Chiscas as people from the mountains who traded in "gold." Both of them claimed—wrongly, no doubt—that the Chiscas were white-skinned, blue-eyed, and red-haired.[146]

Guatari On his return from his second expedition, Pardo went to Guatari, a chiefdom that differed in several respects from the others he had visited. There is no evidence that any of the members of the Soto expedition visited Guatari, and in fact none of the Soto chroniclers mentions its existence. The main town of Guatari was on the Yadkin River in the vicinity of Salisbury, North Carolina. The culture of the people of Guatari was another sixteenth-century derivative of the Uwharrie prehistoric culture already discussed. By and large, the people of Guatari must have been similar to and perhaps in part descended from people of the Uwharrie culture who were forced northward when Indians affiliated with Cofitachequi seized the land along the Pee Dee River.[147]

Pardo learned that the chief of Guatari had no less than thirty-nine chiefs under her control. For the most part, the towns of these chiefs must have been on the Yadkin River and its tributaries. In addition, it is not impossible that the chiefdom of Guatari had power or influence over some towns on the upper Dan River and the headwaters of the Cape Fear River.

The culture of the people of Guatari was probably quite similar to the culture of the people of Upper Saura Town, a protohistoric site on the upper Dan River.[148] These people built houses with circular floor plans some 20 to 30 feet in diameter; Bandera specifically notes that the people of Guatari built circular houses (see figure 18).[149] They were probably like the houses of their Uwharrie predecessors, built by setting poles into the ground and then bending them over to form a dome-shaped structure.

The Indians of Upper Saura Town practiced a mixed economy. They

Figure 18. A circular house floor at the Upper Sauratown site (31Sk1a). It is located on the Dan River in northwestern North Carolina. It dates to about A.D. 1660, but it probably resembles circular houses from an earlier time. (Photograph courtesy Research Laboratories of Anthropology, University of North Carolina.)

continued the hunting and gathering ways of their Uwharrie ancestors, but they also practiced agriculture, cultivating their fields with hoes made with crudely chipped stone blades. They continued some Uwharrie pottery-making practices, such as net impressing. But some new ceramic techniques were adopted, such as surface burnishing on plain pots, corn cob impressing, check stamping, and occasional curvilinear stamping. Peach pits have been found at Upper Saura Town. The Spaniards introduced peaches to the lower South at an early date, and the Indians of the interior began growing them suprisingly quickly.[150] Hence Upper Saura Town would appear to date to the seventeenth century, although most of the basic cultural patterns were probably present at the time of the Pardo expeditions.[151]

The precise location of Guatari Mico's town is not known, but it may have been at a site on the west bank of the Yadkin River opposite the mouth of Swearing Creek. If so, it now lies beneath the water impounded by High

Rock Dam.[152] This was a late prehistoric site at least a quarter of a mile long. In the limited excavation that was done on the site, a few post molds were located, but not enough to reveal houses or palisades. Almost all of the projectile points were small, triangular Uwharrie-type points with concave bases. The pottery is comparable to that excavated at Upper Saura Town. No European artifacts have been found at this site, but such artifacts are extremely infrequent in early protohistoric sites, and so little excavation was done at the site that one is not justified in concluding that because no European artifacts were found there the site predated the Pardo expeditions.

In several respects Guatari was an unusual chiefdom. For one thing, it was ruled by a female chief. And she was not the only woman who held high office. At Otari Pardo was visited both by Guatari Mico, the female chief, and Orata Chiquini, a lesser chief who was also female.[153]

Unfortunately, Bandera does not list the names of the thirty-nine chiefs who were subject to the chief of Guatari. The only names that are known are Orata Chiquini, just mentioned, and Chara Orata and Adini Orata, the two chiefs who met with Pardo at Joara and asked to change their tributary relationship from Guatari to Joara. It is possible that some of the chiefs who met Pardo at Gueça could have been subject to Guatari Mico. It may also be noteworthy that the subsidiary chiefs of Guatari do not appear to have been eager to donate their labor to Pardo's projects; this may indicate that they were less subordinate in their relationship with their chiefs than were the Indians of other chiefdoms Pardo had encountered. It is also possible that these thirty-nine chiefs of Guatari were the headmen of quite small villages.

One of the most unusual aspects of Guatari is that even though it appears to have been a typical Mississippian chiefdom, it lacked substructure mounds. There is neither documentary nor archaeological evidence for their existence on the Upper Yadkin. This raises the possibility that Guatari was a very young chiefdom—one that had not yet set apart its chiefly elite with the ritual apparatus and sumptuary rights typical of older chiefdoms. It is unfortunate that we may never know whether the central town of Guatari possessed a town house or "earthlodge." Such a structure preceded and underlay the Mississippian mound at Town Creek and many other mounds in Georgia, western North Carolina, and eastern Tennessee.[154]

Ironically, it is in the territory of the chiefdom of Guatari that virtually the only discovery of sixteenth-century Spanish artifacts in the Carolina back country has yet been made. Some are possibly artifacts that Pardo gave to the Indians. This discovery was made in the 1880s as a part of the mound exploration project of the Smithsonian Institution. The site—the T. F. Nelson site—is in Caldwell County, near present-day Patterson, on a terrace of the Yadkin River. The artifacts were found in a grave that contained at least ten individuals who were apparently buried in a single episode. This kind of multiple burial sometimes indicates death in a military action or from epidemic disease.[155]

What appears to have been the highest-ranking individual in this grave had a Citico-style gorget beneath his head. This particular style of gorget was closely associated with the chiefdom of Coosa, which is discussed later in the chapter.[156] Near this individual's head were five rolled copper beads, of the type that has been found at several sixteenth-century sites. He also had bracelets on both wrists made of alternating shell and rolled copper beads. But for our purposes the most important artifacts present were four iron implements that lay at his right hand. These included an iron chisel (perhaps one of those given out by Pardo); a celt, evidently made out of a knife or sword blade measuring 1 inch wide and 6 or 7 inches long; and an iron punch or awl that had been fitted to a horn handle.[157]

Quite recently David Moore excavated a burial at the Berry site in Burke County in which he found, along with a number of high-status aboriginal artifacts, a single steel knife blade. It has not yet been determined whether this knife blade is Spanish (and probably early) or English (and probably late).[158] But the fact that this was the *only* European artifact in the burial suggests that it is early. In any case, on the basis of available information it cannot be ruled out that this was one of the knives Pardo was giving out as presents to the Indians.

Guatari appears to have been a young chiefdom whose subjects were neither very obedient nor very devoted to their female chief. Thus, it is puzzling that this place appears to have been so favored by Pardo and his men. Guatari is the place, as we shall see later, that Sebastian Montero picked as the location for his mission. This was also the place where Pardo decided to build one of the most substantial of the string of forts he put up in the interior. And this is where Menéndez planned to establish the center of his noble estate. One possible reason Guatari was so favored by the Spaniards is that the chiefs of Guatari may simply have been the most receptive and compliant of the chiefs whom Pardo met. This would be consistent with their being members of a rather weak chiefdom who were threatened by more powerful and better organized chiefdoms. They may have been anxious to have allies of any sort—even Spaniards.

The Cherokees When Pardo crossed the Blue Ridge Mountains, both going and coming, most of the Indians he met were, with varying degrees of probability, Cherokee-speaking. The first town he came to was Tocae, located in the vicinity of present-day Asheville, North Carolina. It was in Tocae that Bandera first mentions that Guillermo Rufín was using Indian interpreters as intermediaries, although he must have used them before this. Even so, this would seem to indicate that yet another linguistic boundary was crossed when they approached the mountains.

None of the Soto chroniclers mentions the existence of a town in the vicinity of Asheville. They all speak of crossing a mountainous wilderness after departing from Joara.[159] Either they somehow bypassed the town of Tocae or they did not mention it, or else it did not exist in 1540.

Tocae is possibly a Hispanicized and shortened version of *untakiyastiyi*, "where they [the waters] race,"[160] the Cherokee name for the segment of the French Broad River downstream from Asheville.[161] The French Broad River is placid above Asheville, but at Asheville it begins to flow more swiftly and falls some 1,500 feet through a series of cascades as it enters the Tennessee Valley.

It is also possible that *Tocae* is related to the Cherokee place name *dakwai*, which, according to nineteenth-century Cherokees, was located on the French Broad River about 6 miles above Hot Springs (formerly called Warm Springs) in Madison County, North Carolina.[162] The *dakwa* was a monstrous mythological fish that the nineteenth-century Cherokees believed lived at this place in the river. Toqua, an important eighteenth-century Cherokee town on the Little Tennessee River, and Toco, another eighteenth-century Cherokee town, both probably derive their names from Tocae.[163]

An etymology can also be had for *Swannanoa*, the name of the river whose upper reaches Pardo and his men encountered after ascending and crossing the ridge of mountains they came to just after they departed from Joara. They followed a trail along this river to Tocae. *Swannanoa* is an Anglicized version of the Cherokee word *Suwalinunnahi*, "the Suwali trail."[164] The people the Cherokees called *Ani-suwali* or *Ani-suwala* lived to the east of the mountains. They were none other than the people who lived at Soto's Xuala and Pardo's Joara.[165]

In Tocae, Pardo was visited by a most interesting chief, Uastique Orata, who claimed that he had traveled for seventeen days from "the west" in order to meet with Pardo. If this travel time is to be trusted, it is possible that Uastique Orata was from northern Florida. He may have been a member of one of the Timucuan-speaking societies living east of the Aucilla River and north of the Santa Fe River. These were the people whom the Apalachee Indians called "Yustega" and the ones the Saturiwa and Utina called "Houstaqua." It was well known to the people of Saturiwa and Utina that the "Houstaquas" knew where the mountains were and had obtained copper from there.[166]

If Uastique Orata was in fact from northern Florida, a route he could have traveled paralleled the Flint River to the north and from there turned east to the Chattahoochee River at present Atlanta. He could then have proceeded up the Chattahoochee and crossed a watershed to the Keowee-Toxaway River.[167] Continuing up this river, he could have then crossed over a watershed to the headwaters of the French Broad River, which he could have followed to Tocae. This was a distance of about 420 miles, which in seventeen days he could have traveled at the rate of 24.5 miles per day, a not unreasonable rate for an Indian.

Bandera was under the impression that Uastique Orata had traveled all that way to pay obedience to Pardo. But surely he had not. It is possible that he was on a trading expedition. A lesser possibility is that he may have heard that Spaniards were moving about in the interior, and he wanted to learn what they were doing.

On another occasion in Tocae, when Pardo was making his entry into the country, he met with four chiefs whose towns were said to have been located in the neighborhood of Tocae: Enuque Orata, Enxuete Orata, Xenaca Orata, and Atuqui Orata (see figure 17). Xenaca Orata was clearly a Cherokee-speaker because his name closely resembles *Senneca*, a Cherokee town that in the early eighteenth century was located on the upper Seneca River.[168] Atuqui Orata was also probably a Cherokee-speaker, because his name resembles *Taucoe*, a Cherokee town on the Tugaloo River in the eighteenth century.[169] It is striking that both of these were Lower Towns in the eighteenth century. On phonological grounds, Enxuete Orata (possibly the same as eighteenth-century "Chotte") and Enuque Orata might also have been Cherokee-speakers, and it is possible that all four were in the same locations in Pardo's day as they were later. But one cannot rule out the possibility they were on the headwaters of the French Broad or Saluda rivers in Pardo's day and that their inhabitants subsequently moved to the headwaters of the Savannah River.

After departing from Tocae, Pardo visited a second mountain town, Cauchi, which appears to have been more important than Tocae. For one thing, Pardo met with a greater number of chiefs in Cauchi than in Tocae. And a house was built in Cauchi for the Spaniards to use, whereas there is no evidence that a house was built for them in Tocae.

Pardo's travel time places Cauchi in the vicinity of present-day Marshall, North Carolina. No archaeological sites of the protohistoric time period have been found in the vicinity of Marshall, but the problem may be that the valley floor is so narrow in this area that Cauchi is beneath the buildings and streets of Marshall itself. It is also possible that Cauchi was on nearby Ivy Creek, where Pisgah sites are known to occur.

The Soto expedition also passed through this area, but the chroniclers appear to have called Cauchi by a different name—Guasili.[170] In 1540 this was the furthest limit of the influence of the chieftainess of Cofitachequi.[171] Why the name of this town should have been different in 1567 than it was in 1540 is unknown. Soto and his men found little corn at this place. The Indians here gave them about three hundred dogs to eat, although the Indians themselves did not eat them.[172] Even though the supply of food was slight, Soto's men thought themselves fortunate to have found this place to stop and rest.[173] Garcilaso, who is always prone to exaggerate and to transpose towns from one locale to another, says that this town contained three hundred houses, and that the home of the chief was on a high hill (a mound?), around which six men could walk abreast.[174]

Cauchi was the last place where Pardo found that the Indians had built a house for the Spaniards to use on their travels. In part this was because Cauchi Orata was the northernmost chief who had been present at Joara on Pardo's first expedition when he ordered the Indians to build these houses. But in a more profound sense, it was because Cauchi lay on or near a cultural, social, and linguistic frontier. It is quite striking that several of the Cherokee chiefs from the mountains visited Joara, and one even visited Yssa, but there was absolutely no visiting between the Cherokee chiefs and any of the towns of the paramount chiefdom of Coosa in the Tennessee Valley. This implies that the level of real or potential conflict between the Cherokees and the chiefdom of Coosa was greater than that between the Cherokees and Joara Mico and his allies.

At Cauchi, while on his way to the Tennessee Valley, Pardo met with a most interesting group of chiefs: Nequase Orata, Estate Orata, Tacoru Orata, Utaca Orata, and Quetua Orata. Three of these were definitely Cherokee because they were still town names in the eighteenth century and later. *Quetua* was clearly Kittowa, a most important Cherokee town on the Tuckasegee River in the eighteenth century. *Nequase* was clearly Nequasse, a Cherokee town on the upper Little Tennessee River in the eighteenth century, and *Estate* was probably Estatoe, also on the upper Little Tennessee River. It is striking that all three of these were Middle Towns in the eighteenth century. *Tacoru* was probably Cherokee, and *Utaca* was possibly Cherokee.[175] It is possible that these towns were on the Tuckasegee and the headwaters of the Little Tennessee rivers in Pardo's day, but some or all of them could also have been on the upper Pigeon River at that time.

If these chiefs were from the upper Pigeon and Little Tennessee rivers,

The Indians

their meeting with Pardo at Cauchi is something of a puzzle. Tocae would seem to have been a much more likely meeting place. But Pardo was traveling rapidly because he knew that Moyano and his men were in danger of being killed by Indians, so that he only remained in Tocae for four hours. Hence, the five chiefs may simply have missed him at Tocae and caught up with him at Cauchi. It is also possible that they circumvented Tocae (for whatever reason) and reached Cauchi via a trail that went from the upper Pigeon River along Sandymush Creek to the vicinity of Marshall, North Carolina.

It was in Cauchi that Bandera recorded what was virtually his only ethnographic observation. That is, in Cauchi they saw an Indian man who was dressed as a woman and walking in the company of women. When Pardo asked Cauchi Orata for an explanation, he said that the man was his "brother," but because he was not a man for war or for doing the things that men do, he went about as a woman and did a woman's tasks (see figure 19). Pardo asked Bandera to write this down because it showed "how warlike the Indians were." We can only guess at Pardo's meaning. Was he being ironic, implying that this was evidence that the Indians were sissies? Or did he mean what he said—that they were so warlike that a male had the choice of being a man and a warrior, or else a woman?[176]

On his return, Pardo met with additional chiefs at Cauchi. He met with a Canosaqui Orata who was from "another place." This town name resembles that of Canosaca Orata, with whom Pardo met at Cofitachequi. However, it is absurd to think that Canosaca Orata would have traveled all the way to Cauchi to meet with Pardo again. Rather, it is more likely that Canosaqui Orata's town was the same as the "Canosoga" or "Canasagua," which Soto encountered when he was descending the Blue Ridge Mountains while going to Chiaha. Judging from Soto's trail time, Canosaqui was located in the vicinity of present-day Hot Springs, North Carolina, although the town could of course have been in a different location in Pardo's day.[177]

Canosaqui is no doubt the same as *Conasauga*, a place name used by the Cherokees in the eighteenth century and later. Interestingly, this word has a better Muskogean than Iroquoian etymology. In Iroquoian, the ending of the word could be -*aki*, "people of," whereas the first part of the word has no known Iroquoian etymology.[178] If it is Muskogean *kanosaka*, the etymology is probably "ground" + diminutive + plural + place, perhaps meaning "those of the mound."[179]

It is unlikely that a Muskogean chief from the Tennessee Valley would have visited a Cherokee town at the time of the Pardo expedition. Hence, it

Figure 19. Timucuan male transvestites carrying packbaskets of food. Females are shown loading the packbaskets. Engraving by Theodore de Bry after a painting by Jacques Le Moyne. (Photograph courtesy of the Smithsonian Institution.)

is possible that "Canosaqui" was a Muskogean place name that had been adopted by a Cherokee town. This practice was common among the Cherokees in the eighteenth century and later. As we shall presently see, several eighteenth-century Cherokee towns on the Little Tennessee River had names derived from Muskogean place names.

On another occasion at Cauchi, Pardo met with a large contingent of chiefs: Utahaque Orata, Anduque (cf. Enuque) Orata, Enjuete Orata, Guanuguaca Orata, Tucahe Orata, Guaruruquete Orata, and Anxuete Orata. Two of these chiefs may have previously met with Pardo. That is, *Tucahe* may have been the same as *Tocae*, or even *Atuqui*. And Enjuete was probably the same as *Enxuete*. It is remotely possible that *Guanuguaca* was the same as *Cuttagochi*, an eighteenth-century Cherokee town on a tributary of the upper Little Tennessee River.[180] On phonological grounds, all of the others could have been Cherokee-speaking, but their locations cannot be determined, except to say that all were probably located in the mountains.

On yet another occasion at Cauchi, Pardo met a delegation of chiefs who wanted to know where some of them were to pay their tribute. These were Cauchi Orata, Joara Chiquito Orata,[181] Canasahaqui Orata, Arasue Orata,

Guarero Orata, Arande Orata, and two *ynahaes oratas*. The last two, Bandera explains, were like justices. As already discussed, these were probably the same as the Creek *henihas*, who in later times served as mediators in disputes within their towns.

Presumably, the question of where tribute was to be taken was asked by Canosahaqui Orata, Arasue Orata, Guarero Orata, and Arande Orata, whereas Cauchi Orata and Joara Chiquito Orata were there to look after their own interests. Pardo told them that they were to take food to the garrison of Spaniards at Cauchi, but they should take deerskins to "another Guancamu" at Joara, who was "his [i.e., presumably Pardo's] brother." It is possible that "another Guancamu" referred to the Spaniard in command at Joara, or perhaps to Joara Mico. Presumably, most or all of the towns in this dispute were Cherokee towns in the mountains.

This incident has some interesting implications. On one hand, it provides evidence that Joara Mico had some authority over Cauchi and other mountain Cherokees. On the other, it may reflect yet again the chief's displeasure at having to pay tribute to Joara Mico.

On his return, when Pardo departed from Cauchi and went to Tocae, an interesting incident occurred, but its significance can probably never be known. Pardo was visited by Cauchi Orata, whom Pardo had commanded to take "certain captive Indians" to Joara. For performing this service, Pardo gave him a present. But nothing was said about who these Indians were, or where they had come from. One possibility is that they had been enslaved by Moyano when he attacked the Chiscas. Another possibility is that they were young Indians of Satapo and Chiaha whom Pardo was ostensibly taking out to place in a school that was to be built in Havana, but who would surely be held as hostages to ensure the good behavior of their relatives.[182]

It is doubtful that yet-to-be discovered documents will reveal much further information on sixteenth-century Indian towns on the upper French Broad, Catawba, and Nolichucky rivers. But archaeological research might provide some crucial information. At present, there is no evidence for the existence of Qualla phase sites on the upper French Broad River. Since both Tocae and Cauchi appear to have been inhabited by Cherokee-speakers, this can only mean that there must be some undiscovered Qualla sites near Asheville and Marshall, North Carolina, or else some late Pisgah sites in these locations. It is probable that the Swannanoa River had already been abandoned by Cherokee-speakers by the time of the Soto and Pardo expeditions.

Archaeology may also help establish the cultural identity of Joara and the Chiscas. According to Roy Dickens, Marion, North Carolina, where Joara was located, is at the easternmost limit of sites that have significant quantities of Pisgah ceramics. Similar quantities of Pisgah ceramics have been found in sites on the upper Nolichucky River, where the Chiscas probably lived.[183] But in Pardo's day, were the people at Joara and at Chisca the last Cherokees holding out in these areas, or were they new people who had moved in? They could, for example, have been Yuchi-speakers.

If Joara Mico was not a Cherokee-speaker, this would mean that Pardo encountered no micos among the Cherokees. It is true that Pardo only skirted the eastern edge of Cherokee territory, and hence there could have been Cherokee micos elsewhere whom he did not encounter. This may nonetheless lend support to Roy Dickens's hypothesis that the Cherokees were less hierarchically organized than Muskogean-speaking Lamar, Mulberry, and Dallas phase people.[184]

Coosa When Pardo departed from Cauchi and began descending the mountains into the Tennessee Valley, he entered the northern reaches of the territory of the paramount chiefdom of Coosa. The main town of Coosa, the center of the chiefdom, was at the Little Egypt site near Carters, Georgia (see figure 13). But the power and influence of the chief of Coosa extended far to the northeast and to the southwest of the main town. Pardo, in following a trail along the French Broad River, entered the chiefdom of Coosa near its northern frontier. To the south, the influence, if not the power, of the chiefdom of Coosa extended to the vicinity of Childersburg, Alabama.[185]

At the time of the Soto expedition, Indians in a wide area knew that Coosa was a large and powerful chiefdom. Soto first heard of the existence of Coosa when he was in the chiefdom of Ocute, on the upper Oconee River (see figure 2). The people of Ocute told him that Coosa was four days' travel to the northwest, and that it was a rich chiefdom with very large towns.[186] He again heard about Coosa when he arrived at Cofitachequi and was directed toward Chiaha, a chiefdom that was said to be subject to the chief of Coosa.[187]

The outstanding characteristic of the archaeological assemblage at Little Egypt, which is the Barnett phase, is that its pottery consists of about 75 percent grit-tempered Lamar-type pottery and about 25 percent shell-tempered Dallas and Mouse Creek pottery. In addition to the Little Egypt site, Barnett phase materials are present at a cluster of sites just down the Coosawattee River from Little Egypt. They also occur at a site near present-day Rome, Georgia, where the town of Ulibahali was located, and at the Johnstone Farm site and the King site, both of which are on the Coosa River west of Rome.[188] Sites as far south as Childersburg, which were under the power or influence of Coosa, are characterized by the Kymulga phase, yet another variety of Lamar culture. Just as Cofitachequi appears to have had power or influence over Catawba phase and perhaps Pisgah/Qualla phase towns to the north, Coosa had power or influence over Dallas phase towns to the north.

The nature and distribution of native fortifications indicate that Coosa was militarily prepared to defend its territory against outside enemies and that it may have been expansionist, primarily in a northerly direction. There is neither documentary nor archaeological evidence that the Little Egypt site was fortified. Being in the center of the chiefdom, it was relatively safe from suprise attack. But archaeological evidence shows that the King site was fortified, and perhaps its fortifications resembled those that surrounded Ulibahali at the time of the Soto expedition. The palisade around Ulibahali was made of large posts sunk firmly into the earth, with poles as thick as one's arm placed crosswise to the height of a lance, and the whole structure was coated with mud. Small windows for archers were placed at intervals along the palisade (see figure 11).

There is evidence that the chief of Coosa collected tribute from some of the towns that were subject to him, and that if a town failed to pay tribute, he was willing to inflict military punishment on them. Such tributary relationships still existed at the time of the Tristán de Luna expedition. When a party of Luna's soldiers visited Coosa, the chief persuaded them to accompany him on a raid against the Napochies, who lived near present-day Chattanooga, and who were refusing to pay tribute.[189] It is not possible to determine whether the chief of Coosa also ruled Chiaha and other northern towns in such a heavy-handed fashion, but as Pardo was to discover, the chief of Coosa could count on these northern towns for military cooperation.

It is striking that during the time of Pardo's second expedition several Cherokee chiefs visited Joara, and Joara Chiquito Orata visited Cauchi, but no Cherokees visited any of the towns in Coosa, and none from Coosa visited the Cherokees. This implies that the Cherokees and the people of Coosa were at odds with each other, and perhaps they were at war.

The first town of Coosa encountered by Pardo was Tanasqui, situated in a V formed by the junction of two rivers; its land side was fortified by a palisade that had three defensive towers built into it. This appears to have been the first Indian fortification Pardo had seen, because he asked the chief what the purpose of the palisade was. The chief replied simply that it was a defense against his enemies. This experience is consistent with that of Soto, whose chroniclers note that Chiaha was the first fortified town they encountered on their travels, and also similar to the experience of Moyano, who was suprised to find how strongly fortified was the Chisca town he attacked.[190]

There are two possible locations for Tanasqui: at the junction of the Pigeon and French Broad rivers or at the junction of the Nolichucky and French Broad rivers. Excavation has been done at a suspected location of Tanasqui between the Pigeon and the French Broad rivers, but thus far the results are inconclusive.[191]

On phonological grounds, "Tanasqui" could be either a Muskogean or a Cherokee word.[192] As we shall see, it is possible that this place name is the origin of the Cherokee town name *Tennassee*, which in the eighteenth century was located on the Little Tennessee River; if so, it is the namesake of the state of Tennessee.[193] This implies that the town of Tanasqui moved from the French Broad River to the Little Tennessee River at some point after the Pardo expedition.[194]

When Pardo entered the chiefdom of Coosa, Bandera ceased noting down and listing by name the chiefs who assembled at the various towns. In part this was because Pardo, perhaps without knowing it, had gone beyond the area in which he could coerce or intimidate the Indians into obedience. Beyond Cauchi, he found no more houses built for the Spaniards to use, and no more storehouses for food. Bandera's omission is unfortunate, because without the chiefs' names it is difficult to make inferences about the social and political relationships among the various towns.

As already mentioned, the chief of Coosa was a paramount chief—a chief of chiefs. When Pardo reached Zimmerman's Island in the French Broad River, he noted that the place had two names: Chiaha and Olamico. This latter name was applied to the chief as well. Pardo never explains why this place had two names, but the reason can be discerned from the etymologies of the names. Chiaha is one of the more important Creek place names, one that is still in use today. It is possibly the Koasati word *čayha*, "tall," whereas *Olamico* is a Koasati word meaning "chief town" or "head town."[195] As it turned out, Chiaha appears to have been both the name of a particular town, whose headman was Chiaha Orata, and also a name that could be applied to the chiefdom to which this town belonged. But the chief of this larger chiefdom was called Olamico, and the same name was applied to his island town. This usage would appear to be comparable to

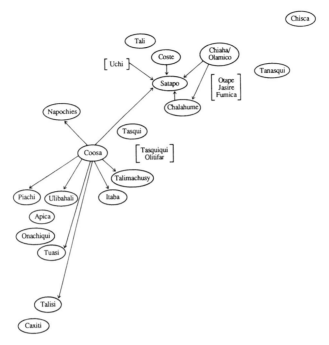

Figure 20. Sociogram of Coosa and its hinterland.

that encountered by Soto at Cofitachequi, whose main town was called *Talomeco*.

At the time of the Soto expedition, Chiaha was the dominant chiefdom in the upper Tennessee Valley, and Chiaha was subject to Coosa (see figure 20). But it is not clear whether the power of Olamico extended over Dallas phase towns to the south such as Satapo and Coste, which are discussed next, or whether these were three chiefdoms that were equal to each other but subject to the paramount chief of Coosa.

As we have already seen, Pardo departed from Olamico and began traveling toward the southwest, going toward Coosa, and ultimately, he believed, toward the Spanish silver mines in Zacatecas.[196] On their first night on a trail that followed along the eastern flank of Chilhowee Mountain, they probably camped near the junction of Cove Creek and Walden Creek. Here they were visited by three chiefs who brought them food: Otape Orata, Jasire Orata, and Fumica Orata. *Otape* appears to be a Muskogean word referring to chestnut trees or to a place where they may be found.[197] *Fumica* is not identifiable; only Muskogean and Yuchi have *f*.

Jasire is not identifiable.[198] It is likely that all three of these towns were located on the Little Pigeon River and its headwaters.

The first town they reached was Chalahume on the upper Little Tennessee River. *Chalahume* appears to be a Muskogean word that may refer to a variety of bass or trout or to fish poison.[199] Chalahume was located at the Dallas phase archaeological site (40BT7) on the Little Tennessee River just below the mouth of Abrams Creek. In 1755 the Cherokee town of *Chilhowee* was located in this vicinity.[200] This indicates that after the Muskogean-speakers moved out of this area in the seventeenth century, Cherokee-speakers moved in, and because /m/ is an infrequent phoneme in the Cherokee language, they substituted *w* for *m*.

It was at the next town Pardo came to, Satapo, that he decided to return to Chiaha because of a conspiracy to attack him. *Satapo* appears to be a Muskogean word meaning "persimmon tree." The town was located at the large Dallas phase site (40MR7) just below the junction of Citico Creek and the Little Tennessee River. In the eighteenth century the Cherokee town of Citico was located in this same area.[201] Again, this is a Muskogean place name that has been modified by Cherokee phonology. Cherokee, lacking bilabial stops, substituted *c* (i.e., *k*) for *p*.[202]

There are several other eighteenth-century Cherokee place names on the Little Tennessee River that have no etymology in Cherokee: Tuskeegee, Tomotley, Tennassee, Tallessee, and Chota. Some Cherokee-speakers have claimed that Tellico is derived from *ade·la e·kwa*, "big bead," or from *ta·li e·kwa* "big two." But both of these may be folk etymologies.[203] Some or all of these are probably Muskogean place names that were adopted by Cherokee-speakers. Thus, the linguistic evidence implies that Cherokee-speakers were latecomers to an area that in the sixteenth century was occupied by Muskogean-speakers.

Upon arriving at Satapo, Pardo almost immediately perceived that he was facing a hostile situation. He was told that in the past some Spaniards had been "through these parts" both on foot and on horseback, and that the Indians subject to Satapo had killed many of them. This is clearly a reference to the Soto expedition, but the substance of these claims is puzzling. None of the Soto chroniclers mentions the loss of life on either side when Soto was in the territory of the paramount chief of Coosa.[204] One possibility is that the people of Satapo may have been exaggerating their military prowess to make themselves appear to be more formidable to Pardo than they really were. Another is that more violence may have occurred than was mentioned by the Soto chroniclers. Still another is that the people of the chiefdom of Coosa (including Satapo) may have aligned themselves with Chief Tascaluza in the suprise attack waged against the Soto expedition at Mabila. Their participation at the battle of Mabila would explain the large number of burials recovered at the King site that showed evidence of having been wounded and killed by European steel weapons.[205] This would be consistent with the testimony, already discussed, of Juan de Ribas in 1600, who appeared to be saying that he had seen Spanish artifacts at Satapo as well as paintings depicting mounted Spanish lancers on the walls of Indians' houses.[206]

When night fell on the day they arrived at Satapo, Pardo posted two sentinels at "two high points" of the village. These high points were probably towers on a palisade surrounding the town. The existence of a palisade at the Citico site has been confirmed by archaeological research. Pardo received another intimation of trouble when the sentinels reported having heard "a very great noise of Indians outside of the place."

The true extent of the trouble became clear that same night when an Indian went to Guillermo Rufín, who was asleep in an Indian house. This Indian, seemingly a member of the chiefdom of Chiaha, had been traveling with Pardo's expedition for two or three days. What is noteworthy is that this man appears to have communicated with Rufín directly, using no intermediaries, and this must mean that the man spoke a Muskogean language that Rufín could understand, or else a lingua franca that both could understand. Rufín learned from him that the chief of Coosa had come to Satapo with many Indians who were his "vassals," and together with Olamico, Satapo, Uchi, and Casque, they planned to ambush Pardo three times on his march southward toward Coosa. Earlier, at Olamico, Pardo had heard that the Chiscas were among the conspirators, although their participation was not mentioned to Rufín.[207] Moreover, the chiefs had agreed among themselves not to give food or anything else to Pardo unless they were given something in return. Thus at the time of the Pardo expedition, the paramount chief of Coosa was still able to count on the military cooperation of Indians as far north as the Little Tennessee and French Broad rivers, and perhaps even from the Chiscas further to the north.

Casque and *Uchi* are mentioned only in connection with this conspiracy. Casque may have been a town subject to Coosa that neither Soto nor Pardo visited. It may be notable that in the late seventeenth century the Tennessee River was known as the *Caskinampo*.[208] This may be derived from Koasati /káski/, "warriors," and /námpon/, "to be so many," that is, "their warriors are many."[209] Other possibilities are that *Casque* may be a garbled version of *Coste*, the town on Bussell Island in the mouth of the Little

Tennessee River that Soto visited. *Coste* is probably the namesake of *Koasati*, a people who lived near the junction of the Coosa and Tallapossa rivers in Alabama in the eighteenth century. Still another possibility is that *Casque* was a garbled version of *Tasqui*, a town that Soto came upon about a day's march before he reached the main town of Coosa.[210]

On their way back to Olamico, Bandera took depositions from three soldiers whose names appear on a supply list of soldiers to whom Pardo gave shoes and sandals the day after he first reached Olamico: Alonso Velas (or Vela), Diego de Morales, and Juan Perez de Ponte de Lima. Those receiving shoes and sandals were Moyano and his soldiers, plus a few others.[211] Hence, some or all of these three men were with Moyano during his stay at Joara and at Olamico, and they appear to have learned to speak at least a few words of the Indian languages spoken there.

Alonso Velas testified to Bandera that when he was at Satapo he talked with an Indian who came along with the company as an interpreter, and who was a native of "Oluga." Either Velas spoke this Indian's language, or else the Indian had some command of Spanish. In any case, they appear to have communicated with each other directly. The Indian told Velas that the paramount chief of Coosa did not intend to let Pardo and his company enter his town, and that he was not going to give Pardo any food unless he was compensated for it. Moreover, this Indian told Velas that it would trouble him very much to go to Coosa because when Soto had come through, he had taken five of his "brothers" with him to Coosa. The Indian believed that the people of Coosa had killed Soto and his men, and that they had enslaved his "brothers."

Regrettably, nothing is known about this Indian of Oluga. It is not clear whether the five men were his actual kinsmen, or whether they were men of his town or chiefdom. It is also not known where Oluga was located, although presumably it was not a town that was subject to Coosa. Hence, Oluga could have been located anywhere in the area between the mountains and the Atlantic coast. In any case, Velas's testimony is further confirmation that Soto had come through the area visited by Pardo, and it may constitute further evidence that subjects of Coosa participated in the battle at Mabila.

Diego de Morales testified that while they were in Satapo he spoke with an Indian who apparently understood some Spanish. The Indian told him that if they went to Coosa, they would not be allowed to enter the town and that if they wanted food they would have to pay for it.

Juan Perez de Ponte de Lima gave the most interesting testimony of the three. Ponte testified about an incident that occurred two days after Pardo first arrived at Olamico. On the night of Sunday, October 9, while he was on guard duty at Olamico, an Indian from Olamico came to him and told him that the chief of Satapo had two sons who "a scundidos [probably *a escondidos*] Españoles que Chafahane." Ponte translated *chafahane* as "he is deceitful." This would seem to indicate that the Indian was speaking partly in Spanish and partly in his own tongue. The Indian further said that Satapo and Coosa "aguacamu E ypeape," which means, according to Ponte, "that they were going to kill him." The word "ypeape" is Koasati *i·bi*, "kill," plus the modal suffix *-a·pi*, thus meaning "as for them, they would kill him." *Chafahane* is Koasati for "he intends deceit."[212]

This shows that by the time of Pardo's second expedition, the soldiers who had been with Moyano were communicating with the Indians at least on a limited basis. And it may indicate that the Indians were learning some Spanish. Even more interestingly, these recorded phrases of Indian words are further proof that the language of Olamico (Chiaha) was predominantly Koasati.

In late May of 1568, Father Juan Rogel visited Guale. He reported that almost all of the Spaniards there spoke the language that was the most widely understood in Florida. A soldier there told Rogel that he had gone into the interior for 200 leagues (undoubtedly with Pardo), and that with this language of Guale he could communicate very well.[213] By December 1569, Father Domingo Agustin Baéz had learned this language, and he was translating sermons and Christian doctrine into it.[214] Baéz is said to have written a grammar of the Guale language, but if he did, it has never been found.[215]

Pardo got no further south than Satapo, but in his shorter report, Bandera summarizes information on towns further south that he obtained from a soldier from Pardo's company who claimed to have gone beyond there. Unfortunately, Bandera does not identify this man, nor does he describe the circumstances that led him beyond the Little Tennessee River. In view of the hostility that was so evident at Satapo, it is unlikely that this man could have been one of the soldiers who were stationed at Olamico or Cauchi. Another possibility is that he was a member of the Luna or Soto expeditions. But if he was a member of the Luna expedition, in all likelihood he would not have known about towns lying between Satapo and Coosa. If he had been a member of the Soto expedition, he would have had some knowledge of these towns, but he would have been a relatively old man by the time of the Pardo expedition. The account of towns to the south of Coosa is so sketchy and inaccurate that it is doubtful if the man in question had been a member of either the Luna or Soto expeditions.

Whatever the source of Bandera's information, it provides a few more details on Coosa. The country between Satapo and Coosa was said to have been thinly inhabited, with only two or three small towns. Two days' travel from Satapo one came upon the towns of Tasqui and Tasquiqui. Soto found the town of Tasqui just south of the Hiwassee River. Tasquiqui is no doubt the namesake of the Creek town "Tuskegee," where a dialect of the Koasati language was spoken.[216] Using the distance Pardo customarily traveled in a day, we can estimate that these towns were about three days from Satapo, although the Indians, who traveled faster than the Spaniards, could have covered the distance in two days. The other town between Tasquiqui and Coosa that was mentioned by name was *Olitifar*, a destroyed town. Bandera gave no explanation for its destruction.

In summary, Coosa was a paramount chiefdom that resembled the paramount chiefdom of Cofitachequi in several interesting respects: (1) The central towns of both of these chiefdoms were characterized by a variety of Lamar/Chicora pottery. (2) But the central towns of both held political sway over towns that were characterized by quite different cultural traditions, that is, Dallas in the case of Coosa, and Uwharrie/Burke/Pisgah in the case of Cofitachequi. (3) Eastern Muskogean languages were probably spoken in the central towns of both chiefdoms. (4) But both had political sway over towns that spoke quite different languages, that is, Koasati in the case of Coosa, and Catawban and Cherokee and possibly Yuchi in the case of Cofitachequi. (5) Both chiefdoms appear to have had a military frontier boundary to the northeast, as evidenced by palisaded towns, and both may have expanded in this direction not many centuries before the arrival of Soto, Luna, and Pardo.

Economic Patterns

From Pardo's transactions with the Indians, several inferences can be made about their economic system. The ease with which Pardo was able to command the Indians to lay up supplies of corn for the Spaniards suggests that an indigenous system of redistribution and/or tribute was already in place. The degree to which most of the Indians gave the appearance of accepting subordination to the Spaniards is also noteworthy. The former subjects of Cofitachequi may have regarded Juan Pardo as a paramount chief. The only exceptions were, first, the Indians of Coosa, who, in their resistance to the Spaniards, demonstrated their subordination to a different master—the paramount chief of Coosa—and, second, the subjects of the chiefs of Guatari, who did not seem eager to work for Pardo, and who may have been less hierarchically organized than were other Indians Pardo met.

When Pardo arrived in Tocae on his return from the Tennessee Valley, he was met by a delegation of four Indians who wanted to know where they were to pay tribute. Pardo ordered the four chiefs, all of whom probably were from towns in the mountains, to take food to Cauchi, where Pardo had stationed eleven men in a small fort he had built. But he said that they should take deerskins to Joara, to "another Guancamu," whom Pardo said was his "brother."

Although this is only a single incident, it implies that tribute in food could be distinguished from tribute in deerskins. It is reminiscent of an incident that occurred after the soldiers from Luna's colony went with the chief of Coosa and his warriors to attack the Napochies, a subject people of Coosa who had rebelled and were refusing to pay their accustomed tribute. After the Napochies were defeated, they promised to again pay tribute to Coosa. They were to pay it three times a year, and it was to consist of game, fruits, chestnuts, and nuts.[217] All of these are forest products. Hence it is possible that in paramount chiefdoms, tribute for the most closely affiliated towns consisted of corn, while the more distant towns paid tribute in the form of skins, nuts, fruits, and so on.

In some of the towns Pardo visited the Indians had built tall cribs or storehouses in which the corn was to be kept. Perhaps these were like the *barbacoas* described by the Gentleman of Elvas, which the members of the Soto expedition first saw at a town of the chiefdom of Toa, on the upper Flint River. He describes the barbacoa as "a house with wooden sides, like a room, raised aloft on four posts, and has a floor of cane."[218] The chiefs and principal men had many large barbacoas, to which their subjects brought tribute in the form of "maize, skins of deer, and blankets of the country."[219] In 1700, John Lawson described similar corn cribs in his journey through the Carolina backcountry:

They make themselves Cribs after a very curious Manner, wherein they secure their Corn from Vermin; which are more frequent in these warm Climates, than Countries more distant from the Sun. These pretty fabricks are commonly supported with eight Feet or Posts, about seven Foot high from the Ground, well daub'd within and without upon Laths, with Loom or Clay, which makes them tight, and fit to keep out the smallest Insect, there being a small Door at the gable end, which is made of the same Composition, and to be remov'd at Pleasure, being no bigger, than that a slender Man may creep in at, cementing the Door up with the same Earth, when they take Corn out of the Crib, and are going from Home, always finding their Granaries in the same Posture they left them; Theft to each other being altogether unpractis'd, never receiving Spoils but from Foreigners.[220]

In a few instances, as we shall see, the Indians apparently stored corn in storage rooms *within* the houses they built for Pardo.

The readiness with which the Indians could supply deerskins to Pardo suggests that they were regularly used as a medium of tribute. This confirms observations made by members of the Soto expedition. Namely, one of the buildings at Cofitachequi contained a great quantity of deerskins (some with designs painted on them), which had no doubt been collected as tribute.[221] Also, Soto was occasionally presented with gifts of deerskins by the Indians, again suggesting that these were especially appropriate gifts given by the less powerful to the more powerful.[222] Both at the mouth of the St. Johns River and at Port Royal Sound, Indians gave gifts of deerskins to Jean Ribaut in 1562.[223]

Some of Pardo's transactions suggest that he participated in a Southeastern trade arrangement that has long been thought to have existed. Namely, conch shells were traded from the coast into the interior, and high-quality deerskins were traded from the interior to the coast. When Pardo was in Tocae, in the mountains, he gave several presents to Uastique Orata, one of which was a large conch shell. On his return journey, Pardo gave painted deerskins he had obtained in the interior to several chiefs who lived on the Coastal Plain.

None of these painted deerskins from the Southeast have, of course, been recovered from archaeological excavations, nor are any known to exist in museum collections. However, a painted skin blanket and shirt are represented in drawings by Philip G. F. von Reck, who depicts their use by Indians along the lower Savannah River in 1736 (see figure 28).[224]

This trade pattern linking interior and coast is confirmed by Father Juan Rogel, who described the Indians in and around Orista. He says that they were very much the traders (*muy mercaderes*), who well understood buying and selling and who took things inland that were not found there to trade for the things that were not found on the coast.[225] Unfortunately, Rogel does not specify what these items were.[226]

On another occasion, Pardo mentions the existence of an indigenous trade in salt, which is confirmed in the Soto chronicles. When Soto reached Cofitachequi, his chroniclers noted that here they were given an abundance of salt of good quality, but there is no indication of where the salt came from.[227] When Pardo was at Gueça, he told the chiefs who had assembled there that if they could not bring in corn or deerskins as tribute, they could bring in salt. The northerly location of Gueça may imply that this salt came from the north. As we have already seen, this is confirmed by the testimony of the two Indian women whom Governor Canço questioned in

1600. It is quite clear that salt was not in good supply everywhere. For example, Soto's men had a great need for salt while they were traveling from Apalachee through Ocute, that is, during the entire time they were traveling through what is now the state of Georgia.[228]

Notes

1. William C. Sturtevant has argued that the documents of the sixteenth-century Spanish explorers of the southeastern United States are too fragmentary and problematical to be of use in reconstructing the social characteristics of the Indians who lived there. To those who read these pages it will be obvious that I have a more optimistic assessment of what may be learned from these documents. William C. Sturtevant, "Tribe and State in the Sixteenth and Twentieth Centuries," in *The Development of Political Organization in Native North America*, ed. Elisabeth Tooker, 1979 Proceedings of the American Ethnological Society (Washington, D.C.: American Ethnological Society, 1983), pp. 3–16.
2. Emmanuel J. Drechsel, "The Question of the *Lingua Franca* Creek," in *1982 Mid-America Linguistics Conference Papers* (Lawrence: Department of Linguistics, University of Kansas, 1983), pp. 388–400.
3. In 1776, Louis Le Clerc Milfort heard a Creek give a long oration in the Lower Creek town of Coweta. Each time the orator paused, the whole audience would say *ka* (possibly a mistranscription of *ya*), which Milfort says means "yes." When the orator finished his speech, the audience said *mado*, "very well." Louis Le Clerc Milfort, *Memoirs or a Quick Glance at My Various Travels and My Sojourn in the Creek Nation*, trans. and ed. Ben C. McCary (Savannah, Ga.: Beehive Press, 1972), pp. 95–100.
4. Randolph J. Widmer, *The Evolution of the Calusa: A Nonagricultural Chiefdom on the Southwest Florida Coast* (Tuscaloosa: University of Alabama Press, 1988); Mark J. Lynott, Thomas W. Boutton, James E. Price, and Dwight E. Nelson, "Stable Carbon Isotopic Evidence for Maize Agriculture in Southeast Missouri and Northeast Arkansas," *American Antiquity* 51(1986):51–65.
5. Bruce D. Smith, "The Archaeology of the Southeastern United States: From Dalton to de Soto, 10,500–500 B.P.," *Advances in World Archaeology* 5(1986):44, 50–51; Vincas P. Steponaitis, "Prehistoric Archaeology in the Southeastern United States, 1970–1985," *Annual Review of Anthropology* 15(1986):383–87; Helen C. Rountree, *The Powhatan Indians of Virginia: Their Traditional Culture* (Norman: University of Oklahoma Press, 1989), pp. 140–52.
6. Robert L. Carneiro, "A Theory of the Origin of the State," *Science* 169(1970):733–38; "The Chiefdom: Precursor of the State," in *The Transi-*

tion to Statehood in the New World, ed. by Grant D. Jones and Robert R. Kautz (Cambridge: Cambridge University Press, 1981), pp. 63–67; "Cross-Currents in the Theory of State Formation," *American Ethnologist* 14(1987):767.

7. David G. Anderson, "Stability and Change in Chiefdom-Level Societies: An Examination of Mississippian Political Evolution on the South Atlantic Slope," paper presented at the 43d Southeastern Archaeological Conference, November 7, 1986.
8. James L. Phillips and James A. Brown, *Archaic Hunters and Gatherers in the American Midwest* (New York: Academic Press, 1983), *passim*. Smith, "Archaeology," pp. 43–53; William H. Marquardt, "The Development of Cultural Complexity in Southwest Florida: Elements of a Critique," *Southeastern Archaeology* 5(1986):63–70.
9. Dan F. Morse and Phyllis A. Morse, *Archaeology of the Central Mississippi Valley* (New York: Academic Press, 1983), pp. 201–35.
10. Bruce Smith argues so strongly against the old theory that Mississippian culture was largely or exclusively spread by the migration of people that he appears to deny that "site-unit intrusion" could have played any role in the Mississippian phenomenon. Bruce Smith, "Mississippian Expansion: Tracing the Historical Development of an Explanatory Model," *Southeastern Archaeology* 3(1984):13–22.
11. David G. Anderson, David J. Hally, and James L. Rudolph, "The Mississippian Occupation of the Savannah River Valley," *Southeastern Archaeology* 5(1986):38; Roy S. Dickens, Jr., "An Ecological-Evolutionary Approach to the Interpretation of Cherokee Cultural Development," MS., 1984.
12. David G. Anderson, "The Mississippian in South Carolina," in *Papers in Honor of Dr. Robert L. Stephenson*, ed. by Albert C. Goodyear and Glen T. Hanson, South Carolina Institute of Archaeology and Anthropology, in press.
13. LAMAR Institute Conference on South Appalachian Mississippian, Macon, Ga., May 1986.
14. Elman R. Service, *Origins of the State and Civilization* (New York: W. W. Norton, 1975), pp. 15–16; Timothy K. Earle, "Chiefdoms in Archaeological and Ethnohistorical Perspective," *Annual Review of Anthropology* 16(1987): 289.
15. Chester B. DePratter, "Late Prehistoric and Early Historic Chiefdoms in the Southeastern United States," Ph.D. dissertation, University of Georgia, 1983, 204–11.
16. Mark Williams, "Piedmont Oconee River," paper presented at the LAMAR Institute Conference on South Appalachian Mississippian, May 9, 1986, Macon, Ga.; Chester B. DePratter and Chris Judge, "Wateree River," paper presented at the LAMAR Institute Conference on South Appalachian Mississippian, May 9, 1986, Macon, Ga.; Anderson et al., "Savannah River," pp. 41–42; Anderson, "Stability and Change."

17. David J. Hally, Charles Hudson, Chester B. DePratter, and Marvin T. Smith, "The Protohistoric Period along the Savannah River," paper presented at the annual meeting of the Southeastern Archaeological Conference, November 7, 1985, Birmingham, Ala.; Anderson et al., "Savannah River," pp. 47–48.
18. Williams, "Piedmont Oconee River."
19. DePratter and Judge, "Wateree River."
20. Charles Hudson, Marvin T. Smith, and Chester B. DePratter, "The Hernando de Soto Expedition: From Apalachee to Chiaha," *Southeastern Archaeology* 3(1984):71–72.
21. Bruce D. Smith, "Variation in Mississippian Settlement Patterns," in *Mississippian Settlement Patterns*, ed. Bruce D. Smith (New York: Academic Press, 1978), pp. 479–503; Christopher S. Peebles and Susan M. Kus, "Some Archaeological Correlates of Ranked Societies," *American Antiquity* 42(1977):421–48. Chester DePratter's "Chiefdoms" is an exception in that it is based both on archaeological evidence and on primary historical documents from the sixteenth and seventeenth centuries.
22. In eighteenth-century English sources, the Catawba chief was *Eractasswa* or *Arataswa*, no doubt a variant of *Orata*. See James H. Merrell, "'Minding the Business of the Nation': Hagler as Catawba Leader," *Ethnohistory* 33(1986): 56, 67.
23. In speaking of the mico of Guale, Jaime Martinez says that the word means "king" or "prince." "Brief Narrative of the Martyrdom of the fathers and brothers of the Society of Jesuits, Slain by the Jacan Indians of Florida," in "The first Jesuit Mission in Florida," comp. Ruben Vargas Ugarte, trans. Aloysius T. Owen, *U.S. Catholic Historical Society Records and Studies*, 1935, vol. 25, p. 129.
24. Bandera II, folio 10 verso to folio 14.
25. Herbert I. Priestly, *The Luna Papers, 1559–1561* (Deland: Florida State Historical Society, 1928), vol. 2, p. 290.
26. Charles Hudson, Marvin Smith, David Hally, Richard Polhemus, and Chester DePratter, "Coosa: A Chiefdom in the Sixteenth-Century Southeastern United States," *American Antiquity* (4)1985:723–37.
27. I am grateful to Paul Hoffman for this explication of sixteenth-century Spanish political terminology. I am grateful to Eugene Lyon for pointing out that as early as 1578–80, the Spaniards were referring to the Chief of Tolomato and Guale as "Mico Mayor." But by this date the Spaniards had been attempting to manipulate native political affairs for more than a decade, and the Mico Mayor of Tolomato/Guale would seem to have been more similar in political stature to Joara Mico and Guatari Mico than to the paramount chief of Coosa.
28. Gentleman of Elvas, *Narratives of De Soto in the Conquest of Florida*, trans. Buckingham Smith (New York: Bradford Club, 1866), pp. 49–50.
29. Rodrigo Ranjel, "Narrative," in *Narratives of the Career of Hernando de*

Soto, trans. Buckingham Smith, ed. Edward G. Bourne (New York: Allerton, 1922), pp. 89–90.
30. Garcilaso de la Vega, *The Florida of the Inca*, ed. John Varner and Jeannette Varner (Austin: University of Texas Press, 1952), pp. 304–8.
31. Elvas, *Narratives*, p. 63.
32. Ranjel, "Narrative," pp. 98–99, 112; Elvas, *Narratives*, p. 62.
33. See DePratter, "Chiefdoms," pp. 111–53.
34. Martinez.
35. DePratter, "Chiefdoms," pp. 189–96.
36. "*ynihaes son como si dixemos justicia o jurados q. mandan al pueblo.*" Bandera II, folio 16 verso. This could also be translated as "Who direct [or command] the people." According to Paul Hoffman there is no precise English equivalent of the Spanish *justicia* or *jurado*.
37. "*Al yniha q. es como alquazil q. manda el pueblo.*"
38. John R. Swanton, *Social Organization and Social Usages of the Indians of the Creek Confederacy*, 42d Annual Report of the Bureau of American Ethnology (Washington, D.C.: GPO, 1928), p. 297.
39. Garcilaso, *Florida*, pp. 297–98.
40. *Mati* has no known meaning in Catawba. *Yssa* is probably Catawban *i·swq̓*, "chief." Robert Rankin, personal communication.
41. Swanton, *Creek Confederacy*, pp. 295–96.
42. Luys Hernandez de Biedma, "Relation," in *Narratives of the Career of Hernando de Soto*, trans. Buckingham Smith, ed. Edward G. Bourne (New York: Allerton, 1922), p. 11.
43. Garcilaso, *Florida*, p. 304.
44. Biedma, "Relation," p. 13.
45. Elvas, *Narratives*, p. 71.
46. Ranjel, "Narrative," p. 91.
47. The Soto chroniclers spelled the name of this chiefdom in various ways: Cofitachequi (Ranjel), Cofitachique (Biedma), Cufitachiqui (Elvas), and Cofachiqui (Garcilaso de le Vega).

   John R. Swanton incorrectly located Cofitachequi on the Savannah River, placing its central town at or near Silver Bluff, and he incorrectly equated Cofitachequi with Kasihta, a Lower Creek town (*Final Report of the United States De Soto Expedition Commission* [Washington, D.C.: Smithsonian Institution Press, 1985], pp. 49, 166–86).

   Wisely, Steven G. Baker rejected both Swanton's location for Cofitachequi and his suggestion Cofitachequi should be equated with Kasihta. Using Leland G. Ferguson's delineation of the "South Appalachian Mississippian" province (Leland G. Ferguson, "South Appalachian Mississippian," Ph.D. dissertation, University of North Carolina, 1971) along with seventeenth-century English documents, Baker correctly placed Cofitachequi on the

Catawba-Wateree River, and he realized that Cofitachequi was a "greater chiefdom" that encompassed several "mini-chiefdoms" located in the area between the mountains and the lower coastal plain (Steven G. Baker, "Cofitachequi: Fair Province of Carolina," M.A. thesis, University of South Carolina, 1974, pp. xi–xii, 145).

Baker incorrectly placed the central town of Cofitachequi too far south, in the High Hills of the Santee area. He also erred in arguing that Cofitachequi was still a "greater chiefdom" as late as 1670 ("Cofitachique," pp. 37, 124). Indeed, it can be argued—as I do—that the paramount chiefdom of Cofitachequi had probably deteriorated by 1540 and certainly by 1566.

48. Biedma, "Relation," pp. 11–12; Ranjel, "Narrative," p. 94.
49. The Square Ground phase on the Lower Ocmulgee and upper Satilla rivers may represent such people. Frankie Snow, "Lower Ocmulgee/Upper Satilla Rivers," paper presented at the LAMAR Institute Conference on South Appalachian Mississippian, May 9, 1986, Macon, Ga.
50. Karen Booker, Charles Hudson, and Robert Rankin, "Multilingualism in Two Paramount Chiefdoms in the Sixteenth-Century Southeastern United States," MS.
51. Ibid. Also see Ranjel, "Narrative," p. 101. Garcilaso spells it *Talomeco* (*Florida*, p. 314).
52. Elvas, *Narratives*, pp. 49–50.
53. Ranjel, "Narrative," p. 101.
54. Vernon J. Knight, Jr., "Mississippian Ritual," Ph.D. dissertation, University of Florida, 1981, p. 47.
55. George E. Stuart, "The Post-Archaic Occupation of Central South Carolina," Ph.D. dissertation, University of North Carolina at Chapel Hill, 1975, *passim*. Time has not been kind to these mounds. The Blanding Mounds were leveled in 1826, and the soil from them was used to fertilize surrounding land. The Boykin site has been almost completely eroded away by the Wateree River. Of the two large mounds at the Mulberry site, Mound B has been much reduced by plowing; over half of Mound A has eroded into the river; and seven of the eight small mounds have been leveled by plowing. Both of the Adamson mounds are still extant, although reduced by plowing and pot hunting.
56. Ibid., pp. 169–72. E. G. Squier and E. H. Davis, *Ancient Mounds of the Mississippi Valley*, Smithsonian Institution Contributions to Knowledge, vol. 1, (Washington, D.C., 1845), pp. 105–8.
57. Anderson, "Mississippian," p. 15.
58. Joffre L. Coe, *The Formative Cultures of the Carolina Piedmont*, American Philosophical Soceity Transactions, n.s., vol. 54, pt. 5 (Philadelphia, 1964).
59. David G. Anderson, Charles E. Cantley, and A. Lee Novick, *The Mattassee Lake Sites: Archaeological Investigations along the Lower Santee River in the Coastal Plain of South Carolina*. U.S. Department of the Interior, Na-

tional Park Service, Interagency Archaeological Services-Atlanta, Special Publication, 1982; Anderson, "Mississippian," p. 8.
60. David G. Anderson, "The Distribution of Prehistoric Ceramics in the Coastal Plain of South Carolina," MS. on file at the Institute of Archaeology and the Charleston Museum, 1975; Anderson, "Mississippian," p. 9.
61. Joffre L. Coe, "The Cultural Sequence of the Carolina Piedmont," in *Archaeology of Eastern United States*, ed. James B. Griffin (Chicago: University of Chicago Press, 1952), pp. 308–11; Jack Hubert Wilson, Jr., "A Study of the Late Prehistoric and Historic Indians of the Carolina and Virginia Piedmont," Ph.D. dissertation, University of North Carolina, Chapel Hill, 1983.
62. Chester DePratter, personal communication.
63. Garcilaso, *Florida*, p. 296. If structures still stood on the Adamson site at the time of the Soto expedition, and if we discount Garcilaso's description of Talimeco, then the Mulberry site could have been the first town Soto reached, and Talimeco could have been at the Adamson site.
64. Ranjel, "Narrative," pp. 101–2.
65. Garcilaso, *Florida*, p. 311.
66. Ibid., p. 314.
67. See Booker et al., "Multilingualism." The Cofitachequi, Santa Elena, Joara, and Coosa (see infra) sociograms have been constructed using the following conventions: Towns whose locations are known with varying degrees of precision, are encircled. Towns whose representatives came from afar to meet with Pardo are bracketed; even though their locations cannot be determined from the documents, I have nonetheless attempted to locate them in terms of broad compass directions with respect to the towns that can be located. A solid line with an arrow indicates that a representative of one town traveled to visit another town. A dotted line with an arrow indicates that one or more visits between towns probably occurred, for example, in instances where Pardo ordered one town to pay tribute to another.
68. Karen Booker, personal communication.
69. Ranjel, "Narrative," p. 100. It is possible that Ranjel's *p* is a mistranscription of *s*, either by Oviedo or by a later copyist. Ranjel calls these cribs *barbacoas*, and he uses the same term for the elevated frames on which bodies of the deceased were laid in the temple at Talimeco.
70. Karen Booker, personal communication.
71. Bandera I.
72. Vowel syncope, especially of an identical vowel, is common in Muskogean. Karen Booker, personal communication.
73. Ibid.
74. Sometimes spelled "Ysaa", and also "Ysa."
75. Booker et al., "Multilingualism," p. 5.
76. According to Mary Haas, *acina* may be a borrowed word from Cherokee. Karen Booker, personal communication.

77. If *Vehidi* was Catawban, it may have been derived from *pida*, "go across," perhaps referring to a ford or a crossing place. Robert Rankin, personal communication.
78. If the word *Vora* had an unrecorded initial vowel, the *v* may represent *w*. If so, it could have been the Catawban word *woru*, "walnut tree." On the other hand, if the *v* represented a *b* or *p*, then it may have been a Muskogean word. Robert Rankin, personal communication.
79. Booker et al., "Multilingualism."
80. Chester DePratter, personal communication.
81. Robert Rankin, personal communication.
82. Ibid.
83. Karen Booker, personal communication.
84. Richard J. Preston, Jr., *North American Trees* (Cambridge, Mass.: MIT Press, 1966), p. 310-11.
85. Robert Rankin, personal communication. *Uca* is possibly related to *oka*, "water," a form occurring in several Muskogean languages.
86. The place name *Cheraw* may derive from *Sarati*. But it could also have been derived from *Chara*, a town that was possibly located on the Upper Yadkin.
87. Karen Booker, personal communication.
88. Felix Zubillaga, ed., *Monumenta Antiquae Floridae (1566-1572)* (Rome: Monumenta Historica Societatis Ieusu, 1946), pp. 315, 321.
89. D. G. Brinton, "Contributions to a Grammar of the Muskogee Language," *Proceedings of the American Philosophical Society* 11(1870):301-9.
90. Karen Booker, personal communication.
91. Robert Rankin, personal communication.
92. Booker et al., "Multilingualism."
93. Ibid.
94. Ibid.
95. Biedma, "Relation," p. 13; Ranjel, "Narrative," p. 96; Elvas, *Narratives*, p. 61.
96. Karen Booker, personal communication.
97. Biedma, "Relation," p. 13; Ranjel "Narrative," p. 96.
98. Ranjel, "Narrative," p. 97.
99. Elvas, *Narratives*, p. 61.
100. "*Tiena aspera y llama de pantanos.*"
101. See Zubillaga, *Floridae*, p. 418-23; cf. Wilson, "Indians of the Carolina and Virginia Piedmont," pp. 584-85.
102. Booker et al., "Multilingualism." A problem with *Orista* is that Muskogean is thought to lack *r*. This means that there was either an *r* dialect at the time of the Pardo expeditions, which has since become extinct, or else the Spaniards wrote *r* when they heard Muskogean *l* in some positions in words.
103. Ibid.
104. Zubillaga, *Floridae*, pp. 471-79.

105. Ibid., pp. 418, 587.
106. Ibid.
107. Ibid., pp. 419-20.
108. Ibid., pp. 423-25.
109. Paul E. Hoffman, *A New Andalucia and a Way to the Orient: A History of the American Southeast during the Sixteenth Century* (Baton Rouge: Louisiana State University Press, in press). Paul Hoffman kindly sent me the tables to his unpublished article, "A Best Text of the Ayllón Indian Name List." All of the place names I cite are from his best text list.
110. Peter Martyr, *De Orbe Novo*, vol. 2, ed. and trans. F. A. MacNutt (New York: 1912), pp. 254-71.
111. That some Southeastern chiefs were significantly taller than their subjects has been documented both historically and archaeologically. For example, Soto and his comrades were astonished at the gigantic size of Chief Tascaluza.
112. Lucy L. Wenhold, trans. and ed., "A 17th Century Letter of Gabriel Diaz Vara Calderon, Bishop of Cuba, Describing the Indians and Indian Missions of Florida," *Smithsonian Miscellaneous Collections*, vol. 95, no. 16 (Washington, D.C.: GPO, 1936), p. 10.
113. It is probable that these towns in the Coastal Plain paid tribute to one or more of Cofitachequi's secondary centers, such as Guiomae or Pasque, and at an earlier time to the town located at the Fort Watson site. And one wonders whether Guale paid tribute to one of Ocute's secondary centers, such as Altamaha. Such an arrangement would explain why the Spaniards on the Guale coast in the seventeenth century referred generally to the Indians in the interior as "Tama."
114. Bandera I.
115. Booker et al., "Multilingualism."
116. Ibid. Other evidence suggests that in the sixteenth century the Yuchis lived in the upper Tennessee Valley.
117. Ranjel, "Narrative," p. 102; Garcilaso, *Florida*, p. 328.
118. Biedma, "Relation," p. 15; Elvas, *Narratives*, p. 67.
119. Elvas, *Narratives*, p. 67.
120. Ranjel, "Narrative," p. 102.
121. Garcilaso, *Florida*, p. 326.
122. Roy S. Dickens, Jr., *Cherokee Prehistory: The Pisgah Phase in the Appalachian Summit Region* (Knoxville: University of Tennessee Press, 1976).
123. Ibid., p. 206.
124. Roy S. Dickens, Jr., "Mississipian Settlement Patterns in the Appalachian Summit Area: The Pisgah and Qualla Phases," in *Mississippian Settlement Patterns*, ed. Bruce D. Smith (New York: Academic Press, 1978), pp. 115-39.
125. Roy S. Dickens, Jr., personal communication.
126. Roy S. Dickens, Jr., "Preliminary Report on Archaeological Investigations at

the Plum Grove Site (40WG17), Washington County, Tennessee," 1980, photocopy, pp. 19-20.
127. Roy S. Dickens, Jr., personal communication.
128. Dickens, "Appalachian Summit Area," pp. 132-33.
129. David Moore, personal communication.
130. H. Trawick Ward, personal communication.
131. William E. Myer, "Indian Trails of the Southeast," 42d Annual Report of the Bureau of American Ethnology, (Washington, D.C.: Government Printing Office, 1928), pp. 771-75.
132. "Inquiry made officially before Don Gonzalo Méndez de Canço, Governor of the Provinces of Florida, upon the situation of La Tama and its riches, and the English settlement." Mary Ross papers, Georgia State Archives, Atlanta, pp. 13, 22.
133. Ranjel, "Narrative," p. 103; Elvas spells it *Xualla* (*Narratives*, p. 67).
134. Karen Booker, personal communication.
135. Nineteenth-century Cherokees told James Mooney that the *Suwali* people were different from themselves, but what they were referring to may only have been a dialectical difference.
136. Robert Rankin, personal communication.
137. Betty Anderson Smith, "Distribution of Eighteenth-Century Cherokee Settlements," in *The Cherokee Nation: A Troubled History*, ed. Duane King (Knoxville: University of Tennessee Press, 1979), pp. 46-60.
138. Robert Rankin, personal communication.
139. Duane King, personal communication.
140. Robert Rankin, personal communication.
141. Ranjel, "Narrative," p. 103.
142. Elvas, *Narratives*, p. 68.
143. Garcilaso, *Florida*, p. 330.
144. Howard H. Earnest, Jr., personal communication; Roy S. Dickens, Jr., "Preliminary Report on Archaeological Investigations of the Plum Grove Site (40Wg17), Washington County, Tennessee," photocopy.
145. Earnest, personal communication.
146. "Inquiry before Don Gonzalo Méndez de Canço," pp. 12, 21-22.
147. Coe, "The Cultural Sequence," pp. 308-9.
148. H. Trawick Ward, "A Review of Archaeology in the North Carolina Piedmont: A Study of Change," in *The Prehistory of North Carolina: An Archaeological Symposium*, ed. Mark A. Mathis and Jeffrey J. Crow (North Carolina Division of Archives and History, 1983), p. 73.
149. Bandera I.
150. Ward, "Review," pp. 75-76. Also see Marvin T. Smith, *Archaeology of Aboriginal Culture Change in the Interior Southeast* (Gainesville: University of Florida Press, 1987), pp. 123-26.
151. H. Trawick Ward, personal communication.

152. Charles D. Howell and Donald C. Dearborn, "The Excavation of an Indian Village near Trading Ford," *Southern Indian Studies* 5(1953):3.
153. John Lawson encountered "Wateree Chickanee" Indians on his 1699–1700 journey through the Carolinas. By this time they had moved from the Yadkin to the Wateree River to the general area previously occupied by the central towns of Cofitachequi. See infra, chapter 5. John Lawson, *A New Voyage to Carolina*, ed. Hugh Talmage Lefler (Chapel Hill: University of North Carolina Press, 1967), p. 38.
154. James L. Rudolph, "Earthlodges and Platform Mounds: Changing Public Architecture in the Southeastern United States," *Southeastern Archaeology* 3(1984):33–45.
155. Smith, *Archaeology of Aboriginal Culture Change*, pp. 60–68, 84–85, 143–45.
156. Charles Hudson, et al., "Coosa," pp. 732–35.
157. Cyrus Thomas, *Report on the Mound Explorations of the Bureau of Ethnology*, 12th Annual Report of the Bureau of American Ethnology (Washington, D.C.: GPO, 1984), pp. 335–39.
158. David Moore, personal communication.
159. Garcilaso, *Florida* pp. 333–34; Elvas, *Narratives* p. 67; Biedma, "Relation," p. 15; Ranjel, "Narrative," pp. 104–6.
160. James Mooney, *Myths of the Cherokee*, 19th Annual Report of the Bureau of American Ethnology (Washington, D.C.: GPO, 1900), p. 543.
161. Contemporary Cherokees use a shortened version of this word to refer to Asheville, North Carolina—*tho khiya sti*. Duane King, personal communication.
162. Duane King favors this interpretation.
163. Duane King, personal communication.
164. Mooney, *Myths*, p. 543.
165. Ibid., p. 552.
166. Charles Hudson and Jerald Milanich, "Hernando de Soto and the Indians of Florida," MS., pp. 76–77.
167. The name of the Toxaway River may be derived from "Tocae."
168. Duane King, personal communication.
169. Smith, "Distribution of Eighteenth-Century Cherokee Settlements," pp. 46–60. "Taucoe" (probably derived from either "Atuqui" or "Tocae") is probably the namesake of present Toccoa, Georgia.
170. Ranjel, "Narrative," p. 106. Variant spellings are "Guasuli" (Biedma, "Relation," p. 15) and "Guaxule" (Elvas, *Narratives*, p. 67; Garcilaso, *Florida*, p. 335).
171. Elvas, *Narratives*, p. 67.
172. Ibid., p. 68.
173. Ranjel, "Narrative," p. 106.
174. Garcilaso, *Florida*, pp. 335–36.

175. Duane King, personal communication.
176. The most complete documentation of the role of transvestites in sixteenth-century Southeastern Indian society is from the French Huguenots who were at Fort Caroline at the mouth of the St. Johns River in Florida. Paul Hulton, *The Work of Jacques Le Moyne de Morgues: A Huguenot Artist in France, Florida, and England* (London: British Museum Publications, 1977), p. 145. From the French one gets the impression that there were *many* transvestites among the Indians. But from Bandera one gets the opposite impression. From Santa Elena to Cauchi they appear to have seen only one transvestite, and this instance was so notable that Pardo asked Bandera to make a written record of it.
177. Ranjel, "Narrative," p. 106; Elvas, *Narratives*, pp. 68–69.
178. Mooney, *Myths*, p. 518.
179. Karen Booker, personal communication.
180. Smith, "Cherokee Towns."
181. "Joara Chiquito Orata" implies than in addition to Joara, there was a Joara the Lesser, a town not mentioned elsewhere.
182. Ugarte, "The First Jesuit Mission in Florida," p. 63 and *passim*.
183. Dickens, *Cherokee Prehistory*, p. 189.
184. Roy S. Dickens, Jr., "An Ecological-Evolutionary Approach to the Interpretation of Cherokee Cultural Development," in *The Conference on Cherokee Prehistory*, comp. David G. Moore (Swannanoa, N.C.: Warren Wilson College, 1986), pp. 81–94.
185. Hudson et al., "Coosa," pp. 723–37; David J. Hally, *Archaeological Investigations of the Little Egypt Site (9Mu102), Murray County, Georgia, 1969 Season*. University of Georgia, Laboratory of Archaeology, Report no. 18, 1979.
186. Elvas, *Narratives*, p. 58.
187. Ibid., p. 65.
188. Hally, *Little Egypt*; David J. Hally, "Introduction to the Symposium: The King Site and Its Investigation." *Southeastern Archaeological Conference Bulletin* 18(1975):48–54; James B. Langford, Jr., and Marvin T. Smith, "The Coosa Province: An Update," paper presented at the Lamar Institute Conference on South Appalachian Mississippian, Macon, Ga., May 10, 1986.
189. Charles Hudson, Marvin Smith, Chester DePratter, and Emilia Kelley, "The Tristán de Luna Expedition, 1559–1561," *Southeastern Archaeology* 8(1989):31–45; Charles Hudson, "A Spanish-Coosa Alliance in Sixteenth-Century North Georgia," *Georgia Historical Quarterly*, 62(1988):599–626.
190. Biedma, "Relation," p. 15; Ranjel, *Narratives*, p. 108.
191. Richard Polhemus, "The Search for Tanasqui," paper presented at the 1985 Southeastern Archaeological Conference meeting, Birmingham, Ala. Polhemus informs me that the site he investigated appears to be the "ancient fortress" described in J. M. W. Brezeale's, *Life As It Is* (Knoxville, 1842), pp. 233–34.

192. Robert Rankin, personal communication.
193. Smith, "Cherokee Settlements."
194. It is possible that Cherokee-speakers in the upper Tennessee Valley or elsewhere could have moved to occupy the site of Tanasqui (i.e., *tanaski*) after it was abandoned by Muskogean-speakers, shortening the name of the place to *tanasi*. Duane King informs me that contemporary Cherokees have shortened *tanasi*, "Tennessee," to *tansi*.
195. Booker et al., "Multilingualism."
196. Charles Hudson, "Juan Pardo's Excursion beyond Chiaha," *Tennessee Anthropologist* 12(1987):74–87.
197. Ibid.
198. Robert Rankin, personal communication.
199. Booker et al., "Multilingualism."
200. Smith, "Cherokee Settlements." Also sometimes spelled "Chihoe."
201. Richard R. Polhemus, "The Early Historic Period in the East Tennessee Valley," MS. This place name is sometimes spelled "Settico." Smith, "Cherokee Settlements."
202. I am indebted to Duane King for this observation.
203. Duane King, personal communication.
204. Ranjel, *Narrative*, pp. 109–110; Garcilaso, *Florida* pp. 341–42.
205. Charles Hudson, Marvin Smith, and Chester DePratter, "The King Site Massacre Victims: An Historical Detectives' Report," in *The King Site: Biocultural Adaptations in Sixteenth-Century Georgia*, ed. Robert L. Blakely, (Athens: University of Georgia Press, 1988), pp. 117–34.
206. Hudson, "Pardo's Excursion."
207. Pardo.
208. John R. Swanton, *Early History of the Creek Indians and Their Neighbors*, Bureau of American Ethnology Bulletin 73 (Washington, D.C.: GPO, 1922), p. 214.
209. Booker et al., "Multilingualism."
210. Chester B. DePratter, Charles H. Hudson, and Marvin T. Smith, "The Hernando de Soto Expedition: From Chiaha to Mabila," in *Alabama and the Borderlands*, ed. R. Reid Badger and Lawrence A. Clayton (University: University of Alabama Press, 1985), p. 116.
211. See "Three New Documents from the Pardo Expeditions," transcribed and translated by Paul E. Hoffman, in part II.
212. Booker et al., "Multilingualism."
213. Zubillaga, *Floridae*, p. 325.
214. Ibid., p. 403.
215. John T. Lanning, *The Spanish Missions of Georgia* (Chapel Hill: The University of North Carolina Press, 1936), p. 43.
216. Hudson, "A Spanish-Coosa Alliance"; Swanton, *Early History*, pp. 207–211; Booker et al., "Multilingualism."
217. Augustín Dávila Padilla, *Historia de la Fundación y discurso de la provincia*

de Santiago de Mexico, de la Orden de Predicadores (Mexico, 1955), p. 216. First published in 1596.
218. Elvas, *Narratives*, p. 52.
219. Ibid.
220. Lawson, *New Voyage*, pp. 23-24.
221. Elvas, *Narratives*, p. 63; Garcilaso, *Florida*, p. 320.
222. Ranjel, "Narrative," p. 99; Elvas, *Narratives*, p. 62. In Soto's day, shawls woven out of mulberry fiber and grass fiber resembling nettles are mentioned several times (e.g., Elvas, *Narratives*, p. 62). None are mentioned in any of the Pardo documents.
223. René Ladonnière, *Three Voyages*, trans. Charles E. Bennett (Gainesville: University of Florida, 1975), pp. 18, 39.
224. Kristian Hvidt, ed., *Von Reck's Voyage* (Savannah, Ga.: Beehive Press, 1980), p. 127.
225. Zubillaga, *Floridae*, p. 400.
226. Rogel further says that the Indians were much given to playing a game of dice (*un juego de dados*), saying that they became gamblers through the influence of Spanish soldiers.
227. Ranjel, "Narrative," p. 99.
228. Elvas, *Narratives*, p. 55.

4
The Foundations of Greater Florida

In order to defend against colonial competition from other European powers in North America, Pedro Menéndez de Avilés founded three forts on the lower Atlantic coast: St. Augustine, San Mateo, and Santa Elena. St. Augustine was strategically located to protect Spanish ships passing through the narrowest part of the Bahama Channel, where they were vulnerable to attack by pirates. San Mateo, on the site of former French Fort Caroline, would control the entrance to the St. Johns River, which Menéndez mistakenly thought might provide water passage to the west coast of Florida. And Santa Elena, which was to be his capital city, was to be the starting point of a road that would lead to the Spanish silver mines in Zacatecas, Mexico.

To the north and west of these three settlements, missionaries were to be sent to establish missions and to evangelize the Indians. Menéndez sent missionaries to Tequesta (Biscayne Bay), Calusa (Charlotte Harbor), Guale, Orista, the Bay of Santa Mariá (Chesapeake Bay), and to Guatari (see figure 1).[1] Beyond this, he wanted missionaries stationed along the road to Zacatecas to Christianize and pacify the Indians and to make the land suitable for additional Spanish settlements. These missions, according to Menéndez's plan, would constitute the outer frontier of what might be called greater Florida, beyond which lay an unexplored terra incognita. Greater Florida would consist of all the territory westward from the Atlantic coast between Chesapeake Bay and peninsular Florida and extending westward in a broad arc to New Mexico and Texas.

What is most noteworthy in all that Pardo and his men did is the degree to which they were dependent on the infrastructure of the Indian societies: The Spaniards relied entirely on the Indians for their means of transportation, shelter, and food. Moreover, it is clear that the fabric of the Indian societies was the stuff of which they intended to build a colony. The Spanish approach to colonization was premodern, in fact, basically medieval. That is, their modus operandi for seizing the land and wealth of pre-state societies was to work through the caciques, whom they regarded as *señores naturales*—natural lords.[2] Through them the Spaniards gained control of the people who paid tribute.

Outfitting the Second Expedition

It is impossible to know precisely how many men were under Pardo's command on his second expedition, but it was in the neighborhood of 150. He was authorized to take 120 men with him, but there is evidence that he only took 110.[3] To this must be added Sergeant Moyano and the 30 men whom Pardo had stationed at Joara on his first expedition, plus the 5 men carrying additional supplies who were sent from Santa Elena in pursuit of Pardo.[4] Everything that Pardo took with him to supply this force had to be carried on a human back. For this reason, the amounts of stores and equipment issued to him by the keeper of the King's Food and Munitions in Fort San Felipe were remarkably small. The issue for the second expedition was as follows:

168	pounds	matchcord for harquebuses
255	pounds	lead balls
309	pounds	matchlock powder
30	pounds	cannon powder[5]
85	pounds	biscuit
48.4	liters	wine
3		linen bags
265	pairs	fiber sandals
178	pairs	shoes.[6]

Although this may have been the entire store of food and wine that Pardo carried with him, it was not his entire store of munitions. In his accounting of how these stores were used, Pardo indicates that they used 189.5 pounds more of matchcord than they were issued. And he indicates that 113 pounds more of lead was expended than was issued. Some but not all of this shortage was probably made up from munitions left over from the first expedition. That is, at the end of the first expedition they had expended 77 fewer pounds of cord than was issued, and 112 fewer pounds of lead than was issued.[7]

There are several differences between the outfitting of the first and second expeditions. For one thing, they took far less food on the second expedition than on the first, presumably because they assumed that they could count on the Indians for supplies of corn, which Pardo had ordered them to lay up for the Spaniards. On the first expedition, they had been issued 793 pounds of biscuits and 10 cheeses, while on the second expedition they were only issued 85 pounds of biscuits and no cheese. On the first expedition, Pardo claimed that he issued his men the wine, biscuits, and cheese only in times of greatest need. On the second expedition they were issued about twice as much wine as on the first expedition—48.4 liters compared with 23.83 liters.

Pardo also must have taken on his second expedition some of the tools that had been issued to him on the first: 2 iron picks (*picos de hierra*), 7 iron shovels (*palas de hierra*), and 5 mattocks (*azadores*). Several of these tools—2 picks, 3 shovels, and 2 mattocks—they left behind at Joara on the first expedition. They carried with them an unspecified supply of carpenter's tools. When Pardo was at Orista, for example, he built a stronghouse out of sawed lumber, and this means that he had taken along at least one saw. And he left behind at Orista a drill (*barrena*) which was presumably to be used for further construction.

They were issued 74 linen sacks on the first expedition and 3 on the second. After using these sacks as containers, they cut some of them up and used the cloth to make game bags and munitions bags. It is likely that some of the pieces of linen that Pardo gave the Indians was cut from these sacks. Beyond this, there is no accounting given for the supply of articles they took with them to give the Indians. It is probable that at least some of these articles were carried by the five men who were sent from Santa Elena in pursuit of Pardo.

Pardo took with him 178 pairs of shoes and 265 pairs of fiber sandals. Some of the sandals were made in Spain and others in Campeche. Pardo probably issued these shoes and sandals as needed, but the day after he arrived at Chiaha he issued 53 pairs of shoes and 45 pairs of sandals at one time, and their issuance was important enough for him to have made a detailed accounting to Bandera. Most of them appear to have gone to men in Moyano's detachment who had been in the interior for almost a year, and who must have been walking on thin soles if they had any soles left at all. Most of the men were each issued a pair of shoes and a pair of sandals.

Sergeant Moyano, however, was given four pairs of shoes, and Corporal Lucas de Canizares was given two pairs of shoes.

Interestingly, Pardo made gifts of shoes and sandals to some of the important Indians at Chiaha. He gave a pair of shoes each to Olamico, the chief of Chiaha, and to a mandador of Chiaha. He gave a pair of shoes and a pair of sandals to a mandador of Olamico and to a heniha of Olamico. This appears to have been a deviation from Pardo's practice of giving gifts to Indians only on the occasion of their accepting Spanish dominion and agreeing to provide the services Pardo required of them.

The only indication of how much weight one of Pardo's men carried into the interior on his back comes from an incident in which one of his men lost a sack of lead harquebus balls weighing 31 pounds when he dropped them into a river. In addition to this burden, the soldier was probably carrying his own personal effects and arms of perhaps 10 pounds or so. Thus, a weight of about 40 to 50 pounds per man would seem to be a conservative estimate.

In addition to relying upon his own men, Pardo also relied upon Indian porters to carry some of his goods. He tested the veracity of the chief of Satapo by asking him to go out and round up some porters to serve him. When the chief came back with nothing but excuses, Pardo was persuaded that indeed, as he had been informed, there was a plot to attack him. On his return, while traveling from Ylasi toward Canos, Pardo split his force and sent twenty of his men and twenty Indian porters south to Guiomae carrying bags of corn. Also, when Pardo reached Coçao, he was met by Orista Orata with thirty Indians and Uscamacu Orata with additional Indians, all of whom were probably to serve as porters on the way to Ahoya and Orista. However, Bandera nowhere tells how many Indian porters actually served them, nor does he say by what means they were impressed into service. The generally peaceful character of Pardo's expedition suggests that he did not treat his Indian porters as harshly as Soto did.

The Road to Zacatecas

On May 25, 1567, in Fort San Felipe at Santa Elena, Menéndez commanded Pardo to find a road to the silver mines at Zacatecas and San Martín. Leaving aside the impossibility of carrying out this order, particularly in the seven months that were allotted, one may well ask why the order was made at all. The reason is that in 1546 an extremely rich deposit of silver was discovered in the territory of the Zacateco Indians who lived on the frontier to the northwest of Mexico City. After this discovery,

Spaniards built the city of Zacatecas in this area, and it became the center of an extraordinary mining venture. Subsequently, other towns, such as San Martín, were founded in the vicinity. Later, as much as one-third of all the silver mined in the New World came from Zacatecas. The consequences of this strike were quickly felt throughout the Spanish economy.

The mines lay in a territory that was not easy for the Spaniards to conquer. Spanish colonial authorities were of the opinion that the Zacatecos and other Indians on the northern frontier of Mexico were ungovernable. They were less dependent on agriculture than Indians living further south, and they were not members of state-level societies. They lived in small communities, and they were highly mobile and difficult to track down in their rugged, barren homeland. They began waging a guerilla war against the Spaniards and their native allies that was both costly and, because the silver mines were so important, profoundly disturbing to the Spaniards. In time, the Spaniards called these wild and ungovernable Indians "Chichimecas," believing that their territory extended far to the north into what is now the United States. The homeland of these Indians was at first as impenetrable for the Spaniards as it had been for hundreds of years to the native state-level societies of central Mexico.[8]

In the 1550s the Chichimecas began attacking Spaniards on the roads leading to Zacatecas. The "land of the Chichimecas" became a synonym for "land of constant war."[9] In 1561 several groups of Chichimecas formed an alliance and attacked a number of Spanish settlements, for a time cutting off the town of Zacatecas from the outside world. This threw the Spaniards into a panic, and afterward they were probably more afraid of the northern Indians than they had need to be.[10] In time, as the Spaniards began getting the upper hand in this conflict, they would make sorties into Chichimeca territory in order to provoke them. When the Chichemecas resisted, it became a "just war," and the Spaniards were entitled to seize Chichimecas and sell them as slaves, a practice that made the Chichimecas even more hostile.[11] In the 1560s the bloody and bitter Chichimeca war was at a stalemate.

Hostilities did not lessen until after 1565, when Viceroy Villamanrique instituted a policy of freeing Chichimeca slaves and distributing gifts of food, clothing, and tools to the people in an effort to win them over.[12] This policy may very well have had an effect on Menéndez's and Pardo's diplomatic, gift-giving approach to the Indians of greater Florida.

But even after the Zacatecos and their allies were subdued, the Indians who lived further north remained hostile and a worry to the Spaniards. This was especially so with Menéndez, who thought that the French Huguenots, whom he considered to be heathen, were natural allies of the Chichimecas, and he was afraid that if the French could find a road to Zacatecas they might join with the Chichimecas and attack the silver mines. This was a palpable fear, and given the importance of these mines in the Spanish economy, it seemed only prudent that if a road to Zacatecas from the north existed, Spaniards should discover it before Frenchmen did.

The importance of silver in sixteenth-century New Spain is nowhere more evident than in the effect that its discovery had on the building of roads. Cattle raising and farming were important in the colonial economy, but they were a poor second to precious metals. After the strike was made in the Zacatecas hinterland, there was an urgent need to build roads to the south. This was the impetus for building the *Camino Real de la Tierra Adentro*, a road built in 1550-55 connecting Mexico City to Zacatecas. Later this road was extended to the Spanish colony in New Mexico, and later still it was extended from there to the Spanish missions in east Texas and Louisiana.[13] If Pardo had succeeded in doing what he was ordered to do on his second expedition, the *Camino Real* would no doubt have had its eastern terminus at what is now Beaufort, South Carolina.

Pardo and his men were not, of course, cutting a new road on their travels: They were traveling along the extensive network of trails used by the Indians. There are no detailed descriptions of the physical appearance of these trails either in the Soto or Pardo documents. About all that can be said is that the trails were visible to travelers, and that the trails were better and more easily traveled in some places than in others.

William Meyer, one of the earliest students of Indian trails in the Southeast, argued that Indian trails followed paths made by animals.[14] Although this theory may explain the origin of some of the trails, it does not explain all of them; in particular, it does not explain the maintenance of the trails for human travel. From incidents that occurred during the Soto expedition, it appears that the trails served social functions, and to that extent they were man-made, or at least man-maintained. This was clearly not the case in the wilderness of Ocute, the no-man's-land between the paramount chiefdoms of Ocute and Cofitachequi. While they were crossing this wilderness the Soto chroniclers complained of not being able to find a trail to follow, particularly after they crossed the Savannah River.[15] Clear trails were particularly important for Soto, who had a contingent of cavalry, because horses do not like to travel through brushy undergrowth. The wilderness of Ocute is especially interesting because a century and a half earlier, when the middle Savannah River was occupied by people, the trails must have been there.[16] This would seem to imply that after the exodus of people from the Savannah River, the trails became overgrown through disuse.

The best trails probably existed between the central towns of chiefdoms and their closest secondary centers (e.g., between Cofitachequi and Gueça), and to a lesser degree between the central towns and their more remote secondary centers (e.g., between Cofitachequi and Ylasi). And trails were probably poorest through the buffer zones that separated antagonistic societies. This was certainly true of the wilderness of Ocute, and the same appears to have been true of the area between Chiaha and the Chiscas. When Soto made ready to send a detachment to the Chiscas to find out whether these people possessed gold, the Indians told him that the way there was so rough that horses could not make the journey. Hence, Soto sent two of his men on foot with Indian guides.[17]

Bandera complains of difficulties in traveling on the trails on only three occasions, and in all three cases the circumstances were unusual. The first occurred when they went from Chiaha to Chalahume and Satapo over a trail that passed by the eastern side of Chilhowee Mountain, and also as they were returning to Chiaha on a trail that ran to the western side of Chilhowee Mountain. In both instances they complain of having to travel by a very rough way (*un camino muy aspero*). As previously discussed, the reason why Pardo followed this route on his way to Coosa is not known. Either the Indians took him by a roundabout way to set him up for an ambush, or else the Spaniards were prospecting beyond Menéndez's orders, and they had reason to think that they might discover precious metals by taking this trail. The fact that a native of Satapo informed them that a better trail to Coosa existed which ran parallel to the French Broad/Tennessee River implies that Pardo did not know about this better road, which is probably the one that Soto took.

The other two instances of difficult travel occurred when they had to make water crossings. In one case, when they were traveling from Ylasi to Cofitachequi, they had to cross the floodplain of Thompson's Creek, which had backed up from the Pee Dee River because of heavy rain. Had it not rained so hard, this probably would have been an easy trail. In the other case they were going from Guiomae to Coçao, with everyone loaded down with sacks and baskets of corn, and they had to cross swamps bordering several rivers and creeks. Three of these swamps were especially difficult, yet Bandera says that none of the men fell or dropped their load into the water. Moreover, on this particular day they were able to travel 7 leagues— 2 leagues more than usual.

The crossing of swamps, creeks, and rivers could be costly. Pardo had to issue 25 pounds of matchcord and 100 pounds of powder to replace supplies that had got wet when they crossed swamps and rivers. The greatest loss came when a soldier crossing a large river lost a sack containing 31 pounds of lead.

Dugout Canoes

When Pardo and his men had to transport themselves and the goods they were carrying across water, they again had to depend on the Indians. Everywhere Pardo and his men traveled, the dominant Indian mode of water transportation was the dugout canoe.[18] These canoes were generally made from large cypress or pine trees. They were hollowed out and shaped by the controlled application of fire. That is, to shape the canoes the Indians would build small fires that they would fan to create more heat and then damp to reduce the heat. They used marine shells to scrape away the charred wood until the vessel had the desired shape. A number of these canoes have been recovered in South Carolina.[19]

The Indians used these dugout canoes to cross bodies of water they could not ford, and they used them for fishing. They also used them to carry necessary supplies, such as corn (see figure 21). The degree to which carrying supplies of necessities in dugouts was important for the operation of a chiefdom such as Cofitachequi is unknown, although it is a matter of some interest. That is, it is possible that the towns of Mississippian chiefdoms were located on rivers and creeks not only to take advantage of the rich agricultural soils and good fishing, but also the water transport.

Pardo used dugout canoes to transport his men across water on several occassions. He used them to carry his men and the corn they were carrying from Ahoya to Orista, and then from Orista to Santa Elena. In fact, one of the concessions Orista Orata had made to Pardo, as a way of persuading him to build a fort at Orista, was to promise him the use of both canoes and Indians to paddle them whenever needed. But this may have been a promise that Orista Orata would later regret. One of the reasons the Indians of Ahoya rebelled was that they were exasperated at having to provide this service on demand for the Spaniards.

Although none of the Pardo documents mentions it, members of the expeditions must have used dugout canoes when they crossed the river going to and from Guiomae. This is confirmed in an incident that occurred when Pardo was returning from his second expedition. He sent Fernán Gomez back from Orista to Guiomae with a socketed axe and an adze. Gomez was to give these tools to the chief of Guiomae and other chiefs

Figure 21. Timucuan Indians transporting food in a dugout canoe. The Indians could paddle these narrow canoes from a standing position. Engraving by Theodore de Bry after a painting by Jacques Le Moyne. (Photograph courtesy of the Smithsonian Institution.)

who were his allies and friends, and he was to order them to construct four canoes that were to be reserved for the Spaniards.

Like Soto before him, Pardo must have used dugout canoes in crossing the Wateree River to reach Cofitachequi. Oddly, he did not send back any instructions to Cofitachequi to build canoes. Perhaps he assumed that plenty of canoes would always be available there. He did order Alonso Bela to go to Ylasi with a large socketed axe to give the chief, with instructions that he and his men were to build three canoes to be reserved for the Spaniards. If these canoes were to be used in crossing Thompson Creek, as seems likely, they would only have been necessary after a great deal of rain had fallen.

It may be significant that the only documentation of dugouts being used to transport corn in the interior was between towns in the central part of Cofitachequi. When going into the interior on his second expedition, Pardo arrived at Guiomae on September 8, 1567. The Indians told him that their corn had already been harvested, but it was not yet cured (*curandose*). They said that after it was cured it would be brought to Guiomae by canoe.

Foundations of Greater Florida

This implies that their corn was grown in bottomland fields along the river, and that canoes were used to transport corn from these fields to the towns. Pardo met with Pasque Orata at Guiomae, and this chief promised to build a corn crib in his town (possibly in the vicinity of the Fort Watson site) in which corn would be kept and transported to Guiomae by canoe when it was needed. From this we can conclude that during the heyday of the chiefdom of Cofitachequi it was possible to transport gifts and tribute in corn from many but not all of the subsidiary towns. Corn could be transported by dugout canoe from such towns as Guiomae and Pasque, but not from Ylasi, on the Pee Dee River, nor even from Gueça, which was located above the falls of the Wateree River.

The narrow dugout canoes were not to the liking of some Spaniards. Francisco Villarreal, a Jesuit missionary in Guale, complained that they were no more than "hollow poles," and that they frequently capsized.[20] Father Antonio Sedeño also criticized their tendency to capsize, even though when he was shipwrecked on the coast of Florida he would surely have perished had not the Indians allowed him and his companions to use their canoes.[21] The Indians, who were habituated to these canoes, could paddle and pole them while standing upright.

Pacifying the Indians

Menéndez ordered Pardo to appease the Indians and to attract them to the service of Christianity and Spain, but at the same time he was to take possession of the land in the name of the king of Spain. Menéndez emphasized that Pardo was to be especially friendly to the caciques and to persuade them to be obedient to his majesty. When Pardo told Father Juan Rogel the story of his two expeditions, he said that he had been obliged to pacify the Indians with much gentleness.[22]

It would be interesting to know how well the Indians understood what they were agreeing to by saying *yaa* after hearing Pardo's speech. The fact that they built houses, corncribs, and laid up supplies of corn indicates that they did more than merely give verbal assent. It is possible that they were familiar with the process of changing one's tributary relationship from one paramount chief to another, and, as already noted, they may have regarded Pardo as simply another paramount chief, although an extraordinary one.

However, Pardo was using more than persuasion and gentleness. He did, after all, come marching in with more than a hundred armed men. Also, on his first expedition he expended 37.5 pounds of powder and on his second expedition 50 pounds of powder firing his harquebuses in the presence of assembled caciques in order to intimidate them. Furthermore, Sergeant Moyano's dog was probably one of the large dogs that the Spaniards used against the Indians, and they would certainly have been intimidated by it. The Indians had seen such dogs before when the Soto expedition had come through.[23]

Pardo's main device for soothing and winning over the Indians was to give them gifts of European manufacture.[24] These were ordinarily presented after he gave his speech and got the Indians' assent. There were five broad categories of gifts: (1) axes, (2) chisels and wedges, (3) knives, (4) jewelry and novelties, and (5) cloth. It is impossible to accurately enumerate the total number of gifts Pardo distributed because on several occasions he gave them to the Indians without making a count of them. This is what happened when he arrived at Guiomae during the return trip of his second expedition. Here he dismissed most of the Indian interpreters who had been traveling with him, and he apparently gave away all his remaining gifts without counting them. His diplomatic tasks were essentially over at this point, and he needed to lighten his load in order to carry as much corn as possible back to Santa Elena. The figures that follow are counts of gifts that are reported and described by Bandera.

The most desirable of the gifts that Pardo distributed were iron cutting tools, and of these the axe was the most desirable of all. This was the gift that was so desired by the Satapo Indian informer who crept under cover of darkness into the house in which Guillermo Rufín was sleeping. The iron tools Pardo gave them were not manufactured in a standardized way. This may account for some of the variability in Bandera's characterization of these artifacts. Beyond this, however, some of the variability may simply be inconsistencies in Bandera's language.

Perhaps the most prestigious of all of Pardo's gifts was a small hafted battle axe (*un hachuela de armas enastada*) that he gave to Joara Mico. Other axes were a large socketed axe (*hacha de ojo grande*), a large Biscayan axe (*hacha biscayna grande*), two socketed axes (*hacha de ojo*), and thirty plain axes (*hacha*). Pardo gave away axes in thirty-five documented instances. These would have been vastly superior woodworking tools compared with the stone axes of the Indians. They were probably also desirable for use in fighting. One of the fundamental weapons of war in the Southeast was the war club, a weapon about the length of a large European hatchet or small axe. The native war club was made of very hard wood, which was worked down to sharp edges; in some instances, celts made of

Figure 22. The murder of the Frenchman Pierre Gambié by Timucuan Indians. This depicts one use the Indians had for iron axes. The Indians built small fires on clay or sand basins in their canoes for warmth and light. Engraving by Theodore de Bry after a painting by Jacques Le Moyne. (Photograph courtesy of the Smithsonian Institution.)

polished stone were set into the war clubs. The axes that Pardo gave the Indians would have fit into indigenous warfare patterns quite neatly (see figure 22).

The second most desirable kind of gifts from the Indian point of view were the iron chisels and wedges that Pardo gave out. This was a highly variable category. Pardo gave away 53 chisels (*escoplo*) and 26 wedges (*cuña*), but also 3 chisels that were like wedges (*escoplo a manera de cuña*), and 8 small wedges that were like long chisels (*cuña pequeña larga a manera de escoplos*). Nor does this exhaust the list. He gave out 7 large wedges (*cuñela* and *cuña grande*), and 29 small wedges (*cuña pequena*). In all, as documented by Bandera, Pardo gave away 126 chisels and wedges.[25]

Several contemporary wedges and chisels that have been recovered from archaeological excavations at Santa Elena may be like the ones Pardo gave away (see figure 23).[26] The metal of which they are made is much the same thickness as iron barrel bands found at Santa Elena, which suggests that

Figure 23. Wedge and chisel from Santa Elena. The wedge (above) is 3¾ inches long; the chisel (below) is 2¾ inches long. (Photograph courtesy South Carolina Institute of Archaeology and Anthropology.)

they were made from the same kind of iron stock as the barrel bands, or else they were made from recycled barrel bands.[27] What the Indians used these wedges and chisels for is unknown. The Spaniards may have wanted the Indians to use them to mine gold, silver, and gems, but most of them probably ended up mounted in war clubs in just the way stone celts were mounted.

Knives must have ranked close in desirability with wedges and chisels (see figure 24). Pardo gave away 32 knives, 30 of which are described as being large (*cuchillo grande*).

Pardo gave away jewelry and novelties and small pieces of cloth, usually along with an iron tool. He gave away 29 necklaces (*gargantilla*), but unfortunately there is no description of the beads of which they were made. Perhaps they were spherical translucent turquoise blue glass beads of a type that has been found in some numbers at Santa Elena and occasionally in the Tennessee Valley (see figure 25).[28] On at least seven occasions he gave away several gilded ball buttons (*botones de atauxia*) (see figure 26). Examples of these buttons have been found both at Santa Elena and in Greene County Tennessee (40GN9), in the general area inhabited by the Chiscas.[29] Pardo gave away six mirrors (*espejo*), none of which have been recovered in the interior by archaeologists.

Foundations of Greater Florida

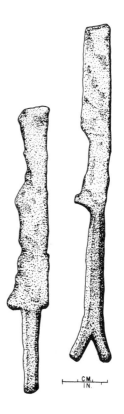

Figure 24. Knives from Santa Elena. The knife on the left was probably mounted in a wood or bone handle. The knife on the right may not have been mounted in a handle. (Drawing courtesy South Carolina Institute of Archaeology and Anthropology.)

The pieces of cloth that Pardo gave the Indians were so small that they were probably used for ornamental purposes. This is certainly the case with the piece of satin (*rraso*) and the pieces of white and red silk ribbon (*unos pasamanos de seda blanca y colorado*)[30] he gave out on one occasion. The same must have been true of the ten pieces (*un poco, un pedaço*) of red taffeta (*tafetan colorado/roxo*), eight pieces of green taffeta (*tafetan verde*), and the several pieces of red cloth (*pano roxo*). Nicholas Burgoigon, whom Sir Francis Drake captured at St. Augustine in 1586, said the Indians used such pieces of cloth to make "baudricks or gyrdles," that is, sashes and belts.[31] The Indians may have been able to make practical use of the two pieces of London cloth (*pano de londres*), each a half a vara long, and the two pieces of linen (*lienco*) each a vara long. The latter were probably obtained from the linen sacks in which the Spaniards were carrying their goods and supplies.

As already discussed, there were several instances in which the Spaniards were not purveyors of exotic goods from beyond the Southeastern

Figure 25. Spherical turquoise blue glass beads from the Hampton site, Rhea County, Tennessee. These heavily eroded beads were excavated from a sixteenth-century context. (Photograph courtesy Frank H. McClung Museum, University of Tennessee, Knoxville.)

world, but were simply middlemen in an indigenous exchange system. This first occurred when Pardo was in the mountains at Tocae. Here he gave Uastique Orata several presents, including a large shell (*un caracol grande*), which was undoubtedly from the coast. Such shells were used as ceremonial drinking cups, and they were greatly valued by the Indians (see figure 27). Other instances occurred as he traveled across the Coastal Plain on his return to Santa Elena. Having given most or all of his remaining European trade goods away at Guiomae, he may have had only Indian artifacts left. He gave away four painted mantles or matchcoats, probably of deerskin, which had come from Indians in the interior (*manta pintada de los yndios de la tierra adentro*), to Aboyaca Orata, Cocao Orata, and their respective mandadores (see figure 28). To the Indians of Ahoya, Orista, and other coastal towns, he gave two large matchcoats of painted catskin (*mantas grandes de gatos pintadas*). These were probably cougar skins whose inside surfaces had been decorated with painted designs. Pardo also gave out breechclouts—the basic garment of Indian men—that had painted

Figure 26. Gilded metal ball button from Santa Elena. Such buttons were widely used in the sixteenth century on front closures and sleeves of garments for males. See, for example, the harquebusier in figure 34. (Photograph courtesy South Carolina Institute of Archaeology and Anthropology.)

designs (*mandiles pintadas de los yndios de la tierra adentro*) and were probably made of deerskin (see figure 29).

It is striking that on his journey into the interior there were several places where Pardo distributed no gifts. Between Cofitachequi and Otari, Pardo visited and gave his customary speech at Tagaya, Gueça, and Aracuchi, but he gave the assembled Indians no gifts. It is possible, as we have seen, that the five men who were sent from Santa Elena were carrying most of the gifts for the Indians, and these men may have caught up with Pardo at Otari. In any case, he was notably more generous with his gifts after reaching the town of Otari.

Pardo's failure to confer a gift at Tagaya may have been an oversight, and perhaps even a diplomatic error. Bandera does not refer to the chief of Tagaya as an Orata. He merely says that he was a chief (cacique), a lord (*señor*). But on the return journey, the chief of Tagaya went to Ylasi to meet with Pardo, and here Bandera refers to him as Tagaya Orata, saying that Pardo gave him a chisel and a wedge, and made him a cacique, because before this time he had not been one, and he had performed many services for the Spaniards.

Figure 27. Conch shell drinking cup from the Hixon site, Hamilton County, Tennessee. (Photograph courtesy Frank H. McClung Museum, University of Tennessee, Knoxville.)

The Houses the Indians Built

On his first expedition Pardo commanded the Indians to build houses and lay up stores of corn for the Spaniards to use in their travel on the road to Zacatecas. Very little information is available on the way these houses were built. Presumably they were like the houses the Indians built for themselves, although perhaps larger. It would seem that the Spaniards intended to use them as inns along the road, about a day's travel apart (see figure 30).

In some places no houses were built by the Indians, nor is there any indication that the Spaniards expected them to build houses. There is, for example, no indication that any were built between Santa Elena and Guiomae. This is another indication that this area was inhabited by very

Figure 28. Indians dressed in matchcoats. The traditional Indian (left), wearing a painted deerskin matchcoat, is armed with a bow and arrow. The "acculturated" Indian (right), wearing an English woolen blanket, is armed with a gun, and he is carrying a bottle of rum or brandy. Painted in 1736 by Philip Georg Friedrich von Reck near Savannah, Georgia. (Photograph courtesy the Royal Library, Copenhagen.)

small populations of Indians. Tocae, near present-day Asheville, North Carolina, is another place where there is no evidence that a house was built. This is a little surprising because the travel from Joara through Swannanoa Gap was difficult, and one would have expected the Spaniards to want a place to rest at Tocae. But perhaps Tocae was simply too small a town to provide accommodations. There is no evidence that the town of Tocae existed at the time of the Soto expedition, or if it did, it was too small for the chroniclers to mention. The last place Pardo found a house built for his use was at Cauchi, in the mountains. Beyond this point, he entered the territory of the paramount chiefdom of Coosa, whose chiefs refused to submit to the Spaniards, so no more houses were built.

Bandera describes many of the houses as being made of wood. But he says that the houses at Cauchi and Yssa were made of wood and earth, and it is probable that all of the houses should have been so described, since there is abundant archaeological evidence that the houses of most Southeastern Indians at this time were made of wattle and daub construction (see figure 31). That is, the typical house was a basketlike structure of inter-

Figure 29. Senkaitschi, the Yuchi "king." He wears a breechclout, leggings and moccasins, and a buffalo skin matchcoat. Painted in 1736 by Philip Georg Friedrich von Reck near Savannah, Georgia. (Photograph courtesy the Royal Library, Copenhagen.)

woven wood and cane that was plastered with a mud and grass covering several inches thick (see figure 32). It was a simple structure that was strong and well insulated against both heat and cold.

Bandera says that the Indians built large houses for them at Guiomae, Canos, Yssa, Guatari, and Ylasi. He does not say that the house that was built for them at Joara was large, but one wonders whether he simply omitted this detail, since Joara was said to have a large corncrib. The house at Gueça was said to have been "somewhat large." The remainder of the houses—that is, those at Tagaya, Aracuchi, Otari, Quinahaqui, Guaquiri, and Cauchi—were presumably small.

Bandera made special mention of the house built for the Spaniards at Guatari. He said that its interior was completely covered with mats.[32] He noted that the Indians of Guatari built their houses with a circular floor plan (see figure 18), and we may presume that this was also true of the house they built for the Spaniards. The fact that Bandera specifies the floor plans at Guatari were circular may imply that the houses they saw elsewhere were rectilinear (see figures 16 and 32).

Evidently there was some variability in the storage rooms that the In-

Foundations of Greater Florida 143

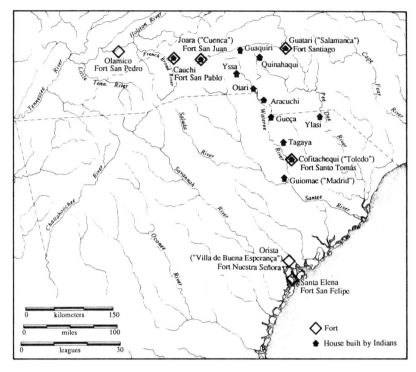

Figure 30. Forts built by Pardo and houses built by the Indians on the "road to Zacatecas."

dians built for storing the corn they collected for the Spaniards. Most of the storage rooms appear to have been corncribs built on posts high above the ground (*una casa alta*, *una camara alta*), which were typical in much of the Southeast (see figure 33). This was clearly the case at Guiomae, where Pardo found a house but no crib, but where he exacted promises from Guiomae Orata and Pasque Orata that they would build cribs. At both Guiomae and Ylasi reference is made to *two* cribs of corn having been collected, and this would seem to imply that these were separate structures. At Cofitachequi, however, the storage room was said to have been *inside* the house that the Indians built.[33] In addition, the house at Aracuchi had an interior corncrib.[34] In passing, it may be noteworthy that at the time of the Soto expedition, the "Lady" of Cofitachequi claimed to have *seven* cribs of corn at Ilapi, which was presumably the same place as Ylasi. If this was so, then the declining number of corncribs may be a measure of the decline of Ylasi and/or the decline of the paramount chiefdom of Cofitachequi during the intervening years.

144   *The Juan Pardo Expeditions*

Figure 31. Wattle and daub construction in a reconstructed house at the Town Creek site, North Carolina. (Author's photograph.)

Figure 32. Reconstructed house at the Town Creek site, North Carolina. (Author's photograph.)

Figure 33. Reconstructed corncrib at the Chucalissa site, Memphis, Tennessee. (Author's photograph.)

The Forts the Spaniards Built

Pardo built five small forts in the interior and one on the coast (see figure 30). On his initial expedition, he built the first of these forts at Joara, a town on which he bestowed the name "Ciudad de Cuenca," in honor of his native city in Spain. Like Joara, Cuenca was situated at the foot of a range of mountains in Spain, and like Joara, Cuenca was surrounded by rivers. It will be recalled that Pardo's initial expedition was cut short because snow had fallen in the mountains, making it impossible to cross to the other side. Pardo and his men spent two weeks in Joara, during which time they built Fort San Juan, so named because they arrived at Joara on the Day of San

Juan, and also because Juan was Pardo's first name. Their justification for building this fort was to prevent the land from remaining a wilderness (*disierto*).

Pardo provisioned each of these tiny forts, and the scant quantities of the supplies he left at each of them is an indication of just how thin the material basis of Menéndez's colonial scheme was. These provisions consisted of gifts for the Indians, tools, and most importantly, weapons and ammunition. Neither of the Spaniards' principal weapons—the crossbow and the harquebus—was notably superior to the bows and arrows of the Indians. Although the crossbow could fire a small missile at high velocity and could be aimed and fired by a person of little skill or strength, an experienced Indian archer could fire an arrow with comparable penetrating power (because of the heavier weight of the arrow), and he could fire several arrows in the time it took a crossbowman to load and fire a single bolt.

The harquebus was the earliest and mechanically the crudest of the handheld firearms. It was fired by touching a piece of lighted matchcord, like a slow-burning fuse, to the powder hole (see figure 34). Some of the har-

Figure 34. A harquebusier with a lighted matchcord. Engraving by Theodore de Bry after a painting by Jacques Le Moyne. (Photograph courtesy of the Smithsonian Institution.)

Foundations of Greater Florida

quebuses used by Pardo's men may have been equipped with devices that held a piece of matchcord and swung it around to the powder hole at the instant the weapon was to be fired. A ball or shot fired from a harquebus had much greater velocity than an arrow or a crossbow bolt, but it was slow to load, not very accurate, and problematic in rainy weather. Because the matchcord had to be kept alight when action might occur, great quantities of it were required.

When Pardo departed from Joara while on his first expedition, he left Moyano in charge of thirty men at Fort San Juan. To defend the fort, Pardo issued Moyano and his men the following supplies:

   150 pounds of harquebus powder
   133 pounds of harquebus matchcord
   135 pounds of lead
     4 crossbows
   240 crossbow bolts.

Pardo also left three shovels, two mattocks, and two picks at Joara.[35] The above is an accounting of supplies by the royal storekeeper. Other supplies belonging to the soldiers were no doubt left at Joara as well. On his second expedition, as he hurried to relieve Moyano at Chiaha, Pardo left Corporal Lucas de Caniçares in charge of Fort San Juan. Bandera does not say how many men remained at Fort San Juan with Caniçares.

When Moyano took twenty men from Fort San Juan and went to attack Chiscas in the upper Tennessee Valley, he proceeded from there to Chiaha (Olamico). Here he found himself in the midst of a vastly superior force of Indians, and he must have recognized the hazard of the situation. He apparently built a small fort in which he and his men could have some protection until Pardo arrived with reinforcements.

All the other forts were built one after another following the military threat at Satapo posed by the allies and tributaries of the paramount chief of Coosa. The degree to which this string of forts had been planned in advance is not clear. Menéndez ordered Pardo to take possession of the land, and it is clear that Pardo considered—or at least claimed—that he had conquered the land all the way to Olamico, and the purpose of the fort at Olamico was to defend what he had conquered. The decision to build this fort was made at Satapo when Pardo met with his ensign, sergeant, and corporals. They agreed that their best course would be to retreat from Satapo to Olamico and to build there a fort at which a detachment would be garrisoned.

The manner in which the fort at Olamico was built is unclear. Bandera says that four days were spent constructing this fort, but the fact is that the full army remained there only two days before continuing on. It is possible that they incorporated into this fort the small structure that Moyano had already built. Or perhaps they got the fort started and then departed, assuming that the soldiers who were left to man it could complete the job in two additional days.

Pardo named the fort at Olamico Fort San Pedro, perhaps in honor of Pedro Menéndez de Avilés, and he manned it with Corporal Marcos Jiminez and twenty-five men. It should be noted that the Indians of Olamico built no house in which the Spaniards could live, and there is no mention of a special corncrib. Also, Pardo did not confer a Spanish place name on Olamico. He supplied the fort with

> 60 pounds of harquebus powder
> 50 pounds of matchcord
> 85 pounds of lead balls
>   a socketed Biscayan axe
>   two dozen chisels and knives to be used as gifts for the Indians
>   a mattock
>   a shovel.

Pardo ordered Jiminez to remain at the fort unless commanded to depart. Jiminez swore that he would do so "under pain of perjury and of infamy and of falling into less value."

On Pardo's return to Santa Elena, the number of men accompanying him dwindled as he traveled along, leaving detachments of men at the forts. Let us estimate that at Olamico his army consisted of 136 men. This figure includes Moyano and his 20 men, with whom Pardo joined forces, and also the 5 men carrying supplies who had come to join them. It does not include the small force he left back at Joara, which we might estimate to have been 11 men. Hence, when Pardo departed from Olamico he had an estimated 110 men under his command.

The next fort to be built was Fort San Pablo at Cauchi, high up in the mountains. The Indians here had built a house for the Spaniards to use. Whether it was situated inside the fortification is unknown. Pardo and his men completed construction of the fort in three to three and a half days, and it was garrisoned with Corporal Pedro Flores and ten men. Flores swore an oath similar to that of Jiminez. Fort San Pablo was supplied with the following:

22 pounds of powder
24 pounds of matchcord
36 pounds of lead balls
two dozen chisels and knives as gifts for the Indians
a socketed Biscayan axe.

The expedition, now numbering ninety-nine, proceeded to Fort San Juan at Joara. Here Pardo and his men rested for two and a half weeks before proceeding any further. When they did depart, Pardo left his ensign, Alberto Escudero de Villamar, in charge. Pardo did not require as stringent an oath from Villamar as he had from Jiminez and Flores. Moreover, Bandera does not say how many men Pardo stationed there under the command of Escudero, but perhaps it was thirty, the same number he left there with Moyano on his first expedition. With eleven men already stationed there, this would have further reduced Pardo's army by nineteen men. The fort was supplied with the following:

85 pounds of powder
68 pounds of matchcord
100 pounds of lead balls
34 pounds of nails (*clabos*)
42 *azolejas*, like chisels[36]
3 iron shovels
2 socketed axes
4 iron wedges
2 mattocks
2 picks.

This is the largest quantity of supplies left at any of the forts. Whether because of its strategic location or because the Spaniards had more time to build a substantial installation, Fort San Juan was clearly to be the principal fort in the interior. Bandera notes that Villamar was charged with conserving the friendship of the chiefs, and he was to visit the other forts in order to govern them. It was on the occasion of accounting for the provisioning of Joara that Bandera wrote a summary of the distances and directions of all the other forts in relation to Joara. Again, this suggests that Joara was to be the metropolis in the interior.

When Pardo departed from Joara, his army had been reduced to about eighty men. They went by way of Yssa and Quinahaqui to Guatari. While he was at Yssa, some Indians from Guatari came to pay him a visit, and he sent them back with two socketed axes and orders to cut wood for the construction of a fort. The construction of the fort at Guatari, renamed "Ciudad de Salamanca," is described by Bandera in more detail than any of the other forts. Because Pardo feared that he and his men might suddenly be summoned to Santa Elena, he first spent five days building two "bastards" (*bastardos*), bastions built of logs and earth. Then, when no summons came, they spent an additional seventeen days building four tall "cavaliers" (*cavalleros*) of thick wood and dirt, and a wall of high poles and dirt. An unknown number of Indians helped them complete the second phase of this construction. Pardo named it Fort Santiago.

Corporal Lucas de Caniçares was left at Fort Santiago in charge of sixteen men. Caniçares swore an oath similar to the ones sworn by Jiminez and Flores, with the added stipulation that the soliders were not to bring Indian women into the fort at night. This may have been included in the oath to satisfy Sebastian Montero, the priest who had been left at Guatari on the first expedition, and who had established a mission there.

The following supplies were left at Fort Santiago:

   34 pounds of powder
   34 pounds of matchcord
   51 pounds of lead balls
      several objects of iron and other things to be used as gifts for the Indians
   2 socketed axes.

Pardo departed from Guatari with approximately sixty-three men, and he proceeded to Aracuchi. Here he split his force, sending five soldiers and some Indians on to Cofitachequi carrying the following supplies:

   20 pounds of harquebus powder in a keg
   17 pounds of matchcord
   34 pounds of lead balls
   2 socketed axes.

Pardo did this in order not to have to carry these munitions to Ylasi with him. With the remaining fifty-five men, Pardo went on to Ylasi and then to Cofitachequi to rendezvous with the others.

Pardo conferred upon Cofitachequi the name "Ciudad de Toledo." He remained there for eighteen days. Bandera does not mention building a fort at Cofitachequi, but they must have built one, because he later refers to Fort Santo Tomás at this place. Unlike the other forts, this one was left with no men and no munitions when Pardo departed. In part, this must have been so that the sixty-three or so men could help carry bags of corn to the coast to relieve the men at Santa Elena. Another reason may have been that he had by this time got word that the Indians of Ahoya were in rebellion and had killed one of his corporals. He expected trouble when he got there.

After arriving at the village of Orista on the coast, which he renamed the "Villa de Buena Esperança" (i.e., the Village of Good Hope), he set about building a fort that was different from the ones in the interior. This was a strong house (*casa fuerte*) built out of sawn lumber. They began construction on February 19 and completed it on March 2, naming it Fuerte de Nuestra Señora. Pardo put Corporal Gaspar Rodriguez in charge of twelve soldiers, and he supplied the fort with

    18 pounds of powder
    18 pounds of matchcord
    18 pounds of lead
    1 socketed axe
    1 large mattock
    1 chisel
    1 drill (*barrena*).

Pardo's force was now down to approximately fifty men.

Aside from the few details on the construction of Fort Santiago at Guatari and Fuerte de Nuestra Señora at Orista, nothing is known at present about the construction of the forts that Pardo built. Presumably they resembled the small, hastily constructed fortifications that Europeans built in other parts of the world.[37] But we can have no sure knowledge until one of these forts is located—if we can be so lucky—and excavated by archaeologists.

Because there was so little food at Santa Elena, Pardo split his force at Villa de Buena Esperança and sent Sergeant Pedro de Hermosa with thirty men back to Fort Santo Tomás at Cofitachequi. Along the way, Hermosa was to leave four soldiers off at Guiomae, renamed the "Villa de Madrid," to oversee the construction of dugout canoes.

Pardo's force was now down to approximately nineteen soldiers. He sent a detachment of three soldiers to Guando—whose location is unknown—with orders to bring back as much corn as possible. It is not clear whether these three men were from the contingent stationed at Fuerte de Nuestra Señora or whether they were from his own force. If they were from his own, this further cut his number to sixteen. Hence, Pardo paddled to the landing at Santa Elena on March 2, 1568, with a scant handful of men. All the others were scattered out in the interior, as much to extract food from the Indians as to hold and defend the territory they had "conquered."

The Missionaries

Menéndez had ordered Pardo to erect a cross and to station a missionary at the town of each of the principal chiefs, but he evidently only had one priest with him on his first expedition, his chaplain (*clerigo*), Father Sebastian Montero.[38] Since Pardo encountered two principal chiefs on his first expedition, Joara Mico and Guatari Mico, Montero presumably had a choice as to where his mission would be. He chose Guatari, but his reason for doing so is not given. Perhaps, as already suggested, he chose Guatari because the chiefs there were more receptive to an alliance with the Spaniards than were those of Joara. Another possibility may be that since Joara was to be the administrative center for the forts that were to be built, fewer soldiers would be stationed at Guatari than at Joara. At this time, many missionaries believed that they could succeed in converting the Indians if only they could keep them from the bad influences and offences of the soldiers.

In the early years of the colony, Menéndez appointed several of the soldiers at remote forts and outposts to serve as lay catechists for the Indians.[39] It is possible that one of these served at Joara, but there is no documentation for it.

In the company of Pardo and his men, Montero arrived at Guatari probably in early February of 1567. He was left there with two boys and as many as four soldiers to evangelize the Indians.[40] Antón Muñoz and Francisco de Apalategui accompanied Guatari Mico to Otari in September of 1567 to meet with Pardo as he was making his second expedition. Montero apparently returned to Santa Elena briefly in September or October, 1567. He must still have been at Guatari when Pardo again reached this place on December 15, although Bandera makes no mention of him.

Presumably Montero returned to Santa Elena in the spring of 1568, after having spent about a year and a half with his flock. From an interrogation conducted in Seville in 1572 it is clear that he had some success with his mission work. Montero initiated this interrogation in an attempt to obtain pay for his services (he was a secular priest) and compensation for money from personal funds he spent on his mission. Five soldiers who had been with Pardo gave testimony at this interrogation: Alvaro de Mendana, Juan

Foundations of Greater Florida

Figure 35. Devils carrying away a naked Indian. A woodcut used by Spanish missionaries to frighten Indians into converting. From *Francisco Pareja's 1613 Confessionario*, edited by Jerald T. Milanich and William C. Sturtevant, translated by Emilio F. Moran. (Courtesy Division of Archives, History and Records Management, Florida Department of State.)

Santos, Francisco de Teran, Pedro de la Sierra, and Hernando Moyano.[41] Mendana and Santos, and perhaps Teran and Sierra, claimed to have been at Guatari with Montero, at least for a time.

They testified that Montero had taught many of the Indians at Guatari four prayers: *Pater Noster*, *Ave Maria*, *Credo*, and *Salve Regina*. The Indians were said to have observed the holy days, and they did not eat meat on Fridays (see figure 35). Moreover, Montero taught the Spanish language to many of the Indians, especially to the caciques. Some of the caciques came to the mission every morning and evening.[42]

To all appearances, Father Montero was more successful in his mission work than were many of the early missionaries who worked among the Southeastern Indians. The Jesuits, for example, were notably unsuccessful. One of the first of these, Father Juan Rogel, arrived in Florida in September 1566. He first worked at a Spanish outpost among the Calusa Indians at Charlotte Harbor. His colleague, Francisco Villarreal, began his work among the Tequesta Indians near present Miami at the same time. After a very short time, both of them returned to St. Augustine convinced that it would be impossible to establish viable missions among the hunter-gatherer Indians of southern Florida.

Rogel next undertook to do mission work at Orista. He arrived at Santa Elena on July 3, 1568.[43] The Indians of Guale and Orista appeared to Rogel to be more promising than the Calusa had been. There was less linguistic diversity at Orista and Guale, and the people were at least part-time farmers. For Rogel, to compare Guale and Orista to Calusa was to compare a politic people to a barbaric people.[44] Father Antonio Sedeño, who worked both at Calusa and at Guale, said that in Calusa there were thirty caciques speaking twenty-four different languages, and all of these consisted of not more than 2,000 people.[45]

A church and a house for the Spaniards was built at Orista, probably by the Indians, and Rogel took up residence there with three Spanish boys who were to learn the language of the Indians. In December of 1569 Rogel saw little to criticize about the people of Orista. They were monogamous (whereas the Guale were polygamous), and they lived orderly lives. The worst thing about them was that they were gamblers and were heavily involved in trading with Indians in the interior.[46] By this time, however, Rogel still had not acquired a sufficient command of the native language to use it effectively.[47]

A year later Rogel was of a different mind about the situation at Orista. He complained that the people all moved into the woods in winter to search for food, and when Rogel tried to follow them, they made it clear to him that they did not wish to be bothered. But the worst thing was that Juan de la Bandera, who was at this time serving as deputy governor of Santa Elena, had gone to Escamacu while a feast was being held, and he forced three or four caciques to take several canoes loaded with corn to Santa Elena. Moreover, Rogel had learned that Bandera was going to send forty of his soldiers inland to live off the Indians because there was not enough food in Santa Elena. Rogel knew that if Bandera did this, the Indians would come to him for protection against the soldiers, and he knew that he would not be able to protect them. And when he could not, he knew that the Indians in their frustration might harm him. Thus, on June 13, 1570, eight or ten days before the soldiers were to be sent to Orista and vicinity, Rogel abandoned his mission and returned to Santa Elena.[48]

It would seem that Rogel had a surer grasp of Indian politics and psychology than did the Jesuit Juan Baptista de Segura and his companions, who attempted to found a mission at Ajacán, on the western side of Chesapeake Bay. This was to be the northernmost outpost in Menéndez's greater Florida. Someone who contributed to the plan for this mission was Don Luis de Velasco, an Indian from the Chesapeake Bay area who had been picked up by one of Menéndez's ships and taken to Spain. Here Don Luis became fluent in Spanish, and he became a nominal Catholic. Father Segura expected Don Luis to serve him in his missionary work as Timothy had served St. Paul.[49] It is likely that Segura had been influenced by the thought and writings of Bartolome de las Casas, who advocated that Indians be brought to Christianity by gentle persuasion rather than by force of arms.[50]

Segura, Don Luis, and eight other members of the mission arrived at Chesapeake Bay in September 1570. They probably disembarked at a landing a few miles up the channel of the James River, and from there moved to a site between the James and York rivers and established their mission. But Don Luis, no Timothy, soon moved away from the mission and refused to come at Segura's bidding. The Spaniards quickly ran short on food and began bartering copper, brass, and eventually axes to the Indians for food. They became so hungry they were forced to go to the woods to look for persimmons to eat.[51] Then, in February 1571, Don Luis led an uprising in which the Indians killed Segura and seven other missionaries, sparing only a boy, Alonso de Olmos, who had been taken there from Santa Elena.[52]

Prospecting for Precious Metals and Gems

Pardo and his men engaged in an activity that Menéndez did not mention in his orders—prospecting for mineral riches. Rumors that precious metals and gems could be found in the area Pardo was exploring had been circulating since the first information on this land had gotten back to Europe. Before 1526, the Indian Francisco ("El Chicorano") told Peter Martyr that pearls and gems could be obtained near Chicora in the land of Duhare and Xapira.[53] This was confirmed by Hernando de Soto, who found a large

Figure 36. Freshwater pearls from the Hixon site, Hamilton County, Tennessee. (Photograph courtesy Frank H. McClung Museum, University of Tennessee, Knoxville.)

quantity of freshwater pearls in a temple at Cofitachequi (see figure 36).[54] Soto exaggerated the importance of these pearls, perhaps because pearls were one of the first treasures the Spaniards discovered in the New World. They were discovered by Christopher Columbus in 1498, on his third voyage, and they were subsequently exploited commercially.[55]

It was also at Cofitachequi that members of the Soto expedition thought they saw traces of gold.[56] An even stronger rumor of gold came when they reached Chiaha, where they heard the same story they had heard in Cofitachequi. Namely, the Chiscas in the mountains mined copper, and another metal of that same color. De Soto sent two men to the Chiscas to investigate, and they came back with the ambiguous message that gold might be found there.⁵⁷

As time went on, however, these bits of information were embellished through telling and retelling to create the "Legend of Chicora," which told of fabulous wealth that could be had in the land explored by Ayllón and Soto. With the publication of Francisco Lopez de Gómera's *General History of the Indies* in 1552, this legend was specifically linked to the Point of Santa Elena, where one entered a land rich in mother of pearl, freshwater pearls, silver, and terrestrial gems. The Legend of Chicora was a factor both in the founding of Charlesfort and Santa Elena.⁵⁸

The French who had been at Fort Caroline were another source of rumors of precious metals and gems in the interior.⁵⁹ They had acquired quite a bit of gold and silver, which the Indians had obtained from wrecked Spanish ships.⁶⁰ But in addition the French were convinced that the Indians found some of their gold in streams that originated in the Appalachian Mountains⁶¹ Jacques Le Moyne drew a picture of how the Indians were supposed to have done this (see figure 37). One of the local Timucuan groups, the Houstaqua, who lived northwest of the St. Johns River and east of the

Figure 37. How the Indians were reputed to have extracted gold from streams in the "Apalatcy Mountains." Engraving by Theodore de Bry after a painting by Jacques Le Moyne. (Photograph courtesy of the Smithsonian Institution.)

Aucilla River, were said to know how to travel to these mountains, and they were known to traffic in copper.[62] These were the same people the Soto chroniclers called "Yustega."[63] And as already mentioned, the chief Uastique Orata, whom Pardo met at Tocae and who had traveled seventeen days from the "west," may have come from this north Florida area.

One indication that Pardo and his men were on the lookout for precious substances came when Pardo led his men from Chiaha to Chalahume and Satapo. On a spur of Chilhowee Mountain, in an area known as The Flats and Lake in the Sky, Pardo picked up a small reddish stone that he gave to Andrés Suarez, who was a "melter of gold and silver." Suarez examined the stone, and although there is no evidence that he tested the ore, he said that it "might be silver ore." Suarez's judgment may have been affected by the fact that in the fabulously rich silver strike the Spaniards had made in the mountains of northern Mexico, two of the ores they found are sometimes called "ruby silver": proustite ($3Ag_2S.As_2S_3$), a scarlet-vermillion mineral; and pyrargyrite ($3\ Ag_2S.Sb_2S_3$), a black to dark grey mineral, reddish in thin pieces by transmitted light.[64]

At this time the Spaniards thought that the mountains of northern Mexico in which they had found these ores extended to the north and that the Appalachians were a part of this same range. But, of course, they are not, and it is extremely unlikely that either proustite or pyrargyrite occurs in the Appalachians. The stone that Pardo found may have been a piece of red jasper, which occurs in association with barite veins in the Appalachians. Another possibility, although a less likely one, is that the stone in question may have been a piece of hematite—iron oxide—which occurs widely in the area through which they were traveling.

There are several indications that one of the moving forces in this prospecting was Sergeant Moyano, whose imagination, one suspects, had been set on fire by the Legend of Chicora. Moyano and his company of thirty men were in residence at Fort San Juan in Joara for three months or more. One clue to Moyano's behavior is that the very next day after Pardo arrived at Joara on his return from the Tennessee Valley, Moyano set out with Andrés Suarez and the two of them traveled to a specific site to examine a "crystal mine." Either they learned from the Indians about the existence of these crystals on the night of November 6, after they arrived at Joara, or what is more likely, Moyano already knew of their existence before he crossed the mountains to attack the Chiscas.

It is possible, even probable, that during his three months' stay at Joara, Moyano made several forays to examine possible locations of precious minerals. Thus, these activities of Pardo and Moyano may be the substance to some of the many reports and stories about Spanish "mining" in western North Carolina.[65] One suspects that most of these stories are modern-day Chicora Legends.

Moyano's foray across the mountains with twenty of his own men along with warriors of Joara to raid the Chiscas may have been motivated as much by the possibility of discovering gold as the desire to attack a chief who had threatened him. When Juan de Ribas told of this exploit thirty-five years later, he said that Moyano had received a "payment in gold."[66] In fact, Moyano was probably not paid in gold, but gold was probably what motivated his action. Also, as already mentioned, Pardo may have chosen the difficult trail from Chiaha to Coosa in order to find a deposit of red stone that might lead him to a deposit of silver ore like that in Zacatecas.

Another indication of Moyano's keen interest in precious substances is the fact that in the years following the Pardo expeditions his name is frequently mentioned in stories about *Los Diamantes*, the fabulous deposit of diamonds.[67] The Indians in and around Joara would have had knowledge of the locations of some of the substances of interest to Moyano and Pardo. For example, Uwharrie pottery was tempered almost exclusively with crushed quartz, and many Southeastern Indians used quartz crystals for divination (see figure 38).[68] Another substance the Indians of Joara would have known about was sheet mica, or muscovite. This mineral is relatively rare, occurring most notably in pegmatite dikes and veins in an area near Boone, North Carolina, southwest into northern Georgia and Alabama. Quite a number of mica mines were opened up in this area in the nineteenth century, and many of them showed evidence of having been worked by prehistoric Indian miners.[69] It may be significant that the largest of these prehistoric mica mines were in Mitchell County, North Carolina, the area through which Moyano took his men to attack the Chiscas on the upper Nolichucky River.

Sheet mica was traded extensively in the eastern United States beginning in the Woodland period, and the trade was still going on in the Mississippian period. The "silver" that the young Indian trading boy Perico promised Soto he would find at Cofitachequi turned out to be slabs of mica.[70] The presence of mica as an exotic material in the territory of Cofitachequi is confirmed both at the Mulberry Mound site and at the Blanding Mound site (38KE17) located about 5 miles northwest of Camden.[71] This mound was leveled in 1826, and its earth fill was spread around to increase the fertility of the adjacent land. Blanding was evidently present when the mound was leveled, and he reports that the uppermost part of the mound contained great quantities of mica, with some pieces 3 or 4 inches across

Figure 38. Quartz crystals from the Toqua site, Monroe County, Tennessee. (Photograph courtesy Frank H. McClung Museum, University of Tennessee, Knoxville.)

(see figure 39).[72] More recently, hand-sized pieces of unworked mica have been found in a house floor at the Mulberry site, one of the principal towns of Cofitachequi at the time of the Soto expedition.[73]

Yet another substance the Indians of Joara would have been familiar with was copper. Free copper did not occur in significant quantities in and around Joara, but it occurred in the country of the Chiscas (see figure 40). From the Indian traders he captured in northern Florida, Hernando de Soto learned that metals were an important component in the indigenous trade.[74]

The way in which the Spaniards explained to the Indians the kinds of substances they were interested in was quite simple. They showed the Indians the precious metals and gems in the jewelry they were wearing and asked the Indians where such materials could be found. This is what Soto and his men did while they were in Apalachee.[75] And this is what Juan Menéndez Marqués did when he visited Chesapeake Bay in 1588.

At 38° latitude, I beheld an Indian with a necklace of gold about his neck; I captured him and took him from over there to where I was. I showed him a chain of gold and he said they had plenty of it and called it *tapisco*; then I showed him a

Foundations of Greater Florida   161

Figure 39. Cut mica discs from Cofitachequi. Gary Tomlin collection. (Photograph courtesy South Carolina Institute of Archaeology and Anthropology.)

candlestick made of brass and the Indian said they had plenty of this and called it *guapacina*; then I showed him copper and he said they also had plenty of this and that it cost nothing and that the Indians made nothing of it and that it was called *ococo*.[76]

It is doubtful that the Indians actually made the distinctions claimed by Menéndez Marqués.

The actual discoveries of precious substances by the Pardo expedition appear to have been minor. Pardo and his men were almost certainly mistaken about the discovery of silver ore. From the descriptions in the Bandera document, it is not possible to be certain about the nature of the crystal found in the two deposits northwest of Yssa, nor of the deposit a league downstream from Yssa. The crystals from the latter deposit were said to have had "sharply faceted edges" and looked different from those to the west of Yssa. The crystals they found were probably corundum, whose value would have ranged from low to very high, depending on the color of the crystals, and quartz, whose value would have been low.

Juan de Ribas, one of Pardo's soldiers, remembered this discovery of crystal many years later, in 1602, when an investigation into the condition

Figure 40. Native copper ax from Long Island, Roane County, Tennessee. (Photograph courtesy Frank H. McClung Museum, University of Tennessee, Knoxville.)

of the Florida colony was held in St. Augustine. Ribas remembered the place where the discovery was made as a high hill that had three or four veins of crystal. One of the veins contained blue "diamonds" and the other contained purple "diamonds." Some of the corundum in the area of their discovery does have a bluish color, and the purple "diamonds" may have been amethyst, which also occurs in this general area. Quite a few of the soldiers collected some of these "tiny diamonds" and later gambled for them. Ribas said that he was sure that they were diamonds because the silversmith pounded on them with a hammer and could not scratch them.[77] This strongly suggests that one of their discoveries may have been corundum, which is next in hardness to diamond.

On November 26, 1567, Bandera registered the first of the crystal mines, promising to exploit it and to give the king his royal due. Moreover, five others were to share with him equally in the mine: Juan Pardo, Esteban de las Alas (governor of Santa Elena), Andrés Suarez, and ensign Albert Escudero de Villamar. The second mine was recorded in the same manner. Then, on December 9, 1567, while they were at Yssa, Suarez and Moyano appeared before Pardo and asked that he decree that no one should dig within a quarter of a league from where the crystal mines were located without the command and consent of all the partners in the mine. The claim to the crystal mine downstream from Yssa was recorded on December 10, 1567, in the names of Moyano, Lucas de Caniçares, Juan Pardo, Esteban de las Alas, Bandera, Escudero de Villamar, and Sergeant Pedro de Hermosa.

It is quite striking that on December 10 at Yssa, Pedro de Hermosa was suddenly promoted from *soldato* to *sargento*. It is equally striking that after this date Bandera refers to Moyano as a "former *sargento*." One wonders whether Moyano's demotion might have come as a consequence of too ardent a pursuit of precious substances.

Notes

1. Eugene Lyon, *The Enterprise of Florida: Pedro Menéndez de Avilés and the Spanish Conquest of 1565–1568* (Gainesville: University Presses of Florida, 1976), pp. 197–205.
2. Charles Gibson, *Spain in America* (New York: Harper & Row, 1966), pp. 149–50.
3. Pedro Menéndez Marqués to Pedro Menéndez de Avilés, Havana, March 28, 1568, AGI, SD 115, fol. 206. I am grateful to Paul Hoffman for this reference.
4. Ibid.
5. No explanation for this cannon powder is given. Paul Hoffman is of the opinion that they intended to use this coarse powder in their harquebuses and that this is an indication of their supply problems at Santa Elena.
6. AGI, Contratación, no. 2, R. 7. I am grateful to Paul Hoffman for this reference.
7. Ibid.
8. P. J. Bakewell, *Silver Mining and Society in Colonial Mexico: Zacatecas 1546–1700* (Cambridge: Cambridge University Press, 1971), *passim*.
9. Ibid., p. 23.
10. Ibid., p. 27.
11. Ibid., p. 32.
12. Ibid., pp. 34–35.
13. Ibid., p. 19.
14. William E. Myer, "Indian Trails of the Southeast," *42nd Annual Report of the Bureau of American Ethnology* (Washington, D.C.: GPO, 1928).
15. Charles Hudson, Marvin T. Smith, and Chester B. DePratter, "The Hernando de Soto Expedition: From Apalachee to Chiaha, "*Southeastern Archaeology* 3(1984):71–72.

16. David Anderson, David J. Hally, and James L. Rudolph, "The Mississippian Occupation of the Savannah River Valley," *Southeastern Archaeology* 5(1986):32–51.
17. Charles Hudson, Marvin Smith, David Hally, Richard Polhemus, and Chester DePratter, "Coosa: A Chiefdom in the Sixteenth-Century Southeastern United States," *American Antiquity* 50(1985):731.
18. Richard P. Kandare, "A Contextual Study of Mississippian Dugout Canoes: A Research Design for the Moundville Phase," M.A. thesis, University of Arkansas, 1983.
19. Ralph Wilbanks, "A Progress Report on the Small-Watercraft Research Project in South Carolina," *The Institute of Archaeology and Anthropology Notebook, The University of South Carolina* 12(1980):17–27.
20. Felix Zubillaga, *Monumenta Antiquae Floridae (1566–1572)* (Rome: Monumenta Historica Soc. Iesu, 1946), p. 421.
21. Father Antonio Sedeño to Father Juan de Polanco, S.J., in *The First Jesuit Mission in Florida*, comp. Ruben Vargas Ugarte, trans. Aloysius J. Owen, U.S. Catholic Historical Society Records and Studies, vol. 25, (New York: 1935), pp. 121–23.
22. Zubillaga, *Floridae*, p. 333.
23. John G. Varner and Jeannette J. Varner, *Dogs of the Conquest* (Norman: University of Oklahoma Press, 1983), *passim*.
24. Chester B. DePratter and Marvin T. Smith, "Sixteenth-Century European Trade in the Southeastern United States: Evidence from the Juan Pardo Expeditions (1566–1568)," in *Spanish Colonial Frontier Research*, ed. Henry F. Dobyns (Albuquerque, N.M.: Center for Anthropological Studies, 1980), pp. 67–77.
25. My counts of the artifacts that Pardo gave the Indians differ from those of DePratter and Smith.
26. Stanley South, Russell K. Skowronek, and Richard E. Johnson, *Spanish Artifacts from Santa Elena* (Columbia: South Carolina Institute of Archaeology and Anthropology, 1988) pp. 179, 183.
27. Richard Polhemus, personal communication.
28. South et al., *Spanish Artifacts from Santa Elena*, pp. 162–63, 425–51. Richard Polhemus is of the opinion that this is probably the bead type that Pardo was distributing.
29. Ibid., pp. 131–35, 412–15. It was Richard Polhemus who noted that the buttons found in Tennessee resemble those found at Santa Elena.
30. Paul Hoffman informs me that this could also have been ornamental appliqués such as the buttons and eyes that are frequently seen on sixteenth-century portraits.
31. Richard Hakluyt, *The Principal Navigations, Voyages, Traffiques & Discoveries* (Glasgow: James MacLehose and Sons, 1904), vol. 20, p. 114.
32. [U] na casa grande de madera nueba y de dentro toda Estreda lo qual. . . .

33. *[U] na casa grande para su mag. y en ella mucho numero de mayz.*
34. *[U] na casa buena nueba de madera y dentro della una camara alta con cierto numero de mayz.*
35. Information on supplies here and in the succeeding pages is based on two documents preserved in the Archive of the Indies (Seville) in Contratación 2929, no. 2, R.7. Both have been translated by Paul Hoffman, and both are included in part II of this book.
36. This is the only occasion on which this particular tool was mentioned.
37. Perhaps they were smaller versions of the first forts that were built at Santa Elena.
38. Michael V. Gannon, "Sebastian Montero, Pioneer American Missionary, 1566–1572," *Catholic Historical Review* 51(1965–66):335–53.
39. Gannon, "Sebastian Montero," p. 337. Michael V. Gannon, *The Cross and the Sand: The Early Catholic Church in Florida, 1513–1870* (Gainesville: University of Florida Press, 1965), p. 29.
40. Pardo stated that he left four soldiers at Guatari, but he made no mention of having left any boys (Pardo, p. 114). Many years later, in the course of an interrogation concerning Montero's performance of his duties, testimony was given that Montero had with him at Guatari his nephew as well as two boys who served him (Gannon, "Montero," p. 340). At Otari, Muñoz and Apaletegui made the somewhat puzzling statement to Bandera that they had been at Guatari for three months at the command of Captain Pardo (Bandera II). But if they were among the original four soldiers, they had to have been there seven months. It is possible that Montero visited Santa Elena in the summer of 1567 and that Pardo sent Muñoz and Apalategui to accompany Montero when he returned to Guatari.
41. Gannon, "Sebastian Montero," p. 349.
42. Ibid., pp. 347–50. It was a small mission, a *doctrina*.
43. Zubillaga, *Floridae*, p. 315.
44. Ibid., p. 332.
45. Ibid., p. 424–25.
46. The Southeastern Indians played several indigenous gambling games. See John Lawson, *A New Voyage to Carolina* (Chapel Hill: University of North Carolina Press, 1967), pp. 34, 35, 43–54, 55.
47. Zubillaga, *Floridae*, pp. 336, 401.
48. Ibid., pp. 474–75. It is striking that in their letters neither Rogel nor any of the other Jesuits refer to Montero. Surely Juan Pardo mentioned Montero when he told Rogel about his travels in the interior.
49. Clifford M. Lewis and Albert J. Loomie, *The Spanish Jesuit Mission in Virginia, 1570–1572* (Chapel Hill: University of North Carolina Press, 1953), p. 26.
50. Ibid., p. 76.
51. Ibid., p. 119.

52. Ibid., pp. 45-48.
53. Peter Martyr, *De Orbe Novo*, ed. and trans. F. A. MacNutt (New York, 1912), vol. 2, pp. 254-71.
54. Gentleman of Elvas, *Narratives of De Soto in the Conquest of Florida*, trans. Buckingham Smith (New York: Bradford Club, 1866), p. 65; Rodrigo Ranjel, "Narrative," in *Narratives of the Career of Hernando de Soto*, trans. Buckingham Smith, ed. Edward G. Bourne (New York: Allerton, 1922), pp. 100-1.
55. Gibson, *Spain in America*, pp. 12-13.
56. Elvas, *Narratives*, p. 72; Ranjel, "Narrative," p. 102.
57. Elvas, *Narratives*, p. 74; Ranjel, "Narrative," p. 110.
58. Paul E. Hoffman, "Legend, Religious Idealism, and Colonies: The Point of Santa Elena in History, 1552-1556," *South Carolina Historical Magazine* 84(1983):59-71.
59. René Laudonnìere, *Three Voyages*, trans. Charles E. Bennett (Gainesville: University Presses of Florida, 1975), pp. 9, 46.
60. Ibid., pp. 76, 96.
61. Ibid., p. 116.
62. Ibid.
63. Luys Hernandez de Biedma, "Relation," in *Narratives of the Career of Hernando de Soto*, trans. Buckingham Smith, ed. Edward Gaylord Bourne (New York: Allerton, 1922), p. 7.
64. Charles Hudson, "Juan Pardo's Excursion Beyond Chiaha," *Tennessee Anthropologist* 12(1987):74-87.
65. Perhaps the most substantial of these reports is a discovery made by geologist Deane F. Kent (personal communication). In 1942 he found a complex sulphide vein (copper, lead, zinc, pyrite, etc.) in a granite gneiss. At the site he found the inscription, "JAN MDLXVII" chiseled into the rock. He also found a small pig of lead with a Spanish coat of arms and a six-sided gun barrel about 18 inches long. The site is east of present-day Hendersonville, North Carolina, about 36 miles from Joara, so that four days or less would have been required for a round-trip prospecting excursion. I have not been able to confirm either the inscription or the artifacts said to have been discovered there.
66. Charles W. Arnade, *Florida on Trial, 1593-1602* (Coral Gables, Fla.: University of Miami Press, 1959), p. 40.
67. Ibid., pp. 35, 38-40.
68. Joffre L. Coe, "The Cultural Sequence of the Carolina Piedmont," in *Archaeology of Eastern United States*, ed. James B. Griffin (Chicago: University of Chicago Press, 1952), pp. 307-8; Charles Hudson, *The Southeastern Indians* (Knoxville: University of Tennessee Press, 1976): pp. 166-69.
69. Leland Ferguson, "Prehistoric Mica Mines in the Southeastern Appalachians," *South Carolina Antiquities* 6(1974):1-9.

70. Garcilaso de la Vega, *The Florida of the Inca*, trans. John G. Varner and Jeannette J. Varner (Austin: University of Texas Press, 1951), p. 311.
71. Kimberly M. Grimes, "Mica Production as Evidence of Craft Specialization at the Mulberry Mound Site," paper presented at the 44th annual meeting of the Southeastern Archaeological Conference, November 14, 1987, Charleston, S.C.; George E. Stuart, "The Post-Archaic Occupation of Central South Carolina," Ph.D. dissertation, University of North Carolina, 1975, pp. 51-52.
72. E. G. Squier and E. H. Davis, *Ancient Monuments of the Mississippi Valley*, Smithsonian Contributions to Knowledge, vol. 1 (Washington, D.C., 1848), p. 106.
73. Grimes, "Mica"; Chester DePratter, "University of South Carolina Department of Anthropology/SCIAA 1985 Archaeological Field School, Mulberry Site (38KE12)," *South Carolina Institute of Archaeology and Anthropology Notebook*, University of South Carolina, 17(1985):33.
74. Garcilaso, *Florida*, pp. 253ff., 33; Ranjel, "Narrative", p. 104.
75. Garcilaso, *Florida*, p. 254.
76. Quoted in Lewis and Loomie, *Jesuit Mission*, p. 207.
77. Arnade, *Florida on Trial*, pp. 38-40.

5
The Failure of Greater Florida

Historians have generally praised Pedro Menéndez de Avilés for having successfully founded a colony in the Southeast where so many had failed before him. But the truth is that Menéndez's success was achieved within a larger failure—a failure of monumental proportions. In part this failure occurred as a consequence of the Spaniards' misconceptions about the nature of the land and the Indians. If Menéndez and Pardo had been the first Spaniards in a totally unknown land, misconceptions might have been expected. But other Spaniards had been there before—Ayllón, Narvaéz, Soto, and Luna. The fact is, however, these were not modern times, nor were these modern men. In the sixteenth century, Europeans still had a poor understanding of the importance of systematically collecting and collating information. The sciences of cartography, geography, history, and anthropology were unborn or in their infancy.

Misconceptions about the Land and the Indians

None of the explorers who had been to the interior of the Southeast before Menéndez had been there to draw an accurate map. Because none penetrated very far into the plains that lay west of the Mississippi River, the vastness of the American continent was unknown. Hence, it is not suprising that Menéndez's conception of the North American continent was inaccurate, or that his conception of east-west distances was grossly inaccurate. In addition to their lack of accurate geographical information, Menéndez and his contemporaries had two concerns that further distorted their conception of North American geography: (1) their interest in protecting the silver mines in Zacatecas, which were so vital to the Spanish economy, and (2) their wish to discover a water passage through the North American continent that would give them easy access to Asia.

In the latter half of the sixteenth century, and for a long time thereafter, many Europeans believed that in the far north of the American continent a "strait of Anian" led from the Atlantic to the Pacific (see figure 41).[1] If only it could be found it would provide easier and more direct nautical access to China and Japan than sailing around Africa. But in addition, Menéndez believed that through an inlet at about 45° latitude and also through Chesapeake Bay at 37° latitude (Bahía de Santa María) one could sail to a large bay far in the interior of North America. From this bay, with two channels to the Atlantic Ocean, it was only a quarter of a league by land to another large bay that led to the Pacific Ocean. By going this way he thought that one could sail quite near the mines in Zacatecas. It was obvious that if Spain's European rivals were to succeed in gaining control of this "large bay," the silver mines would be vulnerable to attack. One justification for the Jesuit mission in Virginia (Ajacán) was to gain a Spanish foothold in this strategic location.

In addition, for reasons already given, Europeans thought that the North American continent was far narrower from east to west than it actually was. This is why Menéndez and his contemporaries thought that the distance from Santa Elena to Zacatecas was only 500 leagues, or about 1,700 miles, whereas the actual distance is more like 2,400 miles.[2] They thought this in spite of the fact that the Soto expedition wandered for four years in the interior of this land. And they thought this in spite of the fact that those who were members of the Soto expedition realized that if they had attempted to cross what is now Texas and northern Mexico they would probably have starved to death. But in Menéndez's plan Santa Elena was to be at the beginning of a road that would traverse this same territory.

Sixteenth-century Spaniards were also mistaken about the nature of the Indian societies of the Southeast. Apparently they thought that the chiefdoms were suitable for colonization in much the same way as the state-level societies of Mexico and South America. Some members of the Soto expedition thought that colonies could be founded at Cofitachequi and at Coosa.[3] And Menéndez and Pardo clearly thought that the chiefdoms could produce goods and services to support a string of inns, forts, and missions.

Given the limited military and economic means available to sixteenth-century Spaniards, there were several reasons why chiefdoms were not as tractable and therefore not as easy to transform into a European colony as

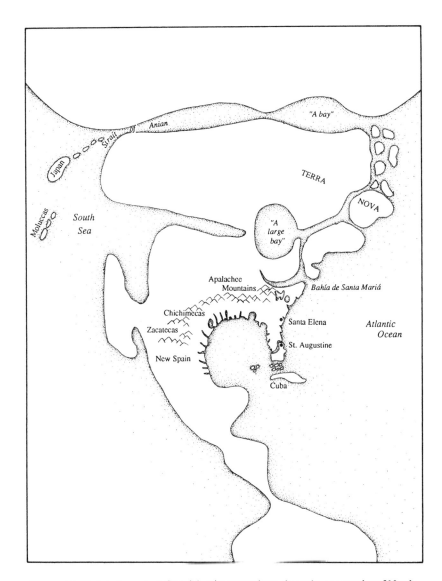

Figure 41. Sixteenth-century Spanish misconceptions about the geography of North America. (Adapted from map in Clifford M. Lewis and Albert H. Loomie, *The Spanish Jesuit Mission in Virginia, 1570–1572* [Chapel Hill: University of North Carolina Press, 1953].)

was a state-level society. Perhaps the most important reason is that the chain of command in the military forces of chiefdoms is far less hierarchical and disciplined than in a state-level society. With such dispersed military leadership, it is difficult to destroy the will to resist. This is the reason why it was so difficult for the Spaniards to conquer the Chichimecas in northern Mexico.[4] This is why it is still difficult for a modern army to wage war against guerillas and insurgents.

And just as the people of the various Southeastern chiefdoms were capable of taking military action on their own, they were also accustomed to seek private redress for personal injury or insult. Then as later, the law of vengeance must have been an important legal principle among them. When they were injured, they sought private revenge rather than justice.[5] And it is probable that Spanish soldiers who considered themselves to be conquerors, gave offense frequently, and they were probably surprised when vengeance was exacted on them. As members of small-scale societies, constantly under the scrutiny of their fellows, the Southeastern Indians fully understood the value of secrecy. They were as adept as any people in the world at concealing their thoughts, motives, and plans of action. And they appreciated the value of surprise attack. The animals they admired most were predators. One was the cougar, who is a master of the lightning attack from concealment, and who moreover attacks his prey singly.

Another aspect of the Southeastern chiefdoms that may have misled the Spaniards is that each chiefdom was ruled by an elite consisting of the lineage of the chief, his or her kinsmen, and perhaps retainers, but this elite was neither as secure in its position nor as powerful as the elites that ruled state-level societies. Because of this insecurity, it is doubtful that the elites of the chiefdoms were as corruptible as the elites of state-level societies. Consequently, they may have been less likely to sell their people out to gain personal advantage from the Spaniards.

It is also the case that chiefdoms are more unstable than state-level societies. That the chiefdoms of the Southeast had limited life spans is amply attested by archaeological evidence. These chiefdoms rose in power, prevailed for a time, and then they went into decline. One reason is that the tributary chiefdoms that were coerced into paying tribute to a paramount chief must in time have become resentful, even vengeful, and only needed an opportunity to rebel. In addition, the problem of succession to the office of chief must have been a problem, so that rival claimants may have led factions in contests for power that would have weakened or destroyed the stability of the chiefdom to which they belonged.[6]

Another factor that made the chiefdoms unsuitable for colonization on a grand scale is that their populations were rather small when compared with state-level societies, and they were fragile once exposed to Old World diseases. Even when the Indians began to die from these diseases, the Spaniards do not seem to have appreciated just how fragile these populations were.

The strategy that worked for the Spaniards in Central Mexico and South America could not have worked in the Southeast. Ironically, Sergeant Moyano may have hit upon the colonial strategy that could have worked for Spain in the Southeast. That is, if he had succeeded in discovering precious substances similar to the silver deposits in Zacatecas, then a population of Spaniards would have been attracted to the mines. Towns would have been built, and then roads, and Spanish might have become the dominant language spoken in the Southeast today. But even if this had come about, a road to Zacatecas would have been a very long time in the making.

The Failure of the Forts

The forts that Pardo built had too few defenders, and they were placed too far inland. Some of the Indians in whose territories the forts were built may have been harboring grievances since the time of the Soto expedition. Indians had long memories when they were insulted or injured. For example, in 1606 Fray Martín Prieto served as a missionary among various Timucuan towns west of St. Augustine. He made some headway in two of these towns, but he encountered stubborn hostility in a third. The reason, he discovered, was that the cacique of this town, who was so old he was physically unable to stand, had as a boy been captive of Hernando de Soto, and he counseled his people to resist the Spaniards. When Prieto went to see this cacique, he became so angry he foamed at the mouth, commanding his principal men to give Prieto a beating and throw him out. This was sixty-six years after the injury he had suffered from Soto.[7]

If Pardo saw signs of hostility among the Indians before he crossed the mountains, there is no documentary evidence of it. But, as we have seen, signs of hostility became all too evident after he crossed the mountains. When Pardo reached Olamico, he heard that several caciques were conspiring to resist him, and the truth of this became clear when he reached Satapo. Moreover, it was in Satapo that Pardo was told that some Spaniards (i.e., Soto) had come through "these parts" previously, both on foot and on horseback, and that the chief of Satapo and his subjects had killed some of them. This is a puzzling claim because the main part of Soto's force got no nearer to Satapo than Coste, which was located on Bussell Island at the mouth of the Little Tennessee River, about 20 miles downstream from Satapo. However, Soto did remain in Coste for six days, and it is possible that he sent a small force of cavalry up along the river to explore the country.[8]

What is even more puzzling is the claim of people of Satapo that their chief and his warriors killed some of Soto's men. The Indians of Coste did put up a show of force when Soto's men were raiding their granaries for food. But none of the Soto chroniclers mentions the loss of life on either side. The most obvious explanations are that either the Indians of Satapo were exaggerating their military prowess, or else more violence occurred than was mentioned by the chroniclers of the Soto expedition. Another, not so obvious explanation is that some of the warriors of Satapo may have been participants in the battle at Mabila, in central Alabama.[9] In this, the greatest battle of the expedition, Soto's soldiers were struck in a surprise attack by the warriors of Chief Tascaluza, whose chiefdom Athahachi (or Atache) was near present-day Montgomery, and Mabila, west of present-day Selma. Fighting alongside Tascaluza's warriors were warriors from several other groups, which unfortunately, are not named. In this battle, Soto suffered heavy losses, while the Indians themselves were devastated.

There are several pieces of supporting evidence for the participation of Coosa and Satapo in the battle of Mabila. One is the testimony of Juan de Ribas, a soldier who had been with Pardo. This testimony was given in 1600, in an inquiry conducted by Gonzalo Méndez de Canço into the situation of the interior of the Southeast.[10] Ribas had gone with Pardo to Satapo, where he claimed to have seen Spanish arms, chain mail, and clothing.[11] He also claimed to have seen painted on the walls of buildings depictions of horses, lances, and the dress of Spaniards.[12] Moreover, Ribas had information from (or heard at second hand from) a Portugese pilot named Almeydo, who Pardo had stationed in the interior to learn an Indian language, who claimed that he had gone to the place where Soto had died. Here the Indians had told him that further on the Spaniards had come and taken away the daughter of a chief on the croup of a horse. They had beaten this girl with a stick, and several months later the girl's father attacked and "killed" Soto.

Although it is impossible that Almeydo could have gone to the place where Soto died, he may very well have gone to a town of Indians (e.g., Satapo, Coste, or some other member town of Coosa) who participated in the battle at Mabila. There is evidence that Soto's men began enslaving Indians in Chiaha, and they continued to do so as they proceeded south-westward through the chiefdom of Coosa. In fact, Soto took the paramount chief and other eminent people of Coosa as hostages, and when he released the paramount chief at Talisi, in northern Alabama, the man went away in tears because Soto had also taken his sister prisoner and would not release her. Since the rule of descent in Coosa was probably matrilineal, the abduction of a female relative of a chief—whether sister or niece—was particularly distressing.

Another piece of confirmation comes from the King site, an archaeological site a few miles down the Coosa River from Rome, Georgia. The King site is thought to have been a town of Ulibahali, which was subject to Coosa, and Soto probably visited it. Although there is no documentary evidence that Soto's men did any fighting in this area, many of the individuals buried there show signs of having been killed and wounded by metal weapons. Moreover, the age and sex characteristics of those who were killed and wounded are consistent with the characteristics of those who were enslaved by Soto.[13] Of all the Spaniards who explored the Southeast in the sixteenth century, the most likely killers were the members of the Soto expedition, and the incident in which the killing may have occurred was the battle at Mabila.[14]

Whether impelled by old or new grievances, the actions of the Indians at Satapo made it clear to the Spaniards that they might expect trouble. Almost inevitably, the Indians attacked the forts and defeated the Spaniards whom Pardo had stationed there. It is not known whether the Indians attacked all of the forts at once or one by one, but there does seem to have been a general uprising. Considering the resistance that Pardo and his men encountered at Satapo, it is probable that the fort at Olamico (Chiaha) was first, or among the first, to be attacked by the Indians. Pardo had to have known that Fort San Pedro at Olamico was terribly vulnerable. This is indicated by the stringent oath he required of the man he left in charge there and also by the fact that immediately after returning to Santa Elena after his second expedition, he sent Pedro Catalan back to Olamico with at least eleven men.[15] If these men succeeded in reaching Fort San Pedro, they brought the total force there up to thirty-seven or so—still pitifully undermanned.

As early as May 1568, news reached Santa Elena that the interior forts had been attacked and overrun by the Indians. The precise number of Spaniards who were killed is unknown, but the loss of life must have been great. Only one man, Juan Martín de Badajoz, is known to have escaped from the interior to return to Santa Elena, although there may have been others.

The Failure of Greater Florida

Writing from Havana on July 25, 1568, Father Juan Rogel mentioned that the forts had fallen.[16] And the cause of the uprising was plain to him. When Rogel first visited Santa Elena, probably in June of 1568, Juan Pardo took him to the village of Uscamacu, where the two of them spent the night. During the night an incident occurred in which one or more Spanish soldiers at nearby Ahoya mistreated the Indians and caused a disturbance. In Rogel's opinion, it was this kind of behavior that had led to the loss of the forts in the interior. If Spanish soldiers treated the Indians so badly this close to where their captain was stationed, what, Rogel asked, must they have done to the Indians 100 to 200 leagues in the interior?[17]

According to Jaime Martínez, one of the causes of the attack against the forts was the Spanish demand for food from the Indians.[18] He claimed that in at least one instance the Indians used trickery to entice the Spaniards to leave their fort, and then a great number of them attacked, killing and torturing the Spaniards with great cruelty. Martínez's main source of information, and perhaps his only source, was Juan Martín de Badajoz, who he says escaped from one of the interior forts and made it back to Santa Elena through more than 200 leagues of brambles and forests. Juan Martín married an Indian woman named Teresa. It would be interesting to know whether Juan and Teresa became man and wife while he was in the interior. His name appears on the list of men to whom Pardo issued shoes and sandals after arriving at Olamico, and this makes it probable that Juan Martín was a member of Moyano's contingent who remained in the interior after Pardo's first expedition. If he took Teresa for his wife while he was in the interior, she was either a native of Joara or Olamico, and more probably the former than the latter.

It is also possible that Teresa was one of the Indian women who were brought out of the interior by Moyano. The names of eight of these appear on a list of Indian slaves who were freed in December 1567.[19] The time when these women were brought out is not given, but they were probably among the slaves and hostages brought out on the return from the second expedition.[20]

When Teresa Martín was interviewed in 1600, she mentioned another cause of hostility. She said that Pardo had promised that he would return in "three or four moons," and when he did not, the soldiers had committed improprieties with Indian women, and this had angered their men.[21] It was to prevent this kind of problem that Pardo had ordered the men of Fort Santiago at Guatari not to take Indian women into the fort at night.

A list of Pardo's men to whom rations were issued at Santa Elena exists, and there are notations alongside some of the names indicating that they were killed by the Indians. For most of these individuals, there is no way of knowing at which forts they were stationed. But the list makes it clear that Pedro Catalan and all of the men who went with him to Chiaha were killed.[22] One cannot help thinking that the uprising must have begun at Chiaha. Interestingly, among those killed in the uprising were six individuals, probably all boys and young men, who had been sent to live among the Indians to learn their languages.

As we have already seen, Father Sebastian Montero survived the uprising and returned to Spain, where in 1572 he petitioned to be paid his salary and to be repaid for his expenses while serving as chaplain in Florida. It is not clear when Montero abandoned his *doctrina* at Guatari. He may have been driven out by the uprising, or he may have somehow been able to remain there.[23]

Hernando Moyano survived the uprising and returned to Spain. But he again returned to Santa Elena, because he still believed that he could become a rich man from the minerals he and his comrades had discovered in the interior. He even persuaded several individuals to return with him to exploit the fabulous gem mines.[24]

The Shrinking of Florida

Regardless of whether it was done in a series of local rebellions or in a large coordinated uprising, the loss of the five interior forts was a serious blow to Menéndez's plan for a Greater Florida, encompassing a vast territory from Chesapeake Bay to what is now New Mexico. But the troubles of the Florida colonists were not over. Indeed, the Indians of the interior were but the first to see what other Indians would see in the next three decades— that Spanish Florida was militarily vulnerable.

The Indians of Guale realized this in 1576. After the Spaniards had executed two of their principal men and cut the ears off another, the Indians of Guale formed an alliance with the Indians of the Santa Elena area, their former enemies. Soon afterward, Sergeant Moyano with twenty-one soldiers went out from Santa Elena to Uscamacu to get food from the Indians.[25] True to character, Moyano may also have hoped to obtain pearls.[26] He and his soldiers spent the night at Uscamacu. At dawn the next day, the Indians launched a surprise attack and Moyano and all the rest were killed, except one Andrés Calderon, who had gone into a thicket to relieve himself just before the attack was launched. Calderon said that the Indians beheaded the slain men, and then danced about brandishing their heads (see figure 42). As the rebellion spread, other Spaniards were killed, and Fort

Figure 42. Timucuan warriors desecrating their enemy dead. The man at the far left is wielding a small war club. The men at the far right are drying scalps over a fire. Engraving by Theodore de Bry after a painting by Jacques Le Moyne. (Photograph courtesy of the Smithsonian Institution.)

San Felipe and Santa Elena had to be abandoned. The survivors fled to St. Augustine.

What the Indians were probably unable to understand was how thin and unsubstantial was the economic base on which Spanish Florida was built. At Santa Elena the farmers had trouble making a go of it because of the thin, poor soil on Parris Island. Because they were on an island, they did not have easy access to the somewhat richer soils on the coast of the mainland. Their food production fell so short that they depended on subsidies from Spanish stores, or else they had to extort food from the Indians. In the worst of times they had to forage for oysters and wild plants. More fundamentally, Pedro Menéndez de Avilés had overextended himself financially, and so had Philip II, who in fact went bankrupt.[27] In the aftermath of the uprising of 1576, Philip II appointed Pedro Menéndez Marqués to serve as governor of Florida. In 1577 he built a new fort—San Marcos—at Santa Elena, and in 1578 new houses and a church were built. Menéndez Marqués also sent out punitive expeditions against the Indians, and an uneasy peace was established. This was broken in 1580, when the Indians again revolted and attacked Fort San Marcos, but they did not succeed in destroying it.[28]

To make matters worse, France and England stepped up their competition with Spain for control of the Atlantic coast of North America. In 1582 Menéndez Marqués received a report that Frenchmen had been seen at Cayagua (Kiawah)—Charleston Harbor. He sent Vicente Gonzalez there in two launches with fifty men to investigate. Gonzalez found the report to have been false, but he was able to ransom two Spaniards and one Frenchman who had been captives of the Kiawah Indians for many years, and he interviewed four local chiefs, one of whom he took to St. Augustine.[29]

The threat from England was far more serious. The Spaniards knew in 1585 that Sir Richard Grenville had planted an English colony somewhere in the general area of Jacán. They did not know its location, but they suspected it was on Chesapeake Bay. Hard on the heels of this disturbing news, Sir Francis Drake attacked in the Caribbean with a very large fleet. The Spaniards at St. Augustine and Santa Elena made ready for Drake as best they could with their poor resources. In June, Drake easily penetrated the defenses of St. Augustine and even succeeded in stealing silver from the royal coffer, ten pieces of artillery, and quantities of supplies.[30] Drake then proceeded toward Santa Elena, but he missed the entrance to the harbor and sailed on to Roanoke Island, where he removed the starving English colonists who lived there, and then he sailed back to England.[31]

After Drake's raid, the Spaniards had to face up to their weaknesses. Pedro Menéndez Marqués recommended that Santa Elena be abandoned and that the defenses of St. Augustine be strengthened. This drastic measure was debated for a time, but in the end this is what was done. In 1587 Fort San Marcos was torn down, the town of Santa Elena put to the torch, and the inhabitants moved to St. Augustine.[32]

In 1588 an incident occurred that revealed just how thoroughly Spain had retreated from her vision of a Greater Florida. Menéndez Marqués again sent Vicente Gonzalez north, this time to discover the location of the English colony. Gonzales searched Chesapeake Bay to no avail. Quite by chance he found the abandoned English colony on Roanoke Island on his return voyage. But what is most noteworthy is that Spain saw fit to spare but a single ship for this important mission.[33]

The Florida colony was in a very bad way when Don Gonzalo Méndez de Canço became governor in June, 1597. Méndez Canço was a gifted naval commander who at the age of fourteen had first served in the fleet of Pedro Menéndez de Avilés. As governor, Méndez Canço wanted to revive

Menéndez's dream of a greater Florida.[34] But instead of expanding inland from Santa Elena, Méndez Canço's plan was to expand inland from Guale to the provinces of "Tama" and Ocute, both of which were on the Oconee River in central Georgia.[35] In fact, from the 1590s onward the Spaniards seldom refer to the Indians whom Pardo had encountered. "La Tama"— presumably derived from the "Altamaha" encountered by Soto—became the general name by which the Spaniards referred to the Indians who lived west of the Savannah River in the interior. On occasion the term was extended to include Indians who lived east of the Savannah River, in the country that Pardo had explored.

After assuming the governorship, Méndez Canço had succeeded in doing little more than to reconnoiter Florida and give presents to the Indians when the Guale Indians lashed out at the Franciscan missionaries in yet another rebellion. The Indians of Guale killed Fathers Miguel de Auñón, Francisco de Verascola, Antonio de Badajoz, Blas de Rodriguez, and Pedro de Corpa. Only Father Francisco de Avila was spared.[36] Méndez Canço put down the rebellion through the drastic expedient of burning the fields of corn the Guale Indians had planted.

In 1598, Méndez Canço reported that there were 225 soldiers in the fort at St. Augustine, plus about four hundred additional Spaniards, Indians, and black servants.[37] The colony was not expanding. By 1602 the difficulties were so great that the king ordered an investigation. There was considerable sentiment, particularly among the Franciscan missionaries, that St. Augustine should be abandoned or drastically diminished, and that the center of the colony should be moved north to Guale, where the soil was more fertile and where there was good land for grazing cattle.[38]

The investigation, conducted by Fernando Valdés, revealed several persistent problems. The harbor at St. Augustine was not as large as the Spaniards would have liked it to be, and the soil was too poor to produce enough food to support a colony. Moreover, there were definitely no deposits of precious stones or metals in the vicinity of St. Augustine. On the other hand, if the center of the colony was moved to Guale or Santa Elena, it would not be as strategically located with respect to ships passing through the Bahama Channel and would not be able to give aid to shipwrecked sailors as effectively as at St. Augustine.

From this investigation it is apparent that the Spaniards' understanding of the geography of the Florida peninsula was still quite limited. The source of the St. Johns River was still unknown. Alonso de las Alas, the business manager of Florida, felt sure that the St. Johns River would provide water passage to the west coast of Florida, whereas Father Baltasar

López, who said that he had walked for 40 leagues in search of a direct water passage, was certain that it would not.[39]

Two of the principal complaints of the Franciscan friars were that Méndez Canço allowed the Indians to disobey their own chiefs, and that he allowed Christian Indians to flee as fugitives to the interior, where they returned to their native customs.[40] As in other parts of Spanish America, an effort was made in Florida to incorporate the caciques into the Spanish colonial apparatus. In some sense, the friars clearly thought that their Indian converts belonged to the colony.

In the end, St. Augustine survived. What did not survive was greater Florida. Méndez Canço was removed as governor, and he was formally condemned for running the affairs of the colony badly. No funds were appropriated for the purpose of exploring and colonizing Tama and Ocute. Spain's interest in the lands to the north waned until the end of the seventeenth century, when the Indians came under the influence of English colonists from Virginia and Carolina and the French from the Mississippi River.

The Decline and Coalescence of the Indians

In the early days of the Florida colony, Menéndez was hopeful that through missionization the Southeastern Indians could be made to serve Spanish colonial interests. But after the interior forts were overrun, and after coastal Indians killed and tortured many Spaniards who were shipwrecked, Menéndez changed his mind. Indeed, he concluded that the best way to handle the Indians of South Florida was to kill and enslave them.[41] But in the course of events, such a drastic measure would not be necessary. If the collision between the Southeastern Indians and Menéndez's soldiers was a setback for Spain, it was a disaster for the Indians. For Spain it was only a disappointment. It forced Menéndez to scale down the size of his intended colony from the Florida of his dreams to almost its present boundaries. For the Indians, it greatly accelerated a population collapse that was already in motion when Menéndez arrived, and it radically changed the structure of their societies. The kinds of societies that Pardo and his men saw on their travels were nowhere to be seen a hundred years later.

The vulnerability of American Indians to germs and viruses from the Old World is well established. The only questions that remain to be answered are what diseases were introduced at what time, and what was the precise shape of the declining population curve?[42] The earliest historical evidence of Old World diseases among Indians in the area that Pardo explored comes from Cofitachequi in 1540, during Soto's visit. If Ponce de León, Ayllón, or Narváez saw evidence of depopulation caused by disease, there is no mention of it. Members of these expeditions are known to have sickened and died, and the Indians may have contracted some of their diseases, but there is no documentary evidence that they did.

It is also possible that the Soto and Luna expeditions introduced diseases among the Indians, but an even more devastating conduit for the pathogens were the towns, outposts, and missions of Spanish Florida. Here, in these tiny communities, Europeans of all sexes and ages were present, and there was a continual coming and going from the Caribbean and from Europe, both of which were stewpots of disease. Moreover, there was also constant, though perhaps small, human traffic between the Spanish towns and missions on the coast and the Indian towns in the interior. This included Indian traders who traveled back and forth, as well as "fugitives"—Indians who fled the regimentation and labor demands of the missions for the native lifeways in the interior.

Writing from Guale in early March 1570, Francisco Villarreal and Antonio Sedeño reported that a disease had broken out among the Indians and that it had killed many of them. The medicines the Jesuits administered to the Indians were ineffective, and when they baptized some of the dying, the Indians of Guale began complaining that the priests were witches and that they were killing the very people whom they were baptizing. Unfortunately, neither Villarreal nor Sedeño describes the symptoms of this disease. Villarreal simply calls it *una enfermedad*—"a disease." But Sedeño characterizes it more seriously as *una como pestilentia*—"like a plague."[43]

The missionaries in Guale do not appear to have recognized the significance of what was happening before their eyes. They merely say that many died. Archaeological evidence from the interior suggests that the Indians died in great numbers and within a relatively short period of time. The population decline in the paramount chiefdom of Coosa in the latter half of the sixteenth century is indicated by a drop in the number of archaeological sites and by the fact that the sites became smaller in size. Moreover, in the territory of the chiefdom of Coosa there is archaeological evidence of the uprooting and coalescence of refugee populations similar to that which, as we shall see, occurred in the area Pardo explored.[44]

After the Pardo expeditions, Spaniards were not anxious to again venture into the interior of the Southeast. In fact, Spaniards seldom ventured into the interior for any purpose. Therefore it is difficult to reconstruct the decline of Cofitachequi, Guatari, Joara, and the other Indian polities Pardo encountered.

Méndez Canço sent two small parties inland to Tama. Documentation from these two ventures suggests that even though they may have been considerably diminished, Tama (i.e., Altamaha) and Ocute were probably in the same locations at the beginning of the seventeenth century as they were at the time of the Soto expedition. In 1597 Méndez Canço sent the soldier Gaspar de Salas along with Fathers Pedro Fernandez de Chozas and Francisco de Velascola to Tama.[45] When interviewed in 1600, Salas recalled that in going into the interior they traveled eight days from Guale to Tama, and on seven of these days they traveled through an uninhabited area. He estimated that this distance was about 50 leagues. Then one day they traveled beyond Tama to the town of Ocute. On their return, they took a different trail, one that only entailed two days of travel through an unpopulated area.[46]

At the time of the Soto expedition, the chiefdom of Altamaha appears to have had its center at the Shinholser archaeological site, about 10 miles south of present-day Milledgeville, Georgia. The chiefdom of Altamaha was tributary to the paramount chief of Ocute, about a day's travel away, and presumably further up the Oconee River.[47] The distance from the Guale area to the Shinholser site is about 160 miles. Using the *legua común* as a unit of distance, this comes out to about 46 leagues, which agrees closely with Salas's estimate.[48] In entering the country, they probably followed a less populated upland trail and returned by a more populated trail that ran near the Oconee-Altamaha River.

Interestingly, when they reached Ocute, the chief wept and implored them not to go any further. He said that Spaniards had come through this place many years before (i.e., the Soto expedition), and that although there were many of them, and many of them were riding horses, the Indians further on had killed some of them.[49] This is reminiscent of what the Indians of Satapo had said, and one wonders whether the chief of Ocute could have been referring to the battle at Mabila.

In 1602 Méndez Canço sent Juan de Lara to investigate rumors that Spanish cavalry from New Mexico had somehow arrived at Tama and Ocute. Lara adds nothing to what Salas reported, and there are serious inconsistencies in his account of his travels. He claimed that he started out with some Indian guides from Tulufina in the Guale area and traveled 60 to 70 leagues to the west before coming to a mountain and a large town named "Olatama." In fact, as we have seen, Tama was more like 46 leagues from Guale, and there were no mountains in Tama. He says that he went 20 leagues further to the north, where he came to the "river of Olatama," whereas in fact Tama was located *on* the Oconee-Altamaha River.[50]

But not even the more credible account of the expedition by Solas,

Chozas, and Velascola reveals much about Tama and Ocute. Most disappointingly, there is no information on population decline nor on social change. One would like to know the degree to which the chiefs and their retainers were set apart from their people at this time. And one would like to know whether Altamaha was still tributary to Ocute. But the documents are silent on these matters.

Nearly a century passed before the door of history opens again on the area that Pardo explored and attempted to colonize. It is not until 1670, with the expeditions of English colonists from Virginia and the Carolinas, that some of the named groups from the Pardo era again enter the documentary record. And it is not until 1700, with the exploration of John Lawson, that sufficient information is available to reconstruct the social characteristics of these people with a comprehensiveness comparable to that of the Pardo period.

In 1670 Governor William Berkeley of Virginia commissioned John Lederer, a German immigrant, to explore the Virginia backcountry.[51] Lederer later claimed to have undertaken three expeditions: (1) from the Pamunkey River to the Blue Ridge Mountains northwest of present-day Charlottesville, Virginia; (2) from the vicinity of present-day Richmond to western Virginia, and from there south and southwest into North Carolina and perhaps as far as northern South Carolina, and from there back to the vicinity of present-day Petersburg by a more easterly route; and (3) a second trip westward to the Blue Ridge Mountains.

That Lederer made the first and third of these expeditions is independently verified. But the second journey is so full of misinformation and suspicious claims that it is doubtful that Lederer actually made this trip in its entirety.[52] However, he did obtain ethnological and geographical information, presumably from an Indian informant who knew the country. Consequently, several scholars have since been persuaded that his second expedition also took place. In traveling southward through the Piedmont he claimed to have visited the Indians of Monakin, Mahock, Sapon, Akenatzy, and Oenock (Eno). Some of these spoke Catawban languages. Traveling beyond these people he claims to have visited several groups whose names are similar to towns visited or encountered by Pardo: Shakory (i.e., Suhere), Watary (i.e., Guatari), Sara (i.e., Chara), Wisacky (i.e., Gueca), and Ushery (i.e., Uchiri or perhaps Yssa). In the 1672 *Discoveries of John Lederer in Three Marches from Virginia*, Lederer locates these names on a map, and most of them (with the exception of Shakory) are in approximately the order that one would have encountered at the time of the Pardo expedition. Lederer is also accurate in saying that the Appalachian Mountains run parallel to the Atlantic Ocean as far as Sara (cf. Chara of the Pardo expedition), beyond which they are known as the Suala (cf. Joara, Xuala) Mountains.[53] Also, he says that at Sara these mountains are low enough so that they can be crossed, and that here they take a westward turn.[54]

More reliable information came from the South Carolinians Henry Woodward and Maurice Mathews, who observed the Indians of the interior at first hand in the late seventeenth century. Henry Woodward traveled to Cofitachequi in 1670 to explore the country and to negotiate a pact with the Indians. According to Woodward, the "Emperor" of the Indians lived at Cofitachequi, and it was at this place that he concluded an agreement with the several "Petty Cassekas" who lived between there and the coast.[55]

In the same year, Maurice Mathews listed the Indians of Carolina he had knowledge of and gave some indication of where they lived. The Keyawah lived near the infant Carolina colony on the south side of the Ashley River. To the south of these were the St. Helena, Ishpow, Wimbee, Edista (i.e., Orista), and Stono. To the west of Charleston and environs were the Kussoo (i.e., Coçao). Others, whose locations were not specifically given were the St. Pa (i.e., Sanapa), Sewee, Santee, Wanniah, Elasie (i.e., Ylasi), Isaw (i.e., Yssa), and Co[f]achicach. Mathews said that some of these groups had four or five "Cassikaes," but contradicting Woodward's report of an "Emperor," he says that the chiefs had scarcely the power of a "Topkin in England," meaning they had very little real power. Moreover, he saw no evidence of any of these being tributaries of others. They visited among themselves, he said, because of intermarrige and poverty, and they never quarreled over who had higher social status. They were, he said, generally "poore and Spanish," and they lived in fear of the Westoes, Indians who lived to the west who were great warriors and were said to be cannibals.[56] In fact, the Westoes were slave catchers in the service of Virginians.[57]

John Lawson's journal of his trip up the Santee-Wateree-Catawbee River in January 1700–1701 is not as comprehensive as Bandera's account of Pardo's second expedition. Lawson does not name every chief he encountered, but on the other hand he had far more interest in the Indians as human beings than Bandera did. By reading Lawson's journal closely, one can discern some of the more significant changes that had occurred since 1566–68.[58]

Having the benefit of hindsight, John Lawson understood far more clearly than did the Spaniards the terrible loss of life that had been caused by Old World diseases. He says of the Sewee Indians, who lived within a

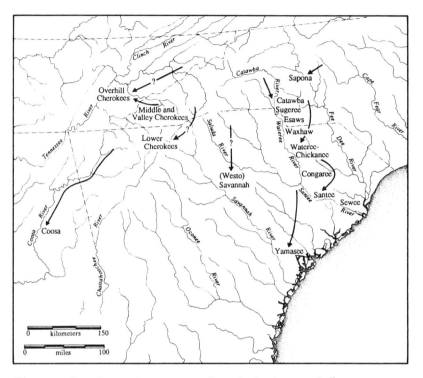

Figure 43. Locations and movements of certain Southeastern Indians ca. 1700.

few miles of the mouth of the Santee River, that they were once a numerous people, but that the Sewee suffered from what always happens when Europeans settle near Indians—they die of diseases, particularly of smallpox (see figure 43). Indian healers, Lawson says, were extraordinarily capable, but they were helpless when confronted with European diseases.[59]

As Lawson proceeded along a trail paralleling the eastern side of the Santee-Wateree River, he encountered the Santee Indians from about present-day Jamestown, South Carolina, to about the junction of the Wateree and Congaree rivers. The Santee, he says, were also known as the "Seretee," and these were probably the same as the *Sarati*, who met with Pardo at Gueça and at Ilasi.[60] Thus, between 1566–68 and 1700–1701, the Sarati appear to have moved southward from the Piedmont or upper Coastal Plain to the middle Coastal Plain.

The "king" or chief of the Santees was reputed to have been the most powerful Indian ruler in the backcountry, even though the Santees were very few in number. It was said that this chief had the authority to sentence any of his people to death. The former "king" of the Santees had been much feared by neighboring societies. When he died, his body had been laid out on a sepulchre on the top of a pyramidal mound. Later his body was defleshed, his bones wrapped into a bundle, and put into a wooden box. Soto's men witnessed similar burial customs in the temple at Cofitachequi in 1540.

It is possible that the Santees retained some vestiges of the structure of a chiefdom, although it is not possible to say whether this was of the paramount chiefdom of Cofitachequi or one of the constituent chiefdoms. Such extravagant claims to social status and power on the part of a chief of such a small society is reminiscent of the Natchez of the lower Mississippi Valley in the early eighteenth century. In both cases it is possible that such claims represented vestiges of a stratified social order that was no longer viable given prevailing conditions.

Lawson next encountered the Congarees, whose number had been much reduced by smallpox. They had their town on an eastern tributary of the lower Wateree River, possibly as far north as present Camden. The chief of the Congarees was called a *Cassetta*, a word that was possibly derived from Spanish *cacique*, again perhaps representing a vestige of chiefdom-level organization. Lawson does not mention either Cofitachequi or Ilasi. Thus, it would seem that in the thirty years since Maurice Mathews had visited this area, the dissolution of the paramount chiefdom of Cofitachequi had finally become complete, and the people who had constituted the center of the chiefdom had coalesced with other peoples.[61]

North of Camden Lawson encountered the "Wateree Chickanee," whose ancestors had been the Guatari of the Pardo era. Lawson says that they spoke a languge (probably Catawban) that was different from the language spoken by the Congarees. They were more numerous than the Congarees, but they lacked English trade goods, most notably guns, so they still had to use bows and arrows. They lived in dark, smoky houses that were less well built than others Lawson had seen. To Lawson, they seemed decidedly "backward."

Just 3 miles distant from the Wateree Chickanees Lawson came to a large Waxhaw town. These people took their name from the town of Gueça of the Pardo era. By 1700 they had probably absorbed other smaller groups of Indians. The Waxhaws apparently occupied the river from somewhere north of present-day Camden to about Sugar Creek.

At the time of Lawson's visit, the area from Sugar Creek upstream to about the mouth of the south fork of the Catawba River (the location of Otari in 1566–68) was held by the Esaws, a name obviously derived from

Yssa. Included among the Esaws were the Sugeree Indians, who must have been descended from the "Suhere," whose chief Pardo met when he arrived at Gueça.

The principal town of the Esaws was that of the "Catawba King."[62] This town appears to have been at or near the site that Otari stood on in the Pardo era. This would seem to imply that the people who had been the Yssa and Cataba on the South Fork of the Catawba River at the time of the Pardo expeditions had moved southward and had occupied the territory in which Otari and Aracuchi had been located. One possible location for this town is the Bell Farm site (31Mk85).

From the Catawba town Lawson traveled for five days until he reached the town (and fort) of Sapona, which appears to have been where Guatari Mico had her town in the Pardo era—that is, on the Yadkin River near present-day Salisbury, North Carolina. Hence, at some time before 1700–1701 the Sapona had moved southward from their original territory to claim the territory once occupied by Guatari Mico.

When the social geography of the Carolinas in Lawson's era is compared with the social geography of the same area at the time of the Pardo expeditions, it is clear that momentous changes had occurred in the interval. The population of the Indians had declined, probably sharply. Many of them had moved in a generally southwardly or westwardly direction (see figure 43). And they were coalescing or combining as refugees in order to form larger and more defensible societies. Lawson observed one such coalascence that the Indians were discussing at the time of his travels: The Saponas were negotiating with two other small groups, the Toteros and Keyauwees. They were discussing whether they ought to live together to strengthen themselves against their enemies, the Iroquois.[63]

The social structure of these early eighteenth-century Indians is not well understood, particularly the position of chief. Certainly there were no paramount chiefs among them, the Congaree chief notwithstanding. And it is possible that even the basis of authority of lesser chiefs had changed. The position of the chiefs must have owed much to the patronage of the English traders. It is significant, for example, that at the "Catawba King's" house Lawson met John Stewart, a Scot from James River in Virginia who had traded there for many years. On his last trip he had brought down seven horse-loads of goods that he had traded to the Indians. This "king" also had several prostitutes in his employ whom he offered to Europeans who passed through his town.[64]

What emerges clearly from Lawson's description is that the Southeastern Indians were being buffeted by winds from a new world order—one in which they had become dependent upon goods that were manufactured in Europe. The social and economic power that Europeans were able to wield in this world far exceeded the power that had been at the disposal of Menéndez or Pardo. Everything indicates that the Indians whom Lawson encountered had been fundamentally transformed by the English trading regime.

By 1700–1701 hardly anything remained in the Carolinas that was identifiably Spanish. The Congerees called their chief *Cassetta*, and this was surely derived from Spanish *cacique*.[65] There is archaeological evidence that the Indians were at this time cultivating peach trees, which had been introduced by the Spaniards. Other than this, it is difficult to see that the Spanish language or culture had any lasting effects on the Indians of the interior.

Anthropologists who began doing field research on Southeastern Indians in the nineteenth century found that the Indians retained no clear memory that their sixteenth-century ancestors had lived in more complicated societies. The only exception may be the Cherokees, who retained a memory into the late nineteenth century of mound construction and, more interestingly, of a class of special people, the *Aní-Kutání*. But opinion was divided among the Cherokees on whether these people were a vestige population of "mound builders" or a clan of Cherokees who had been killed by disease or by violence. What they did agree upon is that the *Aní-Kutání* possessed hereditary privileges, and that they validated their status on religious grounds. There is, in fact, some evidence that this validation was based upon a myth that they were descended from spiritual beings in the Upper World. The Cherokees remembered the *Aní-Kutání* as proud, haughty people who victimized other Cherokees, and because of this the Cherokees exterminated them. This myth of the *Aní-Kutání* may have been the pale vestiges of memory of the once vigorous paramount chiefdoms.[66]

Los Diamantes *and* La Gran Copala

Menéndez and Pardo failed in their attempt to lay the foundations of Greater Florida, a vast empire in the Southeast, but from fragments of the experience of their attempt, legends were created that would make others want to try again. From Pardo's attempt to find precious metals and from his discovery of what appear to have been corundum and quartz deposits, a legend was born in Spanish Florida of *Los Diamantes*, a veritable mountain of diamond, and from other sources a legend was born of *La Gran Copala*, a fabulously rich native society.

Juan de Ribas appears to have been one of the main sources of these stories. When Ribas was interrogated by Méndez Canço in 1600, some thirty-two years after the Pardo expeditions, he said that he went into the interior with Pardo when he was seventeen or eighteen years old. From the information given by Ribas, it is obvious that some of his details are inaccurate, having suffered from the distortion of memory through the intervening years. But it is clear enough that he was a member of Pardo's second expedition (and perhaps also the first expedition), and that he had been stationed at one of the interior forts in order to learn an Indian language. Because he mentions "Juaraz" (i.e., Joara), so frequently, it is likely that if Ribas was stationed anywhere, it was at Joara.[67]

At the time of the Canço inquiry, Ribas was married to Luisa Méndez, an Indian woman from the interior who had been brought out as a girl by Pardo, probably as a hostage. According to Ribas, she was the chieftainess of "Guanaytique."[68] This is a place name not mentioned by Bandera, but it somewhat resembles *Guanaguaca* (or *Guanbuca*), whence a chief came to meet with Pardo when he was at Joara.

Ribas confirms that Moyano had learned from the Indians of the existence of a crystal deposit near Joara. When expedition members went there to investigate, Ribas said they found that it was a high hill, with crystals so hard that they could not break them with mauls and iron wedges. When Ribas was interrogated again two years later, he added that the high hill was located in a naked plain, and that it had three or four veins of crystal, one of which was of blue "diamond" and another of purple "diamond."[69] They named this place *Los Diamantes*. The location of this deposit, probably corundum, may have been near Carpenters Knob in northern Cleveland County, North Carolina. According to Ribas, Moyano took a small point of crystal from *Los Diamantes* to Spain and sold it to a jeweler in Seville for a great deal of money, and afterward the jeweler said that Moyano did not know the value of what he had sold, and that he would not sell it back to him at any price.[70]

Ribas testified that he went with Pardo beyond Joara and Olamico, where they discovered many towns in the mountains. Here they found "mines of gold and silver." This would seem to be based on the signs and rumors of gold that Pardo encountered in the general vicinity of Tanasqui and Chiaha, and the spurious discovery of silver ore on Chilhowee Mountain. Ribas said that the Indians gave Pardo and Moyano many pearls, and this is probably true. But what can be the substance of his saying, as he did, that Pardo and Moyano were given "cups of gold and silver"?

One of the most interesting parts of Ribas's testimony is his story about

La Gran Copala. This story originated with a man named Almeydo, a Portugese pilot whom Pardo left at Cauchi to learn the native language. It is not clear whether Ribas heard this story from Almeydo himself, or from someone to whom Almeydo told the story. But according to Ribas, Almeydo went with his wife (presumably an Indian woman) to the place where Soto had been "killed." Almeydo claimed that at this town he saw articles of silver and pewter, mail, arms, and Spanish clothing, and he saw painted on the walls of Indian houses pictures of horses, lancers, and people dressed as Spaniards.[71]

If we assume that Almeydo told the truth about visiting a town beyond Cauchi, what town could this have been? The only likely possibility—based on details of Almeydo's story given below—is that the town was located on a tributary of the Tennessee River that was navigable by dugout canoe. Olamico would not seem to be a possibility, because other Spaniards were stationed there. But the town could have been Chalahume, Satapo, Coste, or some other town not mentioned by the Soto chroniclers or Bandera.

According to Ribas, Almeydo said that the Indians of this town never tired of talking about a great city, which they called *La Gran Copala*, which was seven or eight days' travel from where they were. They said that one had to travel there in canoes lashed together, and with shields for defense, because one had to go through passes where there were warlike people. If we allow the full eight days, and a rate of travel of 30 or 40 river miles per day, the location of *La Gran Copala* would have been on the Tennessee River somewhere between present-day Guntersville and Florence, Alabama. During the Mississippian period, this area had a very large population, which perhaps exceeded that of the Moundville culture that was located just to the south on the Black Warrior River.[72] There were several large whirlpools in the Tennessee River that make the lashing together of canoes seem reasonable, and there were several places where the river ran alongside steep cliffs, where hostile parties could have had a considerable advantage over people in canoes below, and where shields would have provided welcome protection.[73]

So far, so good. But then Ribas's story becomes more fantastic, probably reflecting distortion on the part of Almeydo's Indian informants, as well as on the part of Almeydo and Ribas. Almeydo claimed that his Indian informants pointed to articles of silver and pewter that they had gotten from the Soto expedition, and they told him that the cooking vessels at *La Gran Copala* were made of similar metal. It goes without saying that such vessels have never been found by Southeastern archaeologists, although the Mississippian people of the Middle Tennessee River possessed considerable quantities of copper, which they used to make articles for their rituals and for personal adornment. It may be relevant that when Ribas's wife, Luisa Méndez, was interviewed by Canço, she said that her own people possessed gold and silver cups that they brought out to display at gatherings and dances, but that she was a young girl at that time and did not understand all that was going on. She said that her people got their gold from the Chiscas who lived in the mountains. What she may have remembered was the display of metal objects in ritual contexts, a practice in the native Southeast for which there is ample archaeological and historical evidence.[74]

Almeydo's story becomes truly fantastic when he says that the houses at *La Gran Copala* were made of stone and were several stories high. Here there can be no substance to Almeydo's story, because there is no archaeological evidence for stone construction in the prehistoric Southeast, anywhere, at any time, and certainly none for structures several stories high. One wonders whether Almeydo and/or Ribas somehow linked a story about a large and powerful Mississippian society—perhaps on the Middle Tennessee and Black Warrior rivers—with the legend of Cíbola, which had deep roots in Spanish mythology as well as in the reports on the American Southwest by Cabeza de Vaca and Fray Marcos.[75] It may, therefore, be no accident that "Copala" and "Cíbola" resemble each other phonetically.

Although Ribas's testimony was affected by the passage of time, and perhaps by the play of his own memory and imagination, it nonetheless had some basis in fact. But as one might expect, as the story was retold by others, it became magnified and Ribas's vague locations became even vaguer.

On the same day he interviewed Ribas, Méndez Canço interviewed Sergeant Francisco Fernández de Ecija, who had been in the service of the Florida colony from the earliest days. He had not been to Tama, he said, nor had he gone into the interior with Pardo, but he had talked many times with both Pardo and Moyano. Moyano had told him that the best entrance to the interior was along the River of Guatari (i.e., the Santee-Wateree-Catawba River), and that about 40 leagues inland there was a crystal mountain, bare of trees. It consisted entirely of diamond. When the Spaniards had tried to break out some crystals with mauls and sharp wedges, he said, it was so hard that the wedges broke into pieces. And he repeated the story that Moyano had taken a crystal and a quantity of pearls to Spain, where he had sold them for a great deal of money.[76]

In 1602, when the inquiry into the fate of St. Augustine was held, the

Spaniards often used "Tama" to refer to the entire interior. Juan López Avilés said that at Tama the Indians wore costumes decorated with gold and silver that they got from the rivers that had their headwaters in the mountains. And he said that he had been told that there was a crystal mountain full of diamonds in this country.[77] Another Spaniard, Francisco López, mentioned the mountain of crystal, and he said that he had spoken with Indians of Tama who possessed gold that they said they got from a naked hill without trees or any other vegetation. Francisco López also told of one Aguilar who was said to have brought out a huge diamond from the crystal mountain, which he took to Spain and sold for two hundred *ducados*.[78]

The legend of fabulously valuable gems from *La Gran Copala* became so widespread that this was one of the avowed motives for Sir Francis Drake's raid on St. Augustine in 1586.[79] Pedro Morales, whom Drake captured, said that *La Gran Copala* was a great city located at sixteen or twenty days' travel northwest of Santa Elena, and that some Spaniards had actually seen it. Morales also claimed to have seen a rich diamond that had come from the mountains that lay to the west of Santa Elena.[80] Here the two elements of Ribas's testimony became fused into a single legend—the Mountain of Diamond was located at *La Gran Copala*.

Another of Drake's captains, Nicholas Burguignon, likewise said the city of *La Gran Copala* lay northwest of Santa Elena. In the mountains near there, one could find crystal, gold, rubies, and diamonds. He said that a Spaniard had brought a diamond from there that was worth 5,000 crowns, and that it was worn by Pedro Menéndez Marqués. He said that to go to these mountains one had to carry hatchets to give to the Indians as well as pickaxes with which to mine the gems. The mountain of crystal shone so brightly during the day, he said, that one could not bear to look at it, and so had to approach it at night.[81]

These extravagant stories of fabulously rich deposits of precious substances must constitute some of the fuel that has fired stories among Anglo-Americans of Spanish mining in the Piedmont and mountains of the Carolinas, Georgia, Alabama, and Tennessee. This includes historical traditions of an ancient settlement near Lincolnton, North Carolina, where there was said to be a dam made of cut stone, a series of low pillars of cut stone, a stone-sided well, a quarry from which the stone had been taken, and a fire pit.[82] If this report is to be believed, this site is quite near the place where Pardo discovered the crystal mines. Other historical traditions include a mine at King's Mountain, a mining shaft on the Valley River in north Georgia, and a mine on Coronaka Creek near Abbeville, South Carolina.[83]

Two questions arise here. First, did the Spaniards actually engage in mining in the interior, and second, if they did engage in mining, when could they have done it? It is extremely unlikely that Spaniards did any significant mining before 1602. During the investigation into the fate of the Florida colony held in 1602, Méndez Canço cast about for every scrap of evidence that might indicate riches could be had in the Southeast. Indeed, his enthusiasm for a greater Spanish effort in Florida probably fueled the legend of *Los Diamantes* and *La Grand Copala*. If any mining had been done before 1602, it is reasonable to think that this investigation would have brought it to light. Yet, the only thing the inquiry uncovered were rumors and stories, and not a shred of material evidence of precious substances.

There are, however, two kinds of evidence that lend some credence to Spanish mining activities: the discovery of old European artifacts by English and American prospectors and miners, and hearsay historical evidence. Such evidence must be evaluated on a case by case basis. One of the best documented instances of European artifacts comes from a mica mine in Macon County, North Carolina. When the Guyer mine was opened up in 1875, a shaft was found that was cleared to a considerable depth. Between depths of 35 and 50 feet, several iron implements were found. These include a socketed axe of an "old pattern which is now [i.e., in 1881] rarely met with." It was a light axe, with the blade and head each about $3\frac{3}{4}$ inches in width, narrowing between them to $2\frac{3}{4}$ inches. A maker's mark appeared on the blade, but it was too worn to be read.

A second tool was a wedge measuring $3\frac{3}{4}$ inches long by $1\frac{1}{2}$ inches wide. And a third tool appears to be the two prongs of a pickaxe that was broken at the socket.[84] Frederick W. Simonds, who reported this discovery, erroneously interpreted the two pieces of the pickaxe as a pair of "gudgeons"—parts of a windlass. It is possible that it was a pickaxe of such an ancient form that it was completely unfamiliar to him. It is not impossible that these were sixteenth- or seventeenth-century Spanish tools, but it is extremely unlikely that they were deposited in a mica mine by Spaniards, for whom mica was a substance of no value. If these were, in fact, Spanish tools, the Indians probably acquired them and used them in their own mica mining.

If the Spaniards did any mining in the area that Pardo explored, they must have done it between 1602 and 1700. Two pieces of historical information, both based on hearsay, argue for this. One is Francis Yeardley's account of men in his employ who in 1654 went by boat to Roanoke Island and then to a Tuscarora hunting quarter in the interior. Yeardley says that the Tuscaroras told them about a Spaniard who had been living in their main town for seven years. He was said to be very rich, with thirty people in his "family," including seven negroes. The Tuscaroras also told them about a place where there was a salt "sea" (probably a saline) where salt could be extracted and where copper could be dug from the ground. One of the Tuscaroras was said to have been wearing gold beads in his ears.[85]

The other incident occurred in 1690, when James Moore traveled from the Carolina coast to the mountains searching for ores and minerals. He later reported that he had found seven different sorts of minerals and ores, which he later sent to England to be tested. At one point in his travels he said that he was told by some Indians that some Spaniards had been at work in mines just 20 miles away. The Indians described to him "bellows and furnaces," and they said that they had killed these Spaniards because they were afraid that they would enslave them and put them to work in the mines.[86] This hearsay evidence could be a somewhat garbled memory of the Pardo expeditions and of the Indians' revolt against the Spaniards.

Greater Florida was a failure. Spain's discovery and initial colonization of the Americas were stupendous achievements. Later, however, Spain's reach exceeded its grasp. After penetrating and exploring the territory that is the southeastern United States, Spain did not have the wherewithal to conquer and exploit it. It would remain for England, with its leaner bureaucracy and mercantilist strategy for dominating the world economy, to conquer Greater Florida and beyond. After founding Charleston in 1670, the English and their Indian mercenaries swept through the outlying Spanish missions of Guale and Apalachee like wolves through a flock of sheep. Henceforth, Spain's control of Florida shrank even further, until at times it controlled only the territory right around St. Augustine.[87] In time, the English and their descendants would find and exploit some of the precious substances that Pardo and Moyano had searched for. Gold would be discovered in western North Carolina, and later an even greater discovery of gold would be made in northern Georgia. And significant quantities of precious stones would also be discovered in both places. But these precious substances were of minor importance when compared with the real wealth of the Southeast—its immense expanses of forest and agricultural land.

Notes

1. Clifford M. Lewis and Albert J. Loomie, *The Spanish Jesuit Mission in Virginia, 1570–1572* (Chapel Hill: University of North Carolina Press, 1953), p. 19.

2. Pedro Menéndez de Avilés to Francis Borgia, Madrid, January 18, 1568. In *The First Jesuit Mission in Florida*, comp. Ruben Vargas Ugarte, trans. Aloysius J. Owen, U.S. Catholic Historical Society Records and Studies, vol. 25 (New York, 1935), p. 74.
3. Rodrigo Ranjel, "Relation," in *Narratives of the Career of Hernando de Soto*, trans. Buckingham Smith, ed. Edward Gaylord Bourne (New York: Allerton, 1922), p. 101; Gentleman of Elvas, *Narratives of De Soto*, trans. Buckingham Smith (New York: Bradford Club, 1866), pp. 64–65.
4. Yucatan and Guatemala were difficult to conquer for the same reason. See Charles Gibson, *Spain in America* (New York: Harper and Row, 1966), p. 29; Inga Clendinnen, *Ambivalent Conquests: Maya and Spaniard in Yucatan, 1517–1570* (Cambridge: Cambridge University Press, 1987), passim.
5. Charles Hudson, *The Southeastern Indians* (Knoxville: University of Tennessee Press, 1976), pp. 229–32.
6. David G. Anderson, "Stability and Change in Chiefdom-Level Societies: An Examination of Mississippian Political Evolution on the South Atlantic Slope," paper presented at the 43d Annual Meeting of the Southeastern Archaeological Conference, Nashville, Tennessee, November 7, 1986.
7. Luís Geronimo de Oré, *The Martyrs of Florida, 1513–1616*, trans. Maynard Geiger, Franciscan Studies no. 18 (New York, 1936), pp. 112–14.
8. Chester DePratter, Charles Hudson, and Marvin Smith, "The Hernando de Soto Expedition: From Chiaha to Mabila," in *Alabama and the Borderlands, from Prehistory to Statehood*, ed. Reid R. Badger and Lawrence A. Clayton (University: University of Alabama Press, 1985), pp. 108–26.
9. Charles Hudson, Marvin Smith, and Chester DePratter, "The Victims of the King Site Massacre: A Historical Detectives' Report," in *The King Site: Continuity and Contact in Sixteenth-Century Georgia*, ed. Robert Blakely (Athens: University of Georgia Press, 1988), pp. 117–34.
10. Gonzalo Méndez de Canço, "Inquiry made officially before Don Gonzalo Méndez de Canço, Governor of the Province of Florida, upon the situation of La Tama and its riches, and the English Settlement," trans. Mary Ross, AGI, Santo Domingo 224, Georgia State Archives, Atlanta.
11. A Nueva Cadiz bead and four Clarkesdale bells have been found at the Citico site (Satapo), and these are thought to have been artifacts carried by the Soto expedition. See Richard Polhemus, "The Early Historic Period in the East Tennessee Valley," MS.; Marvin T. Smith, *Archaeology of Aboriginal Culture Change in the Interior Southeast* (Gainesville: University Presses of Florida/Florida State Museum, 1987).
12. Neither Pardo nor Bandera mentions seeing the paintings reported by Ribas. But there is evidence that in the eighteenth century Southeastern Indians painted pictures of their martial exploits, and this lends some credibility to Ribas's report. See Bernard Romans, *A Concise Natural History of East and*

West Florida (New York, 1775; reprinted by Pelican Publishing Company, New Orleans, 1961), p. 71.
13. Robert Blakely and Bettina Detweiler, "The King Site before, during and after the Spanish Encounter: A Bioarchaeological View," paper presented at the annual meeting of the Society for Historic Archaeology, Savannah, Ga., January 8, 1987.
14. Hudson et al., "King Site Massacre."
15. Noted in "Lista de la gente de guerra a quiem dió ración Tomas Alonso de los Alas," 1566–69, AGI, Contaduria 941. Reference supplied by Paul Hoffman.
16. Felix Zubillaga, *Monumenta Antiquae Floridae, 1566–1572*, Monumenta Historica Societatis Iesu, vol. 69 (Rome, 1946), p. 321.
17. Ibid., pp. 326–27.
18. Jaime Martínez, "Brief Narrative of the Martyrdom of the Fathers and Brothers of the Society of Jesus, Slain by the Jacán Indians of Florida," in *The First Jesuit Mission in Florida*, comp. Ruben Vargas Ugarte, trans. Aloysius J. Owen, U.S. Catholic Historical Society Records and Studies, vol. 25 (New York, 1935), p. 142.
19. They are Teresa, wife of Franciso Camacho; Catalina, servant of Alonso de Olmos; Francisca, servant of Gutierre de Miranda; Luisa, wife of Francisco Gonzales; Isabel, servant of Baltasar de Ciguenza; Juana, servant of Bartolome Martín; Isabel de la Parra, servant of Doña Mayor de Arango; Marina, wife of Juan de Ribas. AGI, Escribania de Camera 154-A, fol. 65–68. Reference supplied by Paul Hoffman.
20. Juan Martín was apparently killed in the Indian uprising of 1576. In addition to his widow, Teresa, he left behind two daughters, Ines and Teresa. Information supplied by Paul Hoffman.
21. Inquiry made officially before Don Gonzalo Méndez Canço, governor of the Provinces of Florida, upon the situation of La Tama and its riches, and the English settlement, trans. Mary Ross, AGI, Seville, Est. 54, Caj. 5, Leg. 9, folio 17, Georgia State Archives, Atlanta, pp. 11–12.
22. Lista de la gente de guerra a quien de racion Tomas Alonso de las Alas, desde Sept. 1566 en adelante, AGI, CD 941. Information supplied by Paul Hoffman.
23. Michael V. Gannon, "Sebastian Montero, Pioneer American Missionary, 1566–1572," *Catholic Historical Review* 5(1965/66):352.
24. Charles W. Arnade, *Florida on Trial, 1593–1602* (Coral Gables, Fla.: University of Miami Press, 1959), p. 35.
25. Eugene Lyon, "Santa Elena: A Brief History of the Colony, 1566–1587," University of South Carolina, Institute of Anthropology and Archaeology, Research Manuscript Series 193 (1984), p. 10.
26. Martinez, "Brief Narrative," p. 142.
27. Lyon, "Santa Elena," pp. 5–9.

28. Ibid., pp. 10–14; Oré, *Martyrs*, pp. 33–41.
29. Paul E. Hoffman, "New Light on Vicente Gonzalez's 1588 Voyage in Search of Raleigh's English Colonies," *The North Carolina Historical Review* 63(1986):204.
30. Verne E. Chatelain, *The Defenses of Spanish Florida, 1565–1763* (Washington, D.C.: Carnegie Institution, 1941), p. 59.
31. Lyon, "Santa Elena," p. 14.
32. Ibid., pp. 15–16.
33. Hoffman, "New Light," pp. 220–23.
34. Arnade, *Florida on Trial*, pp. 3–9.
35. Ibid., pp. 76–77.
36. Oré, *Martyrs*, pp. 66–97.
37. Arnade, *Florida on Trial*, p. 8.
38. Ibid., p. 63.
39. Ibid., pp. 12–18.
40. Ibid., pp. 64–67.
41. Pedro Menéndez de Avilés to his Royal Caesarian Majesty, AGI, Seville, Patronato 257, in *Colonial Records of Spanish Florida*, ed. and trans. Jeannette Connor (Deland, Fla.: 1925–1930), vol. 1, pp. 30–35.
42. Alfred W. Crosby, Jr., *The Columbian Exchange: Biological and Cultural Consequences of 1492* (Westport, Conn.: Greenwood Press, 1972), pp. 35–63; Henry F. Dobyns, *Their Number Become Thinned: Native American Population Dynamics in Eastern North America* (Knoxville: University of Tennessee Press, 1983).
43. Zubillaga, *Floridae*, pp. 418, 423.
44. Marvin T. Smith, *Archaeology of Aboriginal Culture Change in the Interior Southeast* (Gainesville: University Presses of Florida, 1987).
45. Arnade, *Florida on Trial*, pp. 46, 76.
46. Inquiry before Canço, pp. 3–6.
47. Charles Hudson, Marvin T. Smith, and Chester B. DePratter, "The Hernando de Soto Expedition: From Apalachee to Chiaha," *Southeastern Archaeology* 3(1984):70–71.
48. This would have required them to travel 5.75 leagues (about 20 miles) per day. Because they were few in number and had a packhorse, this is not an excessively rapid rate of speed.
49. Inquiry before Canço, p. 5.
50. Arnade, *Florida on Trial*, pp. 42–76.
51. William P. Cumming (ed.), *The Discoveries of John Lederer* (Charlottesville: University of Virginia Press, 1958).
52. Lederer's second expedition is full of ethnological and geographical information that makes no sense. He started out on this expedition with a Major Harris along with twenty Virginians and five Indians. But Lederer parted

ways with them at Mahoc, on the upper James River, and he claims to have gone the rest of the way with a single Indian, a Susquehannock named Jackzetavon.

Time after time, Lederer claims that he was prevented from pushing on just when he was *on the verge* of making great discoveries because he was "frightened." (1) He would have tried to discover minerals such as "*Auripigmentum*" had he not been frightened by the treacherous and hostile Akenatzy Indians. (2) He did not explore west of "Ushery Lake" for fear of being enslaved by Spaniards and of being put to work in their "mines." (3) He did not look for the mines from which the Tuskaroras obtained their copper for fear of being attacked by these selfsame Indians. Such claims must have whetted the appetites of any frontier entrepreneurs who were gullible enough to believe Lederer.

Others have pointed out geographical inconsistencies: palm trees at Oenock; a nonexistent "brackish" lake at Ushery 10 leagues wide; Indian women at Ushery who were fond of peacock feathers and men who were firewalkers, and pieces of silver from the same place; a sandy desert that took twelve days to cross; and so on. It is noteworthy that the most fantastic parts of Lederer's account occur toward the end of his purported journey.

A number of earlier scholars have been skeptical of Lederer's account of his explorations. See, for example, Cyrus Thomas, "Was John Lederer in Either of the Carolinas?" *American Anthropologist* 5(1903):724–27; W. P. Cumming, "Geographical Misconceptions of the Southeast in the Cartography of the Seventeenth and Eighteenth Centuries," *Journal of Southern History* 4(1938):476–92; Lyman Carrier, "The Veracity of John Lederer," *William and Mary Quarterly* 9(1939):435–45; Percy G. Adams, *Travelers and Travel Liars, 1660–1800* (New York: Dover, 1980), pp. 203–10.

Those who have judged Lederer's account of his second expedition to be credible have had much to explain away. Steven Baker, for example, argues that Lederer's "Lake of Ushery" was, in fact, the Wateree Swamp (Steven G. Baker, "Cofitachique: Fair Province of Carolina," M.A. thesis, University of South Carolina, 1974, appendix III). This conclusion led Baker to place the Usherys (i.e., Yssa, Esaws) near the confluence of the Wateree and Congaree rivers, whereas in fact they appear to have always had their towns north of the Fall Line. In addition, Baker concluded that "the Usherys appear to have originally been the same constellation of peoples as the Cofitachiques of the 17th century." This appears to have been a major factor in Baker's locating the center of Cofitachequi near the junction of the Congaree and Santee rivers (i.e., the High Hills of Santee) rather than in the vicinity of present-day Camden.

More recently, Alan V. Briceland has reconstructed a route for Lederer that does not extend westward of the upper tributaries of the Cape Fear River. But

this requires him to locate the Shakory, Watary, Wisacky, and Ushery far to the east of where they were located in the sixteenth century and in the eighteenth century. And he is at pains to explain away the geographical anomalies. Alan V. Briceland, *Westward from Virginia: The Exploration of the Virginia-Carolina Frontier, 1650–1710* (Charlottesville: University of Virginia Press, 1987), pp. 92–123.
53. Cumming *Lederer*, p. 9. Lederer says that *Sasa* or *Sualy* is the way in which *Sara* is pronounced in the "Warrenuncock" dialect. But elsewhere he says that the mountains got their name from the Spaniards. Hence, he may have got some place names from a map containing information from the Soto expedition (p. 28).
54. Ibid., p. 11.
55. Henry Woodward to Sir John Yeamans, Albemarle Pointe in Chyawhaw, September 10, 1670. In *Collections of the South Carolina Historical Society*, vol. 5, 1897, pp. 186–88. Woodward spells it "Chufytachyqj."
56. Maurice Mathews to Lord Ashley, Ashley River, August 30, 1971. In *Collections of the South Carolina Historical Society*, vol. 5, 1897, p. 334.
57. J. Leitch Wright, Jr., *The Only Land They Knew* (New York: Free Press, 1981), pp. 106–7.
58. John Lawson, *A New Voyage to Carolina*, ed. Hugh Talmage Lefler (Chapel Hill: University of North Carolina Press, 1967). Of all the scholars who have attempted to reconstruct Lawson's route through South Carolina, Steven Baker's reconstruction is the most accurate. See Baker, "Cofitachique," appendix II.
59. Lawson, *New Voyage*, p. 17.
60. Ibid., pp. 27–30. It is possible that the Saluda River of South Carolina took its name from "Sarati" and that the Tyger River took its name from "Tagaya."
61. Baker, "Cofitachique," appendix II, pp. 14–15; Lawson, *New Voyage*, pp. 34–35. Some people of the central towns of Cofitachequi may have moved north to coalesce with the Waxhaw (i.e., Gueça). And it is quite possible that some of them moved south to coalesce with others to form the Yemassee (cf. Guiomae, Emae). One wonders whether the Ilapi who lived on the Chattahoochee River in the eighteenth century were descended from the Ylasi or Herapi. One also wonders whether memories of the grand chiefdom of Cofitachequi could have been part of the ideological basis of the Yamassee War of 1715, when many of the descendants of the former chiefdom combined forces in an uprising against the Carolina colonists.
62. Lawson spells this "Kadapaw."
63. Lawson, *A New Voyage*, p. 53.
64. Ibid., pp. 49–50.
65. Ibid., p. 34.
66. James Mooney, "Cherokee Mound-Building," *American Anthropologist*

2(1889):167-71; Raymond D. Fogelson, "Who Were the Aní-Kutánî? An Excursion into Cherokee Historical Thought," *Ethnohistory* 31(1984):255-63.
67. "Inquiry before Canço," pp. 7-8.
68. This was also recorded as "Manaytique."
69. Arnade, *Florida on Trial*, pp. 38-40.
70. Ibid., pp. 38-40.
71. "Inquiry before Canço," pp. 9-10.
72. John A. Walthall, *Prehistoric Indians of the Southeast: Archaeology of Alabama and the Middle South* (University: University of Alabama, Press, 1980), pp. 227-45.
73. I am grateful to Paul Hoffman for pointing out that a similar story was told by Hernando de Escalante Fontaneda, who was shipwrecked among the Indians of Southern Florida about 1545 and held prisoner for seventeen years. He said that 60 leagues to the north of Santa Elena and Guale there were mines of gold and copper near the town of *Otapali* and *Olagatono*. The inhabitants of this place were neither Chichimecas nor people of the River Jordan (i.e., Cofitachequi). He also spoke of a gold mine in the snowy mountains of *Onagatono*, who was the most distant vassal of *Abalachi* (i.e., Appalachee) and *Olagatono*. The people called the chief of this place *mayor y gran señor*, that is, a very great lord. See David O. True, (ed.), *Memoir of Do. d'Escalante Fontaneda Respecting Florida*, trans. Buckingham Smith (Miami: Historical Association of Southern Florida, 1973), pp. 28-31.
74. "Inquiry before Canço," pp. 11-12.
75. Herbert Eugene Bolton, *Coronado: Knight of Pueblos and Plains* (Albuquerque: University of New Mexico Press, 1949), pp. 6-7, 27-29; *Wider Horizons of American History* (New York: Appleton-Century, 1939), pp. 66-68.
76. "Inquiry before Canço", pp. 14-17.
77. Arnade, *Florida on Trial*, pp. 30-31.
78. Ibid., p. 35. Perhaps this man was Luis de Aguilar, whose name appears on a ration list for Pardo's soldiers.
79. Arnade, *Florida on Trial*, p. 41.
80. Richard Hakluyt, *The Principal Navigations, Voyages, Traffiques and Discoveries of the English Nation* (Glasgow: James MacLehose & Sons, 1904), vol. 20, pp. 112-13.
81. Ibid., pp. 113-14.
82. James Mooney, *Myths of the Cherokee*, 19th Annual Report of the Bureau of American Ethnology (Washington, D.C.: GPO, 1900), p. 202.
83. F. A. Sondley, *A History of Buncombe County, North Carolina* (Asheville, N.C.: Advocate Printing, 1930), vol. 1, pp. 140-41.
84. Frederic W. Simonds, "The Discovery of Iron Implements in an Ancient Mine in North Carolina," *American Naturalist* 15(1881):7-11.

85. "Francis Yeardley's Narrative of Excursions into Carolina, 1654," in *Narratives of Early Carolina, 1650–1708*, ed. Alexander S. Salley, Jr. (New York: Charles Scribner's Sons, 1911), pp. 25–29.
86. James Moore to Edward Randolph. Quoted in Samuel Cole Williams, *Early Travels in the Tennessee Country, 1540–1800* (Johnson City, Tennessee: Watauga Press, 1928), p. 93.
87. Charles W. Arnade, "The Failure of Spanish Florida," *The Americas* 16(1959/60):271–81.

Part II
The Pardo Documents

The "Long" Bandera Relation

AGI, Santo Domingo 224

Introduction

This document was first noted in print by Ruidiaz in volume two of his *La Florida*, where he cites it as a 1569 "Autos sobre la relación que dio el Capitan Juan Pardo de la entrada que hizo en Tierras de la Florida," but does not give a source or a transcription.[1] No further reference to the document appeared in print until 1954 when Ketcham discussed his project to translate this document.[2] His draft translation is discussed below.

The next reference to this document, this time with some use made of its contents, appears in Gannon's article about Father Sebastian Montero. Gannon gained access to the copy in the North Carolina Department of Archives and used the information in his discussion of Montero's role as the priest on the Pardo expeditions.[3]

More recently, DePratter, Hudson, and Smith have made extensive use of this document in their study of the Pardo Route. Mary Ross, who examined the Pardo route in the 1920s, did not have access to this document.[4] Nor was it known to Woodbury Lowery.[5] Thus this "long" Bandera account is a new and heretofore little used and never before published source that, as DePratter, Hudson, and Smith have shown, throws considerable light on Pardo's second trip into the interior.

The document here presented is a copy made by notary Juan de la Bandera for Captain Juan Pardo. Its original was the likely source for Pardo's better known and much shorter Relation, as well as for Bandera's "Memoria de los lugares y que tierra es cada lugar de los de las provincias de la Florida por donde el Capitan Juan Pardo entro en 1566 y 1567."[6] That "Memoria" is dated January 23, 1569. The document under study here is dated March 31, 1569. Pardo's Relation has no date but must be after March 1568.

205

The "long" Bandera Relation is preserved in the Archivo General de Indias under the signature Santo Domingo 224.[7] It consists of a cover sheet and 35 folios of text, for a total of some 71 pages. A photostatic copy was made in the 1920s under the supervision of Miss Irene Wright and Dr. W. W. Pierson. An accompanying typewritten note says, "Observe that this is your Banderas document. I believe a translation of this, and a study based upon it and other available material, would be as well worth publication as Priestley's forthcoming book on Tristan de Luna, etc., based on similar, more voluminous material. W." Ketcham determined during a conversation with Pierson that the "W" in question was in fact Irene Wright.[8] Since Priestley's book was published in 1928, this note would date the photostat to ca. 1927.

A microfilm of the photostats in the North Carolina Department of Cultural Resources, Division of Archives and History, Archives and Records Section (Spanish Records, Archivo General de Indias, 1535–1659, Box 1, Banderas Document) was secured by DePratter in 1979. After working with the translation that accompanied the photostat but having some problems with selected words, DePratter and Smith agreed to allow me to review the translation. At my suggestion the present transcription was also prepared and is here published for the first time together with my translation.

Like many manuscripts from Bandera's hand, this document is a mixture of carefully written and hastily, not to say sloppily written text. Except for the first few pages, which contain a very small script, the hand is not overly difficult. It contains a typically mid-sixteenth-century mixture of *cortisano* and italic letters, often used with no apparent consistency in writing the same word in various places in the text.

The greatest difficulty encountered was that of distinguishing "u" and "v," especially where the "u" was used as a vowel within a word. In most cases the sense of the Spanish served as a guide, but in some of the Indian names (e.g., *Cauchi*, note 26) the intended letter was not evident and a careful comparison of the letter form and similar letters that were clearly either "u" or "v" did not serve to show any distinction of value to the transcriber in deciding what Bandera had written. A second difficulty was the writing of "n," "i," "u," and "m." All of them, when next to each other, come out as a series of vertical strokes connected by diagonals of varying degrees of thickness and color. Again, Spanish words did not present a problem because the correct orthography was clear from the context, which indicated what the word was supposed to be. Indian words could not be figured out by this method (e.g., note 37).

The orthography, while generally consistent, follows the canons of the time. It is first of all phonetic, especially for the unfamiliar Indian words. Beyond that it uses the "x" for "j" and "s" in a number of positions; interchanges—relative to modern usage—the "b" and "v"; and seems to contain few "m's," most notably replacing the "m" in *compania* in favor of an apparent "n." Bandera also uses the silent "h" between vowels, a practice common then but largely dropped in modern Spanish.

By contrast, he often omits the initial "h" for words such as *habia*. A capital "R" is frequently used for double "r." Other variants in spelling, which are not clearly slips of the pen, are consistent with sixteenth-century usage but should present no problems to the student able to sound out the word in Spanish as a guide to finding it in a modern dictionary. Where this would not be possible, I have provided a note to explain what I believe the word(s) to be. I have expanded all abbreviations except for the frequently used *dha* and *dho*, which are *dicha* and *dicho*, and *mag.*, which is *magestad*.

Capitalization and punctuation follow no rules. Bandera used capital letters, especially the vowels "A" and "E," more or less at random. The most interesting case is the use of capital "E" at the end of the Indian names *GuiomaE*, *EmaE*, and *TocaE*, in each case to avoid the use of the silent "h" between the vowels (see note 24). Punctuation is shown, if at all, only by the use of slashes, which are indicated in the transcription by the "[slash]" convention. In addition, the slash is occasionally used to separate "o" as an initial vowel from consonants that end the previous word. In preparing the translation I have inserted punctuation and paragraphs where necessary.

Certain lines and words were underlined in the original, probably by a later reader. These underlinings are shown in the transcription and translation with italics. Where the original author used a solidus (/) as a punctuation mark in the manuscript, "[slash]" appears. The solidus is used to indicate line breaks.

Marginal notes are inserted in the transcription and translation at the beginning of the first line that they are opposite. They are enclosed in braces and italicized to distinguish them from the text proper. Similarly, the few emendations written above the line are noted and printed in italics.

Transcription

COVER TEXT

1569
El ten[ient]e govern[ad]or de la florida

FOLIO I
S[an]to Dom[ing]o 1[er]o de Ab[ri]l de 1569

Ju[an] pardo xxxiiii fo[lios]

{*autos p[ara] la Relacion / q[ue] dio el capitan / Ju[an] pardo de la / entrada q[ue] hizo / en tierra / de la / flo/rida / s*} En la muy noble y leal ciudad y fuertes de la punta / de santa helena que Es en las probincias de la florida prime[r]o dia del mes / de Abril año del nascimiento de nuestro Salbador J[e]su Xpto[9] de mill e qui[nient]os / y sesenta y nuebe años en presencia de mi Juan de la vandera escrivano desta dha / ciudad y punta y su distrito y de las testigos di yuso[10]

The "Long" Bandera Relation 207

escriptos parescio el muy magni[fi]co señor Juan / pardo capitan de ynfanteria Española por su mag. contino de su casa lugarteniente del governador de las dhas / probincias de la florida y dixe q[ue] El año proximo pasado de mill E quinientos y sesenta y siete años su m[erce]d con la / mayor p[ar]te de su compania salio desta dha ciudad y fuertes con horden del muy Ill[ustr]e señor pedro menendez de / aviles capitan general de las dhas probincias y adelantado dellas En nombre de su mag. por la tierra de la florida / adentro la bia de nueba España en la qual dha jornada por el dho señor adelantado se le mando q[ue] con todo El /cuydado posible procurase de allanar y quietar a los caciques o yndios de toda la tierra y de atraEllos al s[er]vi[ci]o de / dios y de su mag. y asimismo q[ue] tomase la posesion de toda la dha tierra en su real nonbre para El qual dho / hefeto El dho señor adelantado mando a mi El dho scrivano q[ue] fuese con el dho señor capitan pues hera escrivano des/ta punta y puerto de santa helena y su distrito en la dha jornada y estribiese[11] y asentase lo que se hiziese y ante / mi pasase en forma con las solenidades que se rrequire para lo mostrar a su mag. quando neces[ar]io / fuese todo lo qual le conbiene al dho señor capitan tenello En su poder para dar quenta a su mag. o a / quien En su nonbre se la pidiere por tanto q[ue] me dia y pidio a mi El dho escrivano le de un traslado [e]sc[ri]pto / En limpia firmado y signado En publica forma y en mane[r]a q[ue] haga fe de todo lo que efecto en s[er]vi[ci]o de dios / y Su mag. [que] se hico en la dha jornada lo qual yo El dho escrivano le di como se me pidio juntam[en]te con una aprobacion / que El dho señor adelantado ante mi El dho escrivano hiço de todo lo hecho en la dha jornada con juramento solen/ne q[ue] al dho señor capitan y a otras personas q[ue] se hallaron presentes a toda Ella Rescibio [slash] q[ue] su thenor / de lo suso dha uno en pos de otro disze[12] como se sigue /

{*instrucion*} v. ynstruycion de lo que a de hazer Juan de la vandera en esta jornada q[ue] ba con El capitan Juan pardo de la tierra / adentro y camino de las minas de san martin y çacatecas /

primeramente a qualquier cacique q[ue] llegare a de adv[er]tir al dho capitan [que] trate con los caciques q[ue] si quisieren / s[er] xptianos[13] y dar la ubidiencia al papa y a su mag. y si quisieran rrelixiosos para que les enseñen a ser xptianos / se les enbiaran y si quisieren dar la posesion de la tierra a su mag. se tomara en su real nonbre con todos / solennidades y de todo Esto sacaran testimonios en forma para a su tiempo entregallos a su mag. /

llegado a la tierra de las cacatecas y minas de san martin al primer pueblo de xptianos El capitan / os a de enbiar a la audiencia rreal de la nueba galicia y al birrey de la nueba España para dar les abiso / de su jornada y de lo demas q[ue] le paresciere conviniente al s[er]vi[ci]o de su mag. El qual camino hareis con la / dilixencia y fidelidad que de bos se confia y nuestro señor os de prospero biaje deste fuerte de san pfelipe / y punta de santa helena veynte y ocho de mayo de mill e quinientos y sesenta y siete años pedro menendez /

{*ynstrucion*} v. ynstruycion de lo q[ue] bos el Capitan Juan pardo debeis de hazer

Este presente biaxe q[ue] bais la tierra adentro a / procurar El amistad con los caciques para que vengan a conocimiento de nuestra santa fee catolica / y a ubidiencia de su mag. /

primeramente partireis deste puerto de santa helena y fuerte de san pfelipe principio de septienbre / deste presente año con hasta ciento y veynte soldados arcabuzeros y ballesteros y llebareis El camino / q[ue] os paresziere mas conbiniente y derecho para hacia las cacatecas y minas de san m[art]yn procurando / hazer toda amistad con los caciques que en el camino y lados ubiere y en todos deteniendo dos y tres y q[uat]ro / dias los q[ue] os paresciere / En los lugares mas comodos para que los caciques del derredor os bayan a ver FOLIO I VERSO y hazed con ellos toda amistad p[ro]curando de los atraher a la hubidiencia de su mag. y que queden y / den su palabra si fueren rreligiosos para dezilles las cosas de dios nuestro señor y como an de ser / xptianos los trataran muy bien y a donde quiera que llegaredes donde ubiere cacique principal dexal / cruz y xptianos q[ue] les enseñen la dotrina xptiana /

p[ro]curareis que la gente que llebaredes biba xptianamente y en toda buena disciplina y q[ue] se huelguen / y rregociexen y tomen plazer y en el camino no los trabaxar haziendo alto en los lugares y p[ar]tes q[ue] os paresciere para que se rrefresquen y descansen para q[ue] se puedan bien conservar y ten/gan salud asi para la yda como para la buelta /

en llegando a qualquier pueblo de xptianos de las minas de san martin o cacatecas hareis alto / y dareis abiso al birrey de la nueba España y audienca de la nueba galicia de v[uest]ra llegada y jor/nada que abeis hecho y si fuere nescesario den algun rrelixioso doxto y de buen Espiritu y bida / para que a la buelta a Este fuerte lo dexesis en la parte que os paresciere con alguna gente para que Es/tando alli procurae atraer los yndios al s[er]vi[ci]o de Dios nuestro señor y a ubidiencia de su mag. lo hareis /

p[ro]curaeis de Estar de buelta en todo El mes de março si fuere posible en este fuerte de san felipe para / si de caso a el binieren el berano benidero franceses los poder Resistir y defender /

y a la buelta que bolbieredes si os paresciere bolber por el camino que fuistes lo hareis y si no sera / por el camino que os paresciere /

la gente que llebaredes y la que hallaredes y os entregaren que se aya ydo sin horden mia destas p[ro]bincias / los traereys con bos y si alguno se os fuere o ausentare o quisiere yr o ausentar lo castigareis conforme / al delito q[ue] para Ello vos doy poder cunplido sigun q[ue] de su mag. lo tengo /

a todos y qualesquier just[ici]as de su mag. donde quiera que llegaredes de parte de su mag. pido y ree/quiero y de la mia pido por m[erce]d bos den todo El fabor y ayuda que les pidieredes y ubieredes me/nester y nuestro señor por su bondad bos de prospero bixae y subceso fecho en san phelipe y punta / de santa helena veynte y cinco de mayo de mill e quinientos y sesenta y siete años /

en todas las partes que llegaredes tomareis la posesion de las tierras y sitios q[ue] los caciques os dieren / en nonbre de su mag. y por ante escribano que dello de testimonio los quales testimonios traereis / signados y autorizados en forma [slash] pedro menendez por mandado de su señoria juan de zuiniga /

{*Jornada*} v. en el nonbre de nuestro señor J[e]su Xpto y de la gloriosa sienpre virgen sin mancilla santa maria su ben/ditisima madre a honor y rreberencia de los quales y en su santo nonbre y para s[er]vi[ci]o de su santidad y del /rrey don felipe n[uest]ro señor [slash] Sepan todos los que Esta escriptura bieren como Estando en la ciudad y punta / de santa helena que Es En las probincias de la florida lunes primero del mes de septienbre del año del / nascimiento de nuestro salbador J[e]su Xpto de mill E quinientos y sesenta y siete años en pre/sencia de mi Juan de la vandera escribano destas probincias y do los t[estig]os di yuso esc[ri]pto parescio / El dho señor capitan y dixo q[ue] por quanto oy dho dia El se a de partir dende la dha ciudad a hazer / cierta jornada por la tierra de la florida adentro con horden del muy Ill[ustr]e señor pedro menendez / de aviles adelantado destas probincias En nonbre de su mag. atrayendo y allanando los / yndios de la dha florida hasta llegar a qualquier parte de la nueba España y a llebar en la dha su con/pania a guillermo Rufin frances ynterpetre de mucha tierra de la dha florida al qual su m[erce]d / en mi presencia y de los dhos t[estig]os le tomo juramento sobre la señal de la Cruz por dios y por santa maria / en forma dibida de derecho para que El suso dho como persona que Es lengua de los dhos yndios y los entiende / de hordinario todo sienpre puess que Esta jornada se haze en s[er]vi[ci]o de dios y de su mag. trate y diga verdad FOLIO 2 de todas las cosas que se trataren y conzertaren con los dhos yndios para q[ue] de todas Ellos se de quenta muy por extenso a su mag. y q[ue] / si ansi lo hiziese dios nuestro señor le ayudase En este mundo al cuerpo y en el otro a la anima donde mas a de durar y lo / contrario haziendo que El se lo demanda se mal y caramente como aquel q[ue] jura y se perjura jurando su santo nonbre en bano / de q[ue] yo El dho escribano doy fee y al dho Juramento El dho guillermo Rufin dixo prometiendo como prometio de deszir v[er]dad si juro E / amen y lo firmo de su nonbre a lo qual que dho Es fueron presentes por testigos El señor alverto Escudero de villamar alferez / del dho señor capitan Juan pardo y pedro de hermosa y pedro gutierrez pacheco y pedro de Olibares soldados de la dha compania / y otros muchos soldados della [slash] Juan pardo guillermo Rufin juan de la bandera escribano /

otro si El dho señor capitan mando a mi El dho escribano Estriba y asiente como desde la dha ciudad hasta El lugar que se dieze / guiomaE[14] que ay quarenta leguas su m[erce]d no quiere poner por Estenso los lugares y caciques dellos que ay en las dhas qua/renta leguas por ser la dha tierra aspero y llana de pantanos y demas desto por Estar como los dhos caciques E yndios / {*ojo*} de las dhas quarenta leguas Estan muy subjetos y ubidientes al servicio de su mag. y tanbien por le parescer como / le a parescido q[ue] Es como cosa y Jurisdiccion subjeta a la dha ciudad de santa helena de donde cada y quando que algo / se les enbia a mandar a los dhos caciques y yndios lo hazen con mucho cuydado y presteza de q[ue] yo El dho escrivano doy fee / a lo que qual dho Es fueron presentes por testigos los dhos señor alberto

Escudero de Villamar alferez y pedro de / hermosa y pedro gutierrez pacheco y pedro de olibares soldados de la dha conpania Juan pardo Juan de la / vandera escribano /

En el nonbre de n[uest]ro senor J[e]su Xpto y de la gloriosa sienpre virgen sin mancilla su bendita madre El *primer* / dia del mes de dizienbre del año del nascimiento de nuestro salbador J[e]su xpto de mill E quinientos E sesenta y seis/ años Estando en la ciudad y punta de santa helena que Es en las probincias de la florida los muy magni[fi]cos señores Estebano / de las alas governardor en las dhas probincias y Juan pardo contino y capitan de ynfanteria Española lugar/theniente del dho señor Estebano de las alas El dho señor capitan con horden del muy Ill[ustr]e señor pedro menendez / de aviles capitan general y adelantado en las dhas probincias En nonbre de su mag. salio juntam[en]te con alverto / Escudero de Villamar su alferez y hernando mayano[15] su sargento con los soldados de su conpania por la tierra de / la florida adentro a subjetar y allanar los caciques E yndios de la dha tierra para q[ue] Estubiesen debaxo del dominio / y ubidiencia de su santidad y del Rey don phelipe nuestro señor y con esta horden y presupuesto hizo la dha jor/nada subjetando y atrayendo a los dhos caciques E yndios a la dha ubidiencia hasta llegar como llego a un lugar / q[ue] se disze *Joara* en donde su m[erce]d con la dha su conpania hiço un fuerte llamado por nonbre *san juan* de / donde hasta la dha ciudad y punta de santa helena *ay ciento y beynte* leguas y estando haziendo el dho fuerte le / llego un yndio con una carta esc[ri]pta y firmada del dho señor Estebano de las alas governador en la qual le / hazia saver q[ue] En la dha ciudad y punta de santa helena tenia nescesidad de ser socorrido con su persona y conpania / p[ar]a q[ue] conbenia al s[er]vi[ci]o de su mag. y por el dho señor capitan vista la dha carta y la gran nescesidad q[ue] abia de dar buelta / con la dha su conpania a la dha ciudad y punta de santa helena dexo En el dho fuerte de san Juan al dho hernando mo/yano sargento con treynta honbres de la dha su conpania En guarnicion del y luego dio buelta con el dho su alfereza / y la demas gente de la dha su conpania a la dha ciudad y punta de santa helena En donde llego a siete dias del mes de marco / del año siguiente de mill E quinientos y sesenta y siete años y en ella Estubo con la dha su conpania hasta prime[r]o / dia del mes de septienbre del dho año de mill E quinientos y sesenta y siete años de donde el dho dia salio prosiguiendo la / dha jornada con horden q[ue] le dio El dho señor adelantado El mes de mayo proximo pasado del dho año de sesenta y siete años / {FmaeE / horata / s[eñ]or / grande} y prosiguiendo la dha jornada como es dha llego a un lugar que se disze *guiomaE* en donde El cacique q[ue] se disze EmaE / orata tenia hecha una casa grande para su mag. la qual abia hecho rrespeto de lo q[ue] por el dho señor capitan En nonbre de / su mag se la abia mandado y agora oy lunes ocho dias del dho mes de septienbre del dho año q[ue] fue quando el dho señor ca/pitan juntamente con la dha su conpania llego segunda bez al dha llugar llamado guiomaE En donde por su m[erce]d bista / la dha casa hiço llamar por guillermo Rufin ynterpetre de la dha lengua q[ue] traya consigo en la dha su conpania / al dho EmaE orata al qual por el dho ynterpetre En presencia de mi El dho Juan de la Vandera escrivano se le declara E / dixo q[ue] al s[er]vi[ci]o de dios n[uest]ro señor y a la ubidi[enci]a que se debia a su santidad y a su mag. conbenia q[ue] El y los yndios q[ue] le heran sub/jetos se bolbiesen xptianos y demas desto q[ue] juntase cierto num[er]o de mayz E hiziese hazer una casa En donde se pusiese / al qual dho mayz no se llegase si no no fuese con lic[enci]a de su mag. o de quien El poder que El dho señor capitan tenia tubiese / El qual dho EmaE orata dixo y por el dho ynterpetre se declara q[ue] En quanto a lo q[ue] se deszia de q[ue] se bolbiese xptiano El y los yndios q[ue] le heran subjetos / que el hera contento de lo hazer cada y quando q[ue] El dho señor capitan quisiese y Q[ue] En quanto al hazer juntar y allegar El dho mayz y hazer la dha casa para / lo thener q[ue] El mayz ya lo tiene junto y la casa E q[ue] lo a de thener e En curandose El dha mayz la hara y dello no sacara El ny otri[16] por el cantidad ninguna / si no fuere con la dha lic[enci]a por satisfacion de q[ue] cumple lo q[ue] por el dho señor capitan En nombre de su mag se le manda pidio lic[enci]a al dho señor capitan / para q[ue] del dho mayz q[ue] tiene junto para su mag. pueda sacar una canoa dello para dar de comer al dho señor capitan y a la dha su gente

FOLIO 2 VERSO otro si se le hizo El parlamento q[ue] Esta declarado *a pasque* [slash] orata [slash] cacique [slash] q[ue] Estaba pres[en]te El qual en presen[ci]a de mi El dho escrivano dixo y declara por el / dho guillermo ynterpetre q[ue] Estaba presto de hazer todo lo q[ue] por el dho señor capitan en nonbre de su mag. se le manda y q[ue] tanbien tiene junto El mayz q[ue] se / le a declarado para lo qual hara una casa En que lo tenga quando Este curado a lo qual El ni otro por el no se llegara si no fuere con la dha licencia y en satisfacion / desto pidio licencia al dho señor capitan para del dho mayz sacar una canoa para dar de comer al dho señor capitan y a su gente y por el dho señor capitan / visto lo suso dho y q[ue] conbenia al servicio de Su Mag. dar la dha licencia a los dhos caciques para sacar El dho mayz pues q[ue] hera para El sustento de la g[en]te de su con/pania se la dio y demas desto en presen[ci]a de mi el dho escrivano les dio a cada uno de los dhos caciques una hacha y ciertos botones de autauxia[17] con q[ue] los dhos caciq[ue]s / quedaron muy contentos a todo lo qual q[ue] dho Es fueron presentes por t[estig]os el dho alverto Escudero de villamar alferez y pedro de hermosa y pedro gutierrez pa/checo y pedro de olibares soldados de la dha conpania y el dho señor capitan lo firmo de su nonbre y el dho guillermo rrufin lengua y yo el dho escrivano Juan pardo / guillermo Rufin Juan de la vandera escrivano /

E despues desto En presen[ci]a de mi el dho escrivano El dho señor capitan Juan pardo prosiguiendo en la dha jornada En honze dias del dho mes de septienbre del dho año de / mill e quinientos E sesenta y siete años llego a un lugar q[ue] le disze *canos* en el qual hallo una casa grande con cierto numero de mayz q[ue] por mandado de su magestad / abian hecho como se lo mando El dho señor capitan [slash] canos [slash] orata [slash] ylasi [slash] orata [slash] sanapa[18] [slash] orata [slash] unuguqua [slash] orata [slash] Vora[19] orata [slash] ysaa orata [slash] catapa orata [slash] / {*aqui se junta/ron muchos se/ñores como la / letra lo declara*} vehidi

212   *The Juan Pardo Expeditions*

orata [slash] otari orata [slash] uraca orata [slash] achini orata [slash] ayo orata [slash] canosaca [slash] orata [slash] caciques muy principales sin otros muchos q[ue] Estan subjetos y de/baxo del dominio de algunos de los suso dhos los quales suso dhos hallo El dho señor capitan juntos En el dho lugar llamado canos y asi a todos juntos en presen[ci]a / de mi El dho escrivano y de los t[estig]os di yuso esc[ri]pto los hiço llamar por guillermo rrufin ynterpetre de la lengua dellos y por el se les declaro como la dha casa q[ue] te/nian hecha juntamente con el dho mayz ya sabian q[ue] antes de agora se les abia declarado y dho por el dho señor capitan como la abian de hazer para su mag. / y como conbenia q[ue] se bolbiesen xptianos y estubiesen debaxo de su dominio y q[ue] desto no abian de descrepar un punto y que agora por su m[erce]d visto q[ue] cun/plen con Efeto lo suso dho es menester q[ue] En presen[ci]a de mi El dho escrivano lo rratifiquen y aprueban los quales En presen[ci]a de mi El dho escribano por el dho yn/terpetre declaron y dixeron q[ue] seran muy contentos de hazer lo q[ue] por el dho señor capitan en nonbre de su mag. se les a mandado y declarado y / como tales q[ue]dando como q[ue]daron debaxo del dho dominio hizieron El yaa [slash] de que yo El dho escrivano doy fee y prometieron de no deshazer ni sacar nin/guna {*yaa [slash] quiere dezir / soy contento de / hazer lo q[ue] se me / manda*} cantidad del dho mayz para en caso alguna si no fuere con licencia de su mag. o de quien su poder tubiere fueron t[estig]os de lo q[ue] dho Es y de como El / dho señor capitan les dio a cada uno de los dhos caciques una hacha y un poco de tafetan colorado y ciertos botones de atauxia en señal de buena A/mistad con q[ue] los suso dhos quedaron muy contentos de q[ue] yo El dho escrivano doy fee / El señor alverto Escudero de villamar alferez y pedro de hermosa / y pedro guiterrez pacheco y pedro de olibares soldados de la dha conpania /

otro si en este dia mes E año suso dho ante mi El dho escrivano / y en presen[ci]a del dho señor capitan parescio El dho ylassi [slash] orata [slash] cacique arriba nonbrado y por el dho ynterpetre declaro y dixe q[ue] no ostante q[ue] El ayudo / a hazer la casa del dho lugar llamado canos que El mayz que El de su parte a de dar y casa en q[ue] lo tenga q[ue] El la tiene hecha y junto El dho mayz para su / mag. en su lugar a lo qual prometio de no allegar El ni otri por el si no fuere con lic[enci]a de su mag. o de quien su poder ubiere [slash] t[estig]os El dho señor / Alferez y los suso dhos Juan pardo guillermo rrufin Juan d la vandera escrivano /

E despues de esto en presen[ci]a de mi El dho Juan de la vandera escrivano El dho señor capitan Juan pardo prosiguiendo la dha jornada En treze dias / del dho mes de septienbre del dho año de mill E quinientos y sesenta y siete años llego a un lugar q se disze *tagaya* y el cacique de el se disze Tagaya [slash] El qual / en el dho lugar tenia hecha una casa nueba buena para su magestad y como El dho señor capitan bido la dha casa en mi presen[ci]a hizo llamar ante si / {*tagaya caciq[ue] s[eñ]or*} al dho tagaya por el dho guillermo Rufin interprete y por el se hico El parlamento acostunbrado El qual rrespondio por la dha lengua q[ue] El esta / presto de hazer lo q[ue] por su m[erce]d se le manda de q[ue] yo El dho escrivano doy fee a lo qual fueron presentes por t[estig]os El señor alvertos de Villamar alferez / y pedro de hermosa y pedro gutierrez pacheco y pedro de olibares soldados de la dha conpania Ju[an] pardo guillermo rrufin Juan de la vandera escrivano /

E despues desto en presencia de mi Juan de la vandera escrivano El dho señor capitan Juan pardo prosiguiendo la dha / jornada en catorze dias del dho mes de septienbre del dho año de mill E quinientos y sesenta y siete anos / llego con la dha su compania a un lugar que se disze [slash] gueca [slash] en el qual Estaba el cacique del que se disze gueca / orata [slash] El qual no ostante que Es de los caciques que hizieron la casa de canos se le hallo hecha en el dho su lugar / una casa nueba de madera un poco grande y por el dho señor capitan visto En su presen[ci]a y de mi El dho / escrivano por guillermo Rufin ynterpetre q[ue] presente estaba El dho caciq[ue] dixo y declaro q[ue] la dha / casa la abia hecho para El s[er]vicio de Su Mag. y dicho Esto por la dha lengua se le declaro E dixo El par/lamento acostumbrado y el dho cacique Respondio q[ue] hera muy contento de lo ansi cumplir y demas / desto Estaban En el dho lugar unharca[20] orata [slash] y herape orata [slash] y suhere [slash] orata [slash] y suya [slash] orata [slash] / y uniaca [slash] orata [slash] y sarati orata [slash] y ohebere [slash] orata [slash] todos caciques de otros lugares diferentes / de los suso dhos a los quales se les dixo E declara por la dha lengua q[ue] conbenia asi a Ellos / como a todos los demas que tubiesen quenta y cuydado de cada uno dellos acudir / a su mag. con la cantidad de mayz q[ue] cada uno conforme a su posibilidad podia lo qual pusiesen / y llebasen a una casa ques para su magestad Esta hecha En el lugar de canos / en donde Esta cierta cantidad de mayz que otros caciques an / dado y en defeto de no thener El dho mayz El que no lo tu/biese que fuese obligado de acudir a la dha casa con algunos gamuças / y en defeto de no las thener acudan por la horden y en la parte / suso dha con la cantidad de sal que cada uno pudiese los quales / oydo y entendido lo suso dho por el dho guillermo Rufin / ynterpetre rrespondieron todos a una boz yaa [slash] / que quiere deszir que Estaban prestos de lo cumplir / como se les abia declarado y mandado de que yo El dho escrivano / doy fee a lo qual fueron presentes por testigos el señor alv[er]to / Escudero de Villamar alferez y pedro gutierrez pacheco y pedo de / hermosa y pedro de olibares soldados de la dha compania y Juan pardo [slash] gui/llermo Rufin [slash] Juan de la vandera escrivano /

FOLIO 3 Despues desto en presencia de mi El dho Juan de la vandera escrivano / El dho señor capitan Juan pardo prosiguiendo la dha jornada en quinze / dias del dho mes de septienbre del dho año de mill E quinientos E / sesenta y siete años llego con la dha su compania a un lugar que se / disze Aracuchi En el qual El dho señor capitan hallo hecha una casa buena / nueba de madera y dentro della una camara alta con cierto / numero de mayz y vista por el dho señor capitan por el dho / guillermo rrufin ynterpetre hico llamar al cacique del dho lugar / que se deszia Aracuchi orata [slash] al qual por la dha lengua se le dixo / y declaro como la dha casa con el dho mayz hera para su magestad / y demas desto se le hiço El parlamento Acostumbrado El qual / dho cacique por el dho ynterpetre declaro que para el mismo hefeto / abia hecho la dha casa y juntado El dho mayz y en quanto a lo de/mas del dho parlamento en señal de ubidiencia hiço El yaa [slash] que quiere / deszir que Esta presto de lo cumplir y de no deshazer ni quitar nin/guna cantidad del dho mayz sy no fuere con licencia de su magestad / o de quien su poder ubiere de lo qual yo El dho escrivano doy fee testigos / que fueron palsentes [sic] a lo que dho Es El señor alverto Escudero de villa/mar alferez y pedro de hermosa y pedro gutierrez pacheco y pedro de / olivares soldados de la dha compania Juan pardo guillermo rrufin / Juan de la vandera escrivano /

despues desto en presencia de mi El dho Juan de la vandera scri[van]o / El dho señor capitan Juan Pardo prosiguiendo la dha jornada En / diez y siete dias del dho mes de septienbre del dho año de mill / E quinientos E sesenta y siete años llego con la dha su compania a un / lugar que se disze otari [slash?] En el qual El dho señor capitan hallo una casa / de madera nueba quel cacique del que se disze otari orata tenia / hecha para su magestad y vista por el dho señor capitan hiço llamar / por guillermo rrufin ynterpetre al dho cacique al qual se le hico El / parlamento acostumbrado y el dho cacique rrespondio y dixo yaa que / quiere dezir que esta presto de lo cumplir y prometio de acudir con / lo que tubiese y pudiese a su magestad de lo qual yo El dho scriv[an]o / doy fee testigos que fueron presentes a lo que dho Es el señor alverto FOLIO 3 VERSO Escudero de villamar alferez y pedro de hermosa y pedro gutierrez / pacheco y pedro de olivares soldados de la dha conpania Juan pardo / guillermo Rufin Juan de la vandera escrivano /

otro si en este dho dia mes E año suso dho Estando El dho señor capitan / Juan pardo con su conpania En el dho lugar llamado otari binieron / guatari meco y orata chiquini cacicas de un lugar que se disze / {meco Es gran / señor ora/ta chiquini / menos s[eñ]or} guatari En donde esta un clerigo y dos mochachos Españoles Ense/ñando la dotrina a los yndios de la tierra las quales por el dho gui/llermo Rufin ynterpetre dixeron que Ellas venian a dezir / que tenian hecha una casa en el dho lugar para su magestad / sigun que por el dho señor capitan antes de agora les hera / mandado hazer y a saver lo que de aqui adelante abian de hazer / En servicio de su magestad y por el dho señor capitan visto lo suso / dho por el dho ynterpetre les fue hecho El parlamento acostum/brado las quales abiendo lo Entendido dixeron yaa [slash] que quiere / deszir que estan puestas de lo cumplir de lo qual yo El dho scriv[an]o / doy fee testigos los dhos señor alverto escudero de Villamar al/ferez y pedro de hermosa y pedro gutierrez pacheco y pedro de o/livares soldados de la dha compania /

otro si por el dho guillermo rrufin ynterpetre las dhas cacicas / declararon que Ellas tienen subjecion sobre treynta y nuebe ca/ciques los quales les ayudaron a hazer la dha casa y ellas como cabecas / dellos prometieron de cumplir lo que los dhos treynta y nuebe ca/ciques en servicio de su magestad debian cumplir y asi prometieron / desde oy dho dia de allegar y tener allegado para lo que a su magestad / cumpla Juntamente con los dhos caciques dos camaras de mayz / de que yo El dho escrivano doy fee [slash] asimismo para mas satisfacion / de como la dha casa esta hecha para El dho hefeto porque El dho senor / capitan de presente y yo El dho escrivano por yr de camino pro/siguiendo la dha jornada no la podimos ver y yo El dho escrivano / En su presencia del dho señor capitan tome E Recivi juramento / en forma de derecho de anton munoz y francisco de apalategui FOLIO 4 Soldados de la dha conpania que de tres meses a Esta parte por / mandado del dho señor capitan an Estado En el dho lugar / llamado guatarimico hasta oy dho dia que binieren a la dha conpa/nia para que dixesen y declarasen si hera verdad que las dhas cacicas / tenian hecha la dha casa los quales so cargo del dho juramento / dixeron y declararon que las dhas cacicas juntamente con los / dhos caciques Em presencia de los suso dhos abian hecho la dha casa / grande de madera nueba y de dentro toda Esterada lo qual / Es verdad para El juramento que hecho tienen y lo firmaron / de sus nonbres y yo El dho escrivano doy fee como el dho señor / capitan dio a las dhas cacicas a cada una dellas una hacha y asimis/mo a los demas caciques que antes de agora Estan declarados / a los que se entiende principales a cada uno una hacha y a los de/mas subjetos a Ellos A algunos un Escoplo y a otros botones / de atauxia y algun tafetan colorado testigos los dhos / Juan pardo [slash] anton munoz fran[cis]co de apalategui guillermo / Rufin Juan de la vandera es-crivano /

despues desto En presencia de mi Juan de la vandera escrivano El / dho señor capitan prosiguiendo la dha Jornada en veynte / dias del dho mes de septienbre del dho año de myll E quinientos / y sesenta y siete años llego con la dha su conpania a un lugar / que se disze quinahaqui en el qual hallo El dho señor capitan al / cacique del que se disze quinahaqui orata [slash] El qual tenia hecha / una casa de madera nueba para su magestad y por el dho señor / capitan vista la dha casa aloxo en ella con la dha su conpania E hico / llamar a guillermo Rufin ynterpetre por el qual En nonbre / de Su mag. se le hico al dho cacique El parlamento acostumbrado / por el qual dho cacique entendido lo que por el dho guillermo Ru/fin se le abia declarado Respondio yaa que quiere deszir que Esta / presto de lo Cumplir y por el dho señor capitan visto la ubidiencia / del dho cacique le dio dos Escoplos y un cuchillo delante de mi / El dho escrivano y testigos de que El dho cacique quedo muy contento FOLIO 4 VERSO y prometio de no si[21] deshazer la dha casa ni El dho mayz sin licencia de su mag. / o de quien su poder ubiere antes por la dha lengua prometio y se / obligo de acudir con el dho mayz en el lugar llamado otari de que / yo El dho escrivano doy fee a lo qual fueron presentes por testigos El / señor alverto Escudero de villamar alferez y pedro de hermosa / y pedro gutierrez pacheco y pedro de olivares soldados de la dha / conpania Juan pardo guillermo Rufin juan de la vandera scriv[an]o /

Otro si En el dho lugar quinahaqui bineiron a ver y a bisitar al dho / señor capitan ytaa [slash] orata [slash] cacique muy principal [slash] y cataba / orata asimismo cacique muy principal [slash] y uchiri orata ca/cique los quales En presencia de my El dho escrivano por el dho / guillermo Rufin ynterpetre que estaba presente dixeron E / declararon que Ellos benian a saver y entender del dho señor ca/pitan que que hera lo que les manda que hiziesen En servicio de / su magestad y por el dho señor capitan vista la ubidiencia de / los dhos caciques por el dho guillermo Rufin ynterpetre se les / dixo E declaro El dho parlamento acostumbrado y por ellos

/ savido oydo y entendido El dho parlamento rrespondieron y / dixeron [slash] yaa [slash] que quiere desde que Estan prestos de lo Cumplir / y asimismo los dhos ysaa [sic] y cataba oratas prometieron y se / obligaron cada uno por lo que asi toca de llebar y acudir con el dho / mayz al dho lugar llamado otari lo qual no desharan sin licencia / de su mag. o de quien su poder ubiere y asimismo El dho uchiri / orata prometio y se obligo que El dho mayz con que El tiene de / acudir lo pondra dentro de un lugar llamado canos lo qual no / deshara sin licencia de su magestad o de quien su poder ubiere de / que yo El dho escrivano doy fee a lo qual fueron presentes por t[estig]os / El dho señor alvertos de villamar alferez y pedro de hermosa / y pedro gutierrez pacheco y pedro de olivares soldados de la dha / conpania [slash] Juan pardo [slash] guillermo Rufin Ju[an] de la vandera /

Despues desto En presencia de mi El dho Juan de la vandera / escrivano El dho señor capitan prosiguiendo la dha jornada / En veynte y un dias del dho mes de septienbre del dho año de FOLIO 5 mill e quinientos y sesenta E siete años llego con la dha su conpania / A un lugar que se disze guaquiri En el qual El dho señor capitan / hallo una casa hecha de madera nueba que el cacique del tenia hecha / para El servicio de su magestad y por el dho señor capitan bista / hico llamar por el dho guillermo Rufin ynterpetre al / cacique que se disze guaquiri orata [slash] al qual se le hiço y dixo por / El dho ynterpetre El parlamento acostumbrado y por el / dho cacique oydo y entendido dixo yaa [slash] que quiere desdir / que Esta presto de lo cumplir y que El mayz que por el dho s[eñ]or / capitan se le manda juntar que esta presto de lo juntar / En la dha casa y no llegar a Ello sin licencia de su magestad / o de quien su poder ubiere y por el dho señor capitan visto / la ubidiencia del dho cacique le dio dos cunas[22] con que El dho / cacique quedo conthento de que yo El dho escrivano doy fee / testigos El dho señor alferez y los demas suso dhos gui/llermo Rufin [slash] Juan pardo [slash] Juan de la vandera escrivano /

En la muy noble y leal ciudad de quenca y fuerte de san juan que Es en un lugar que se disze Joara que Esta al pie / de la sierra que descubrio El año presente de mill e / quinientos y sesenta y siete años El muy magnificio senor / Juan pardo contino y capitan de ynfanteria Española por / su mag. veynte E quatro dias del mes de septienbre del / año del señor de mill e quinientos y sesenta y siete años / en presencia de mi Juan de la vandera escrivano El dho dia El / dho señor capitan Juan pardo prosiguiendo la dha Jornada / llego al dho lugar llamado Ioara al qual puso nonbre la / ciudad de quenca porque su m[erce]d Es natural de la ciudad de / quenca que Esta En el rreyno de España y al pie de una sierra / zercada de rrios como lo Esta El dho lugar llamado Joara / y asimismo su m[erce]d puso nonbre al dho fuerte llamado San / Juan que su m[erce]d antes de agora hico hazer con la dha[23] FOLIO 5 VERSO su conpania en nonbre de su magestad san Juan porque El dho / señor Capitan se llama Juan y tanbien porque El dia de san juan / Apostol y envangelista proximo pasado deste dho pre/sente año llego con la dha su compania al dho lugar llamado / Joara que es ciento y veynte leguas de la ciudad y punta de / santa helena que Es En las probincias de la florida y tan/bien porque su m[erce]d bido que la dho sierra Estava llena de / niebe que no se podia pasar y porque al servicio de su mag. conbenia / que su m[erce]d diese buelta a la dha ciudad de santa helena / y lo [que] an dado y trabaxado hasta El dho lugar llamado joara / no quedase disierto hiço hazer con la dha su conpania El / dho fuerte llamado san Juan en el qual porque quedase / con la fuerca que conbenia dexo dentro del a hernando mo/yano su sargento con treynta honbres de la dha su conpania / y ansi dio buelta a la dha ciudad de santa helena con toda / la demas gente a donde Estubo hasta primero dia del / mes de septienbre que comenco sigunda vez a proseguir / la dha Jornada y prosiguiendo por sus jornadas con todas / llego El dia dho al dho lugar llamado Joara En donde En el / hallo hecha una casa nueba de madera con una camara grande / alta llena de mayz que El cacique del dho lugar que se / llama Joada [*sic*] mico tenia hecha por mandado del dho señor ca/pitan para El servico de su magestad al qual dho cacique El / dho señor capitan hiço llamar por guillermo Rufin yn/terpetre y benido En presencia de mi El dho escrivano por el / dho guillermo Rufin se le declaro como al s[er]vicio de su mag. / conbenia que de hordinario tubiese y acudiese con numero / de mayz a la dha casa para El sustento de cierto numero de / soldados que su m[erce]d de presente dexa En el dho fuerte llamado / san juan y demas de lo suso dho se le hico El parlamento acos- FOLIO 6 tumbrado y el dho cacique oydo y entendido lo suso dho por el dho / guillermo Rufin ynterpetre rrespondio diziendo yaa [slash] que / quiere deszir que Esta presto de lo cumplir y por el dho señor / capitan visto la ubidiencia del dho cacique demas de una / hachuela de armas Enastada que antes de agora le abia / dado En mi presencia Agora le dio una hacha y para El y otros / caciques sus subditos ocho cuñas pequeñas largas a manera / de Escoplos y ocho cuchillos grandes y un pedaco de rraso y otro / de tafetan colorado con que El dho cacique y los demas / fueron muy contentos de que yo El dho escrivano doy fee / otro si por el dho señor capitan visto como al servicio de / su magestad conbenia que El dho fuerte quedase con la fuerca / de gente que hera menester y tanbien alguna municion / y los bastimentos nescesarios dexo en el dho fuerte por / alcayde del a lucas de canicares cabo de Esquadra de su con/pania persona en quien concurren las calidades de derecho / Al qual le mando y encargo lo que al servicio de su magestad / conbiene y juntamente con el para la dha fuerca ciertos / soldados y cierto numero de mayz con ciertas municiones las / que avastaban para los dhos soldados de que yo El dho escrivano / asimismo doy fee a todo lo qual que dho Es fueron presentes / por testigos El señor alverto Escudero de villamar alferez y / pedro de hermosa y pedro gutierrez pacheco y pedro de olibares / soldados de la dha compania Juan pardo guillermo Rufin / Juan de la vandera escrivano /

Despues desto En presencia de mi El dho Juan de la vandera / escrivano El dho senor capitan Juan pardo prosiguiendo la / dha jornada En primero dia del mes de otubre del dho año / de mill E quinientos y sesenta y siete años llego con la dha / su conpania a un lugar que Esta desecabo de la sierra que se FOLIO 6 VERSO dieze tocaE[24] [slash] En el qual El dho señor capitan se detubo cantidad / de quatro oras y en el Entretanto que Estubo alli hico llamar / por guillermo Rufin ynterpetre y otras

218 *The Juan Pardo Expeditions*

lenguas al cacique / del dho lugar que se disze tocaE orata [slash] Al qual por mandado / del dho señor capitan por el dho guillermo Rufin y las de/mas lenguas se le declaro y dixo El parlamento acostumbrado / y por el dho cacique oydo y entendido rrespondio yaa [slash] que / quiere deszir que Esta presto de cumplir lo que se le manda / agora y en todo tiempo y por el dho señor capitan visto la u/bidiencia del dho cacique le dio una cuña pequeña y un cuchillo / grande y un poco de tafetan verde con que El suso dho que/do muy contento de que yo el dho escrivano doy fee [slash] otro si / bino Al dho lugar un cacique que se deszia [slash] uastique [slash] orata [slash] de / diez y siete jornadas del dho lugar de la parte del poniente / y en presencia de mi El dho escrivano por el dho señor capitan / fue E mandado al dho guillermo Rufin ynterpetre le de/clarase al dho cacique lo que debia de hazer En servicio de su / mag. y por el dho cacique Entendido por las dhas lenguas Res/pondio que hera muy conthento de cumplir lo que se le man/daba porque para El dho hefecto abia venido de diez y siete / jornadas de alli por q͞: cumplir y hazer lo que al servicio de su / magestad conbenia Agora y en todo tiempo y a cada cosa hico / El yaa [slash] y por el dho señor capitan visto la dha ubidiencia / le dio una cuña pequeña y un cuchillo grande y un caracol / grande de la mar y un poco de tafetan verde con que El dho / cacique fue muy conthento de que yo El dho escrivano doy fee asi/mismo binieron al dho lugar otros quatro caciques de los / derredores del dho lugar que se llamaban [slash] Enuque [slash] orata / y enxuete [slash] orata [slash] y xenaca [slash] orata [slash] y atuqui [slash] orata [slash] los / quales En presencia de mi El dho escrivano parescieron ante[25] FOLIO 7 dho señor capitan y por el dho guillermo Rufin y las demas / lenguas dixeron que benian a ver lo que por el dho señor / capitan se les mandaba y por el dho señor capitan visto lo / suso dho mando Al dho guillermo Rufin y las demas lenguas / que declarasen a los dhos caciques lo que conbenia que / hiziesen En servicio de su magestad que Es El parlamento / acostumbrado y todos quatro juntos hizieron El yaa [slash] dan/do a entender que Estan prestos de lo cumplir y por el dho / señor capitan visto la ubidiencia de los suso dhos les dio sie/te cuñas pequeñas y un pedaco de tafetan verde con que / los dhos caciques fueron muy contentos de que yo El / dho escrivano doy fee a todo lo qual que dho Es fueron pre/sentes por testigos El señor Alverto Escudero de Villa/mar alferez y pedro de hermosa y pedro de olibares y luis / ximenez soldados de la dha compania Juan pardo guillermo / Rufin Iuan de la vandera escrivano /

despues desto En presencia de mi El dho Juan de la van/dera escrivano El dho señor capitan Juan pardo pro/siguiendo la dha Jornada En dos dias del dho mes de otubre / del dho año de mill E quinientos E sesenta E siete años / llego a un lugar que se disze cauchi[26] En el qual hallo hecha una / casa nueba de madera y tierra que El cacique del dho lugar / que se disze cauchi orata tenia hecha para su mag. por / que asi se le abia mandado por el dho señor capitan antes / de agora Estando El dho señor capitan En el lugar que se / disze Joara a donde El dho cacique fue a rreconocimiento / y por el dho señor capitan visto como El dho cacique abia / hecho y cumplido lo que por su m[erce]d antes de agora se la / abia mandado lo

The "Long" Bandera Relation 219

hiço llamar por el dho guillermo Rufin FOLIO 7 VERSO ynterpetre y por el y las demas lenguas se le declaro y dixo El / parlamento acostumbrado y el dho cacique rrespondio di/ziendo yaa [slash] y por el dho señor capitan visto la dha ubidiencia / le dio una cuña pequeña y un cuchillo grande y un poco de / tafetan verde y colorado con que El suso dho fue muy con/tento de que yo El dho escrivano doy fee otro si Este dho dia / En presencia de mi El dho escrivano antel dho señor capitan / binieron a la dha subjecion y rreconocimiento otros cinco / {ojo} caciques que se dezian neguase orata [slash] y estate orata [slash] y ta/coru orata [slash] y utaca orata [slash] y quetua orata a los quales / por el dho guillermo Rufin ynterpetre y las demas / lenguas se les declaro y dixo lo que para El servicio de dios / y de su mag. debian de hazer y ellos todos cinco juntos / dixeron ya [sic] [slash] que quiere deszir que Estavan prestos de lo cun/plir agora y en todo tiempo y por el dho señor capitan vista / la dha ubidienca les dio ciertas cuñas pequeñas y un / poco de tafetan verde y colorado y unos botones de ata/xia[27] con que los suso dhos fueron muy contentos de que / yo El dho escrivano doy fee fueron testigos de lo que dho Es / El señor alberto Escudero de Villamar alferez y los / dhos pedro de hermosa y pedro de olivares y luis ximenez / soldados de la dha compania [slash] Juan pardo [slash] guillermo Rufin / Juan de la vandera escrivano /

despues de lo suso dho En tres dias del dho mes de otubre / del dho año de mill E quinientos y sesenta y siete años / Estando El dho señor capitan en el dho lugar llamado ca/uchi bido a un yndio andar Entre las yndias con un man/dil[e] delante como Ellas lo traen y hazer lo que Ellas hazian / y visto Esto El dho señor capitan mando llamar al / dho guillermo Rufin lengua y a las demas lenguas [slash] y asi FOLIO 8 llamados El dho señor capitan delante de muchos soldados / de su compania les dixo que preguntasen que como andaba / aquel yndio entre las yndias y con mandile delante / como Ellas y las dhas lenguas preguntaron lo suso dho / al cacique del dho pueblo y el dho cacique Respondio por / las dhas lenguas que el dho yndio hera su hermano y que / por no ser honbre para la guerra ni hazer cosas de honbre {ojo} an/daba de aquella manera como muger y hazia todo lo que / a una muger les Es dado hazer y por el dho señor capitan / savido lo suso dho mando a mi Juan de la vandera escrivano / lo Estribiese En la forma suso dha para que se sepa y entienda / quan belicosos son los yndios destas probincias de la florida / y para lo dar por fee y testimonio cada y quando que me / sea pedido de todo lo qual que dho Es yo El dho escrivano doy fee / porque paso todo lo que dho Es en mi presencia y de muchos / soldados de la dha conpania Juan de la vandera escrivano /

Despues desto En presencia de mi El dho Juan de la / vandera escrivano El dho señor capitan Juan pardo / prosiguiendo la dha jornada En seis dias del dho mes de / otubre del dho año de mill E quinientos y sesenta E siete / años llego a un lugar que se disze tanasqui El qual / Estaba situado En cierta parte de tierra fuerte a / manera de ysla zercada de agua porque asi lo Estaba El / *dho lugar zercado de dos rrios caudalosos los quales El* / uno con el otro se juntan a un cabo de la dha ysla que es / a donde Estaba El sitio del dho lugar y por el camino que benia / El dho señor capitan que Es El que biene a dar al dho lugar / por el qual para entrar En el

El dho señor capitan con la / dha su conpania paso a pie uno de los dhos rrios que hera FOLIO 8 VERSO harto trabaxoso porque daba al onbligo[28] y antes mas que menos / y por la p[ar]te que El dho señor capitan con la dha su compania / paso El dho rrio para llegar al dho lugar hasta llegar a El / abia buen trecho y por aquella parte El cacique E yndios del / *dho lugar tenian hecha una muralla con tres torres para su* / defensa en el qual como El dho señor capitan con la dha su / compania llego por guillermo Rufin ynterpetre y otras lenguas / hiço llamar Al cacique del dho lugar que se disze tanasqui / orata al qual su m[erce]d pregunto por las dhas lenguas que por q[ue] / causa thenia hecha muralla por aquella parte de donde su m[erce]d / entro mas que por otra parte a la qual pregunta El dho ca/*cique rrespondio que para defensa de sus enemigos los quales si* / *biniesen a le hazer dano no tenian por donde le entrar al dho su* / pueblo si no hera por aquella parte y por el dho señor ca/pitan vista la buena Razon que El dho cacique le dio man/do al dho guillermo Rufin y las demas lenguas que dixesen / al dho cacique lo que en servicio de dios nuestro señor y de su / mag. se debia de hazer los quales dhas lenguas dixeron al dho / cacique El parlamento acostumbrado y el dho cacique Res/pondio haziendo El yaa [slash] por donde dio a entender que hera muy / conthento de hazer e cumplir lo que se le mandaba y por el / dho señor capitan visto la ubidiencia del dho cacique le dio / una cuñiela y media bara de paño de londres y una bara de / lienco y a tres yndios principales del dho cacique a cada uno / una cuña pequeña con que El dho cacique E los dhos yndios / quedaron muy conthentos de que yo El dho escrivano doy fee / fueron testigos a lo que dho Es El señor alverto Escudero de / villamar alferez y pedro de hermosa y pedro de olibares / y luis ximenez soldados de la dha compania Juan pardo [slash] gui/llermo Rufin Juan de la vandera escrivano /

FOLIO 9 despues desto En presencia de mi Juan de la vandera / escrivano El dho señor capitan Juan pardo prosiguiendo la / dha jornada En siete dias del dho mes de otubre del / dho año de mill E quinientos E sesenta E siete años llego / *a un lugar que se disze chiaha que para Entrar dentro del* / *El dho señor capitan con la dha su compania paso tres rrios* / caudalosos y entro dentro del dho lugar El qual Esta en un / buen sitio fuerte porque Es ysla zercada de rrios caudalosos / y en el dho lugar Estubo con la dha su compania ocho dias / porque hera lugar grande y en el muchos yndios y porque / del cacique y los yndios fuese bien rrecivido y como El dho s[eñ]or / capitan estubo en el dho lugar luego yncontinente / por guillermo rrufin ynterpetre hiço llamar al dho cacique / que se disze ola mico[29] al qual por el dho guillermo Rufin / y otras lenguas por mandado del dho señor capitan se le dixo / muy por estenso lo que en servi[ci]o de dios y su mag. debia de / hazer El qual dho cacique entendido lo suso dho hico El / ya [sic] [slash] dando a entender que hera muy contento de hazer lo que / se le mandaba y por el dho señor capitan visto la ubidiencia / del dho cacique le dio una hacha y un pedaço de paño de londres / y una bara de lienco y un poco de tafetan colorado con que / El dho cacique quedo muy conthento y demas desto dio / a tres yndios principales del dho lugar a cada uno una / cuña pequeña y ciertos botones de atauxia de que yo / El dho escrivano doy fee a lo qual que dho Es fueron presentes por / testigos El señor alberto Escudero de villamar alferez y / hernando moyano de morales Sargento y pedro de hermosa / y pedro de olibares soldados de la dha compania Juan pardo / guillermo Rufin Juan de la Vandera escrivano /

despues desto En presencia de mi Juan de la Vandera   FOLIO 9 VERSO escrivano El dho señor capitan Juan pardo prosiguiendo la dha jor/nada en treze dias del mes de otubre del dho ano de mill / e quinientos y sesenta E siete años salio del dho lugar lla/mado olameco y este dia llego con la dha su compania cinco leguas / del dho lugar que fue en campaña a do[30] hiço noche y durmio y es/tando alli binieron a traher de comer para la dha con/pania tres caciques que se dezian [slash] otape [slash] orata [slash] jasire / orata [slash] fumica [slash] orata [slash] y un yndio principal con ellos / los quales dieron a entender por guillermo Rufin / ynterpetre que benian a saver lo que El dho señor ca/pitan En nonbre de su mag. les queria mandar que hiziesen / y por el dho señor capitan Juan pardo bisto lo suso dho / les mando hazer y dezir El parlamento acostumbrado / El qual se les dixo y por los dhos caciques Entendido hi/zieron El ya [sic] En que dieron a entender que heran muy con/tentos de hazer lo que se les mandaba y por el señor / capitan visto la ubidiencia de los suso dhos les dio a / los tres caciques A cada uno una cuña pequeña y un pe/daco de tafetan verde y unos pasamanos de seda blanca / y colorada y al yndio principal que benia con ellos una / cuñuela con que los suso dhos quedaron muy contentos / de que yo El dho escrivano doy fee fueron testigos a lo suso / dho El señor alberto Escudero de Villamar alferez y el / señor hernando moyano de morales sargento y pedro de / hermosa y pedro de olivares soldados de la dha compania / Juan pardo guillermo Rufin Juan de la vandera escrivano /

despues desto En presencia de mi Juan de la vandera escrivano / El dho señor capitan Juan pardo prosiguiendo la dha / jornada en catorze dias del dho mes de otubre del / dho año de mill E quinientos E sesenta E siete años partio / la dha p[ar]te donde abia dormido y este dia fue con la dha su con/pania otras cinco leguas mas adelante por un camino   FOLIO 10 muy aspero en donde subiendo una montana muy alta / En la otra parte dormio y estando En lo alto della / *hallo una piedra pequeña vermeja y hallada hiço llamar* / ante si a andres xuarez fundidor de oro y plata / al qual se le mostro la dha piedra y el suso dho la / bio y vista dixo que El humo[31] que la dha piedra mostraba / daba a entender q[ue] sea plata de que yo El dho escrivano / doy fee a lo qual fueron presentes por testigos El señor / alverto Escudero de billamar alferez y el señor her/nando moyano sargento y pedro de hermosa y pedro / de olivares soldados a dha compania Juan pardo / guillermo Rufin Juan de la vandera escrivano /

Despues desto En presencia de mi Juan de la vandera / escrivano El dho señor capitan Juan pardo prosiguien/do la dha Joranda en quinze dias del dho mes de o/tubre del dho año de mill E quinientos E sesenta e sie/te años llego con la dha su compania a un lugar que se llama / chalahume [slash] en donde hico alto y durmio aquella noche / y luego yncontinente hico llamar al cacique del dho / lugar que se deszia chalahume orata [slash] al qual por / guillermo Rufin ynterpetre se le dixo El parlamento / acostumbrado y el dho cacique hiço El ya [sic] [slash] en que dio a en/tender que Estava presto de lo cumplir y por el dho s[eñ]or / capitan visto la ubidiencia del suso dho le dio una hacha / y una gargantilla con que El dho cacique quedo muy con/tento de que yo El dho escrivano doy fee a lo qual fueron / presentes por testigos El señor alverto Escudero de / villamar alferez y el señor hernando moyano sar/gento de la dha compania [slash] Ju[an] pardo guillermo Rufin [slash] Ju[an] de la vandera escriv[an]o /

FOLIO 10 VERSO despues desto En presencia de mi Juan de la vandera escriv[an]o / El dho señor capitan Juan pardo prosiguiendo la / dha jornada en diez y seis dias del dho mes de otubre / del dho año de mill E quinientos E sesenta y siete / años llego a un lugar que se disze satapo En donde fue / bien Recivido y este dia hiço alto y se detubo En el dho / lugar y luego yncontinente hiço llamar al dho cacique / del dho lugar que se disze Satapo orata [slash] al qual por gui/llermo Rufin ynterpetre se le dixo El parlamento a/costumbrado y por el dho cacique Entendido hiço El yaa [slash] dan/do a entender que Estaba presto de lo cumplir como se / le mandaba y por el dho señor capitan visto la ubidiencia / *que el dho cacique mostro le dio una hacha y un Espejo y una* / gargantilla Respeto de le conthenter porque alli El dho / *señor capitan fue ynformado que otros muchos Españoles* / *que antes de agora an benido por estas partes ansi de a* / pie como de a cavallo y el dho cacique y los yndios que / le son subjetos an muerto los dhos Españoles y demas / desto El dho señor capitan dio a un mandador y dos yndios / principales del dho lugar a cada uno una cuña pequeña / y una gargantilla con que asi El dho cacique E yndios que/daron muy conthentos de que yo El dho escrivano doy fee testigos / que fueron presentes a lo que dho Es El señor alverto Escu/dero de billamar alferez y el señor hernando moyano / sargento y pedro de hermosa y pedro de olibares sol/dados de la dha conpania [slash] Juan pardo [slash] guillermo Rufin / Juan de la vandera escrivano /

despues desto Este dho dia diez y seis dias deste dho / mes E año dho a la tarde ya que começaba a / Escurecer Estando puestas las centinelas que se acos- FOLIO 11 tumbran poner para la guardia y anparo de la dha conpania dos / dellas que Estavan puestas En dos partes altas del / dho lugar dieron abiso como abian oydo muy gran rremor / de yndios fuera del dho lugar y por el señor capitan En/tendido lo suso dho mando y dio abiso a la dha su compania / que estubiese alerta porque le abia parescido mal / cierta Junta y cantidad de yndios que Estavan fuera / del dho lugar junto a El a tal ora y ansi la dha conpania / Estubo muy alerta toda la noche de que yo El dho escriv[an]o doy fee /

otro si la dha noche ya que seria la media noche poco mas o me/*nos Estando guillermo Rufin ynterpetre de la dha conpania* / durmiendo dendro de un bohio bino a El un yndio del / dho lugar que lo conocia porque abia dos u tres dias que / benia con la dha compania de atras El qual dixo Al dho gui/llermo Rufin lengua que se hacia que el dho señor capitan / le diese una hacha que El discubriria y diria cierta traycion / que los yndios del dho lugar y cacique del y los caciques / e yndios de cosa y huchi y de casque y de olameco que / hasta alli abian ydo con la dha conpania tenian hor/denada El qual dho guillermo Rufin Juntamente / con el dho yndio fue a donde estaba El dho señor ca/pitan y le dixo lo que el dho yndio le abia dho y por el / señor capitan entendido lo suso dho dio una hacha al / dho yndio y entonces El dho yndio dio a Entender / y dixo como cusa que aquella noche abia llegado alli / con muchos yndios sus basallos y olameco [slash] y uchi [slash] y casque / caciques con todos los yndios que alli tenian y el ca/cique de aquel lugar con todos sus yndios Estaban / acostumbrados de matar muchos Españoles que / antes de agora abian benido por Estas partes y que / asimismo ni menos abran tratado y conzertado de / entre todos Ellos hazer tres enboscadas y se poner por[32] FOLIO 11 VERSO la p[ar]te donde El dho señor capitan con la dha su compania / abia de pasar derecho a cosa y demas desto que El / dho cosa y los otros caciques antes de agora abian tra/tado de no dar de comer ni acoxer[33] al dho señor capitan / con la dha su compania En ningun lugar de los suyos / sy no se lo pagaba [slash] y por el dho señor capitan entendido lo su/so dho Agradecio mucho al dho yndio lo que le abia a/bisado y como la dha su compania Estaba alerta por a/quella noche disimulo y otro dia siguiente ya que hera / salido El sol hiço llamar al cacique del dho lugar al qual / su m[erce]d pidio que le diese ciertos yndios que heran me/nester para que llebasen ciertas cargas y el dho ca/cique disimuladamente hico que yba a buscar los dhos / yndios y a cabo de rrato bino y no traxo ninguno dan/do las disculpas que le parescio por donde El dho señor / capitan entendio y bio que lo que El dho yndio le abia abisado / hera verdad porque ni los yndios que hasta alli abian / ydo con su m[erce]d ni los del dho lugar ninguno parescio / mas que las mugeres y niños del dho pueblo y asi hico pa/rescer y llamar ante si al señor alberto Escudero / de Villamar alferez y a hernando moyano sargento / y a pedro de hermosa y a Juan de la vandera y a mar/cos ximenez y a pedro florez y a Juan de Salazar y a / miguel de haro y a gaspar rrodriguez cabos de Esquadras de / la dha compania y estando asi todos Juntos su m[erce]d / les dixo como abia savido que El cacique de aquel lugar / y olameco que hasta alli abia ydo con el y cosa y otros / caciques con sus yndios se abian confiderado de le matar / a El y su compania y para Esto le tenian puestas cier/tas envoscadas en el camino por donde abia de pasar / que El dho señor alferez y sargento y cabos de Esquadra FOLIO 12 mirasen si seria bueno para lo que conbenia al serbiçio de / su magestad pasar adelante o desde alli dar la / buelta y con esto se aparto y ~~con esto~~ por el dho señor / alferez y sargento y cabos desquadras Entendido lo / que dho Es cada uno de por si dixo aquello que le pa/rescia por donde ni se conformaron en lo uno ni en lo otro / syno antes se tornaron a Juntar con el dho señor ca/pitan y cada uno dio El parescer que bien le Estubo y al cabo de / todo Esto El dho señor capitan Juntamente con los dhos se/ñores alferez y sargento y los cabos de Esquadras se binieron a / conformar en que al servicio de dios y de su magestad con/benia que desde alli se diese buelta derecho a olamico que / hera tres Jornadas de alli por otro camino diferente / del que se abia ydo hasta alli y que en el dho olamico se hiziese / un fuerte y que En el quedase El numero de gente y muni/cion que al dho señor capitan le paresciese y asi con este pa/reszer El dho señor capitan juntamente con la dha / su compania dio buelta derecho a olamico por caminos / y montañas muy asperas por traher sigura la dha su conpa/nia en el qual camino durmio dos noches en campaña que / se pasaron con harto trabaxo y en diez y nuebe dias del dho / mes llego a un lugar que se disze chiaha en donde hiço lla/mar al cacique del dho lugar que se disze chiaha orata / al qual se le dixo El parlamento acostumbrado y el dho / cacique hico El ya [sic] dando a entender que Estaba presto de / cumplir lo que se le mandaba y asi El dho señor capitan / dio a dho cacique una hacha con que quedo muy conthento / fueron presentes por testigos a todo lo que dho Es pedro de / olivares y luis ximenez y Juan garcia de madrid solda/dos y otros muchos soldados de la dha conpania y firmaron / lo El dho señor capitan y alferez y sargento y los cabos desquadra[34]   FOLIO 12 VERSO que sabian Estribir [slash] Juan Pardo [slash] alberto Escudero de Billamar / fernando moyano de morales [slash] pedro de hermosa juan de sa/lazar [slash] guillermo Rufin [slash] Juan de la vandera escrivano /

Despues de lo suso en diez y siete dias del dho mes de o/tubre del dho año de mill E quinientos y sesenta E sie/te años en presencia de mi Juan de la vandera escrivano y / de los testigos di yuso esc[ri]ptos parescio El muy magni[fic]o señor / capitan Juan pardo y me pidio E rrequirio le diese por fee / y testimonio como ayer que se contaron diez y seis dias / del presente mes llego con la dha su conpania a un lugar / que se disze satapo con yntento y boluntad de proseguir / la dha Jornada y fenecerla como por el muy Ill[ustr]e señor / pedro menendez de aviles Adelantado de las probincias de / la florida en nonbre de su magestad se le abia mandado y como / por saver y entender que los caciques E yndios de los lugares / por donde abia de pasar desde alli adelante Estaban / yndignados de lo matar a El y a la gente de su conpania / y que ya que Esto no fuese que en ninguno de los dhos lu/gares le dexarian Entrar ni le daria de comer si no lo / pagava y que demas desto Este dho dia por el camino por / donde abia de pasar le Estaban aguardando En tres partes / mucha cantidad de yndios para le matar y destruyr la dha / su conpania y como por estas causas y por otras muchas / que heran muy notorias en la dha su conpania su m[erce]d Es/te dho dia daba buelta desde El dho lugar derecho a la ciudad / de santa helena con ynthento de hazer y edificar en un lugar / que disze olamico que por otro nonbre se llamaba chiaha / {ojo} un fuerte en donde pretendia dexar numero de gente para / allanar y tener siguro lo q[ue] hasta alli En servicio de su / mag. a conquistado y lo mas que desde alli pudiese por/que ansi conbenia al servicio de su mag. y yo El dho escrivano   FOLIO 13 entendido lo que por el dho señor capitan se me pidio para mas / me zertificar de lo suso dho y le dar por testimonio lo que su / m[erce]d me pide con licencia espresa que para ello me dio y poder / bastante tome e rrescibi juramento a algunos sol/dados de la dha compania que abian hablado con algunos yn/dios y entendido lo que se me pide en nonbre de los quales / Es el siguiente Juan pardo juan de la vandera escrivano /

luego yncontinente yo El dho juan de la vandera escrivano me/diante El dho poder y licencia que arriba Esta esc[ri]pto tome / y rrecibi juramento en forma debida de derecho de al[ons]o / velas soldado de la dha conpania El qual abiendo jurado E / syendo preguntado azerca de lo suso dho dixo que lo que / Este testigo sabe es que estando en un lugar que se disze / satapo que es a donde a estado la dha conpania començo / a hablar y hablo con un yndio que en la dha conpania se trae por / lengua que es natural de oluga[35] al qual Este testigo le pre/gunto que que sentia o entendia que trataban o habla/ban los yndios del dho lugar de setapo [*sic*] y como El dho yndio / lengua entendio a Este testigo le rrespondio que los dhos / yndios deszian que un cacique grande que se llamaba cosa / por donde abia de pasar El dho señor capitan no consintiria / que entrase en su pueblo la conpania ni les darian de comer / sy no lo pagaban y que antes que el dho señor capitan lle/gase a cosa abian de salir muchos yndios al dho señor capitan / de guerra y no de paz y que al dho yndio le pesaria mucho / de yr porque en el dho lugar de cosa y otros al rrededor / tenia el dho yndio cinco hermanos cautibos los quales ca/utibaron yendo lo dhos sus hermanos en otra conpania que El capitan se dezia soto y ansi los dhos sus hermanos / quedaron por esclabos porque El dho capitan y la gente se FOLIO 13 VERSO perdio porque los dhos yndios de cosa y de otros lugares / al rrededor los mataron y esto Es lo que este testigo sabe / y el dho yndio le dixo para El juramento que hecho tiene E no / lo firmo porque dixo no saver Escribir declaro ser de hedad / de beynte E seis años poco mas o menos Juan de la vandera /

{T[estig]o} otro si yo El dho Juan de la vandera escrivano para mas zer/tificacion de lo suso dho tome E recivi juramento En / forma debida de derecho de diego de morales soldado de la / dha compania El qual aviendo jurado E siendo pregun/tado zerca de lo suso dho dixo que lo que sabe Es que hablando / Este testigo con un yndio En el dho lugar de satapo por/que El dho yndio le entendia algunas cosas de las que el hablaba / le dixo a este testigo que si El señor capitan abia de yr / por cosa con su conpania que no le tenian de Recibir En el / dho lugar ni le darian de comer para la dha su gente si no lo / pagaba muy bien y que esto le oyo y no otra cosa lo qual Es ver/dad y lo que save para el juramento de hiço y no lo firmo por/que dixo que no sabia escrivir y que Es de hedad de beynte y / quatro años poco mas o menos Juan de la vandera escrivano /

{T[estig]o} Otro si Este dho dia diez y siete dias del dho mes de otubre del / dho año de mill E quinientos y sesenta y siete años yo El / dho Juan de la vandera escrivano para mas zertificacion / de lo que dho Es tome E Rescibi juramento En forma de/vida de derecho de Juan perez de ponte de lima soldado de la / dha compania El qual abiendo jurado En forma E siendo pre/guntado dixo que lo que save es que estando este testigo El / domingo En la noche proximo pasado que se contaron diez / y nueve dias del presente mes en un lugar que se disze chiaha / a donde Este dia El dho señor capitan llego con la dha su / compania syendo de guardia haziendo la primera llego a el / un yndio de los del dho lugar y començo a hablar con este t[estig]o FOLIO 14 de suerte que le entendio que satapo que tenia dos muchachos / Ascundidos[36] Españoles que chafahane que quiere deszir q[ue] / Es vellaco y que El y cosa aguacamu[37] [slash] E ypeape[38] [slash] que quiere / dezir que lo abian de matar y esto Es lo que sabe y en/tendio y no otra cosa para El juramento de hiço y no fir/mo porque dixo que no sabia y dixo ser de hedad de beynte / y ocho años poco mas o menos Juan de la vandera escriv[an]o /

otro si yo El dho escrivano doy fee que conforme lo que los dhos / testigos an dho Es cosa cierta la causa porque El dho señor / capitan dio buelta desde el dho lugar de satapo y por/que la dha noche que el dho señor capitan Estubo En el dho lugar / con la dha su compania yo El dho escrivano Estube de guardia / con un Esquadra como cabo de Esquadra de la dha conpania y bide / la junta de yndios que estaba fuera del dho lugar andando / Rondando por el y sigun la dha Junta que otro dia siguiente / no parescio ningun yndio en el dho lugar digo y doy fee que lo que / El dho señor capitan me pide Es cosa cierta y pasa asi como se / me pide y ansi yo El dho escrivano se lo di por testimonio syendo testigos / al ver pedir y dar del El señor alverto Escudero de villamar / alferez y hernando moyano de morales sargento y pedro de / hermosa y pedro de olibares soldados de la dha conpania / Juan de la vandera escrivano /

otro si en este dho dia diez y siete dias del dho mes de otubre / del dho año de mill E quinientos E sesenta E siete años / El dho señor capitan Juan pardo pidio y mando a mi El dho / {ojo} escribano asentase como [inserted above line: ante] su m[erce]d bino guillermo Rufin / lengua y con el un yndio natural de satapo los quales / hablando de algunas cosas tocantes a la jornada que el dho / señor capitan yba a hazer El dho yndio por la dha lengua / dio a entender que harto mejor se haria la jornada yendo / por un rrio que pasaba por Junto a olameco que por otro[39] FO-LIO 14 VERSO nonbre se llama chiaha [slash] que no yendo por aquel camino derecho / a cosa porque mediante esto y las demas cosas arriba de/claradas su m[erce]d daba la buelta que Estaba acordado que / diese [slash] y yo El dho escrivano lo escribi como me mando y lo firme y a/simismo lo firmo el dho señor capitan y el dho guillermo / Rufin lengua [slash] Juan pardo [slash] guillermo Rufin [slash] Juan de la / vandera escrivano /

Despues de lo suso dho en veynte dias del dho mes de otubre / del dho año de mill E quinientos E sesenta E siete / años en presencia de mi Juan de la vandera escrivano El dho / señor capitan prosiguiendo En el dar de la dha buelta con la / dha su conpania llego al dho lugar llamado olamico que por / otro nonbre se llama chiaha [slash] en el qual luego como llego Este dia / dio traca en el hazer de el fuerte que estaba acordado que se hi/ziese en el dho olameco y asi se començo a hazer luego y en fin / de quatro dias estubo acabado y como se acabo El dho señor / capitan dexo en el dho fuerte a marcos ximenez cabo de Es/quadra de su compania y con el veynte E cinco honbres soldados / de la dha compania para que estubiesen y guardasen El dho fuer/te y encomendo y mando al dho marcos ximenez tubiese / quenta con lo que al servi[ci]o de su magestad se debia y que no saliese / ni se apartase del dho fuerte sy no fuese con licencia de su mag. / o de quien en su nonbre se lo mandase y para mas siguridad / de lo suso dho en presencia de mi el dho escrivano le tomo Ju/ramento solemne sobre que guardaria lo suso dho y el dho / marcos ximenez absolbiendo al dho juramento prome/tio de lo ansi cumplir so pena de perjuro E ynfame y de / caher en caso de menos valer y por el dho señor capitan / visto lo suso dho entendiendo como entendio que hera / menester que en el dho fuerte dexase alguna municion / para defensa de la dha gente syendo nescesario entrego y dio / al dho marcos ximenez una hacha de ojo bizcayna y un   FOLIO 15 azadon y una pala y sesenta libras de polbora y cinquenta l[ib]ras / de cuerda y ochenta y cinco libras de balas de que yo El dho escrivano / doy fee y asimismo doy fee de como en mi presencia El dho / señor capitan por dexar gratos y conthentos al cacique y / mandador del dho lugar demas de otras herramientas y cosas / que les abia dado este dho dia dio al dho cacique una hacha biz/cayna grande y al dho mandador y a un yndio principal del / dho pueblo una cuña grande a cada uno y al dho marcos ximenez / cabo de Esquadra dos dozenas de escoplos y cuchillos para que tu/biese con que contentar a algunos yndios que seria me/nester a todo lo qual que dho es fueron presentes por testigos / El señor alverto Escudero de Villamar alferez y hernando / moyano de morales sargento y pedro de hermosa y pedro / de olivares soldados de la dha compania [slash] Juan pardo [slash] gui/ llermo Rufin [slash] Juan de la vandera escrivano /

Despues de lo suso dho en veynte y dos dias del dho mes de / {#} otubre del dho año de mill E quinientos y sesenta / y siete años El dho señor capitan Juan pardo prosiguiendo / en el dar de la dha buelta este dho dia salio con la dha su con/pania del dho lugar llamado olameco y en veynte E siete / dias del dho mes llego a un lugar que se disze *cauchi* y en el / luego como llego dio traça de hazer un fuerte al qual puso / por nonbre san pablo El qual dio horden y mando que se hi/ziese Respeto de mejor conserbar El amistad de los / yndios de la dha tierra el qual se hiço y acabo a treynta dias / del dho mes de otubre y por el dho señor capitan visto / entendiendo como entendio que al servicio de su mag. / conbenia que en el dho fuerte dexase una dozena de / honbres de los soldados de su conpania y con ellos una per/sona que los rrixiese governase y administrase en las cosas   FOLIO 15 VERSO que al s[er]vicio de dios y de su mag. conbenian y para que tanbien / conserbase El amistad de los dhos yndios y asi dexo a / pedro flores cabo desquadra de la dha su conpania y con el a diez / soldados della a los quales su m[erce]d les mando que obedesziesen / y tubiesen por superior de alli adelante al dho pedro flo/res [slash] Y al dho pedro flores le encomendo y mando tubiese gran / quenta y cuydado de guardar El dho fuerte y de no salir del / syn licencia de su magestad o de quien en su nonbre se le man/dase y tanbien con el gobierno de los dhos soldados y conserva/cion del amistad de los dhos yndios y para todo Esto le dio su / poder cumplido bastante como su m[erce]d de su magestad lo / tiene y despues de aver se lo dado en presencia de mi el dho / escrivano sobre una señal de cruz que hiço con su mano de/recha tomo E Recibio juramento por dios E por santa ma/ria sigun derecho al dho pedro flores cabo de Esquadra sobre / que fiel y lealmente guardaria el dho fuerte y cum/pliria las cosas que le heran mandadas y no las quebrantaria / ni saldria del dho fuerte syn licencia de su magestad o de / quien en su nonbre se lo mandase y el dho p[edr]o flores absol/biendo al dho juramento dixo si juro e amen E prome/tio de lo ansi cumplir so pena de perjuro E ynfame E de / caher en caso de menos baler y por el dho señor capitan visto / lo suso dho y entendiendo como entendio que El dho pedro flo/rez para el servicio del dho fuerte tenia nescesidad / de algunas municiones por si fuesen menester le dexo / treynta E seis libras de balas de plomo y beynte E dos l[ibr]as / de polbora y veynte E quatro libras de cuerda y un hacha / bizcayna de ojo de que yo El dho escrivano doy fee y asimesmo / le dexo dos dozenas de escoplos y cuchillos para que diese / a algunos yndios que seria menester a lo qual que dho Es / fueron presentes por testigos El señor alverto Escudero de   FOLIO 16 villamar alferez y hernando moyano de morales sar/gento y pedro de hermosa y pedro de olivares sol/dados de la dha conpania Juan pardo guillermo Rufin / Juan de la vandera escrivano /

otro si en veynte y siete dias del dho mes de otubre del / dho año de mill E quinientos y sesenta y siete años / que es quando el dho señor Capitan llego de buelta / al dho lugar llamado cauchi dio al mandador del dho lugar / y a un hermano suyo porque El cacique no Estaba alli / una cuña grande y una gargantilla y un Escoplo con que / estubieron muy conthentos y a un cacique que hera de / otro lugar que se deszia canos-a-Aqui[40] orata una cuña / y un escoplo y una gargantilla con que ansimismo Es/tubo muy conthento [slash] y luego En veynte E nuebe dias / del dho mes Estando El dho señor capitan con la dha su / conpania En el dho cauchi binieron a le ver y bisitar / cinco [sic] siete] caciques que se llamaban utahaque orata [slash] anduque / orata Enjuete orata [slash] guanuguaca[41] [slash] orata [slash] tucahe orata / guaruruquete orata [slash] anxuete orata [slash] a los quales / por su m[erce]d visto la ubidiencia con que binieron les dio cinco Escoplos y dos cuñas y dos gargantillas y un pedaço de tafetan / verde y se les dixo por guillermo Rufin ynterpetre / y se les declaro El parlamento acostumbrado los quales / todos hizieron El ya [sic] En que dieron a entender que estaban / prestos de lo cumplir de que yo El dho escrivano doy fee a lo / qual que dho Es fueron presentes por testigos El señor alverto / Escudero de villamar alferez y hernando moyano sargento / y pedro de hermosa y pedro de olivares soldados de la / dha conpania Juan pardo [slash] guillermo Rufin [slash] Juan / de la vandera escrivano /

FOLIO 16 VERSO Despues de lo suso dho En primero dia del mes de nob[iemb]re / del dho año de mill E quinientos E sesenta E siete / años El dho señor capitan juan pardo prosiguiendo / en el dar de la dha buelta llego a un lugar que se disze / tocahe[42] en donde el dho señor capitan fue bien rrecibido / por el mandador y principales del dho pueblo a los quales / El dho señor capitan dio dos escoplos y dos gargantillas y un pedaço de tafetan Roxo con que los suso dhos Es/tubieron muy conthentos y otro dia siguiente es/tando El dho señor capitan descansando con la dha su / conpania En el dho tocahee [sic] bino alli cauchi orata / que havia llebado ciertas yndias cautibas a Joara por / mandado del dho señor capitan al qual su m[erce]d dio / una cuña grande con que El dho cacique estubo muy con/tento y se bolbio al dho su lugar de que yo El dho escrivano / doy fee a lo qual que dho Es fueron presentes por testigos / El señor alverto Escudero de villamar alferez y her/nando moyano sargento y pedro de hermosa y pedro de / olivares soldados de la dha conpania [slash] Juan pardo gui/llermo Rufin Juan de la vandera escrivano /

despues de lo suso dho En veynte E nuebe dias del dho / mes de otubre del dho año de mill E quinientos / y sesenta E siete años que es quando El dho señor ca/pitan Estubo En el lugar llamado cauchi en presencia / de mi Juan de la vandera

The "Long" Bandera Relation   229

escrivano y de los testigos di yuso / esc[ri]ptos binieron antel dho señor capitan [slash] cauchi orata / canasahaqui[43] orata [slash] arasue orata [slash] guarero orata / Joara chiquito orata [slash] arande orata [slash] y dos ynahaes / {*ynihaEs / son como si / dixesemos / Just[ici]a o Jura/dos q[ue] mandan / al pueblo*} oratas y por guillermo Rufin ynterpetree [*sic*] dixeron y / declararon que El dho señor capitan les declarase   FOLIO 17 a donde mandaba que Acudiesen con el tributo / que Estaban obligados por el dho señor capitan enten/dido por el dho guillermo Rufin les declaro y dixo que / ya sabian como su m[erce]d dexaba alli un honbre principal / de los de su conpania con otros diez soldados della que / tubiesen cuydado de que con las cosas de comer que Es/taban obligados de acudir a su m[erce]d que acudiesen desde / oy dho dia en adelante al dho honbre principal y sol/dados que dexaba En el dho cauchi y que en quanto a lo / que tocaba a camucas[44] que acudiesen a Joara a otro / guançamu que hera hermano suyo que Estaba alli por/que desto se serbiria su magestad y por los dhos caciques / entendido lo suso dho todos juntos hizieron El ya [*sic*] / de que yo El dho escrivano doy fee a lo qual que dho Es fueron / presentes por testigos el señor alverto Escudero de villa/mar alferez y hernando moyano sargento y pedro de / hermosa y pedro de olivares soldados de la dha conpania /

despues de lo suso dho En tres dias del mes de nob[iemb]re / del dho año de mill E quinientos y sesenta E siete años / en presencia de mi Juan de la vandera escrivano y de los / testigos di yuso esc[ri]ptos El dho señor capitan Juan pardo pro/siguiendo en el dar de la dha buelta Este dia se partio del / lugar de tocaE y llego con la dha su compania cinco leguas / del dho lugar en campaña do hiço alto y durmio y estando / alli binieron cinco caciques de los derredores a traher bas/timento a la dha compania a los quales El dho señor ca/pitan dio cinco cuñas pequeñas con que fueron muy con/tentos y no se les dixo azerca del rreconocimiento Cosa / ninguna porque antes deste dia se les dixo y estan con el / mismo Reconocimiento que otros fueron testigos el señor / alverto Escudero de billamar alferez y hernando moyano / sargento y pedro de hermosa y pedro de olibares soldados   FOLIO 17 VERSO de la dha conpania Juan pardo Juan de la vandera escrivano /

Despues de lo suso dho En presencia de mi Juan de la / vandera escrivano y de los testigos di yuso esc[ri]ptos El dho se[eñ]or / capitan prosiguiendo En el dar de la dha buelta En qua/tro dias del dho mes de nobienbre del dho año de mill E qui[niento]s / y sesenta E siete años partio con la dha su conpania de la / parte que la noche pasada abia dormido y este dia llego otras / cinco leguas mas adelante junto a un arroyo en una ca/ñada a do hiço alto y durmio y estando alli bino un cacique / que se disze atuqui El qual traxo bastimento para / la dha conpania y por el dho señor capitan visto le dio / un pedaço de tafetan verde con que Estubo muy con/tento a lo qual que dho Es fueron testigos El señor / alverto Escudero de billamar alferez y hernando mo/yano sargento y pedro de hermosa y pedro de olibares / soldados de la dha conpania Juan pardo juan de la / vandera escrivano /

despues de lo suso dho En presencia de mi Juan de la van/dera escrivano y de los testigos di yuso esc[ri]ptos El dho señor / capitan Juan pardeo prosiguiendo en el dar de la dha / buelta En cinco dias del dho mes de nobienbre del dho año / de mill E quinientos E sesenta E siete años salio con la / dha su conpania de la p[ar]te y lugar donde abia dormido la noche / pasada y este dia llego con la dha su conpania otras quatro / leguas mas adelante donde hiço alto y aquella noche / durmio la dha conpania y estando alli bino joara mico ca/cique a traher vastimento fueron testigos El señor / alverto Escudero de billamar alferez y hernando moyano / sargento y pedro de hermosa y pedro de olivares sol/dados de la dha conpania Juan Pardo Juan de la vandera escriv[an]o /

despues de lo suso dho En presencia de mi Juan de la / vandera escrivano y de los testigos di yuso esc[ri]ptos El dho señor FOLIO 18 capitan Juan pardo prosiguiendo En el dar de la dha buelta / en seis dias del mes de nobienbre del dho año de mill / E quinientos E sesenta E siete años llego con la dha / su conpania a la ciudad de quenca y fuerte de san Juan / que en lengua yndiana se dize Joara a donde hiço alto / y estubo veynte dias Respeto de que la gente de la dha / su conpania benia cansada y mal Reparada porque tu/biese lugar de descansar y se rreparar y este dia El dho s[eñ]or / capitan dio a un yndio principal de la dha ciudad hijo de / la cacica bieja della una cuña grande y a la dha cacica un Es/coplo y un Espejo con que los suso dhos Estubieron muy con/tentos y estando En la dha ciudad en diez y seis dias del / dho mes de nobienbre del dho año binieron a Ella a hazer / El ya [sic] que es ubidiencia cinco caciques que se llamaban / quinahaqui orata los dos catapes oratas [slash] guaquira⁴⁵ / orata [slash] y ysa chiquito orata con algunos yndios / principales los quales traxeron que comer a la dha / gente y por el dho señor Capitan visto les dio siete / cuñas pequeñas y tres gargantillas y unos pedaços de / paño Roxo con que los suso dhos fueron muy contentos / y estando como Es dho en la dha ciudad en veynte dias del / dho mes de nobienbre del dho año binieron a Ella diez y ocho / caciques a hazer El ya [sic] que Es ubidiencia como antes de / agora lo suelen hazer y a traher mayz y otros bastimentos / para la dha gente que se llamaban atuqui orata osu/guen⁴⁶ orata [slash] y sus mandadores [slash] y aubesan⁴⁷ orata [slash] guenpu/ret orata [slash] ustehuque orata [slash] Tres pundahaques oratas / tocaE orata [slash] guanbuca orata [slash] ansuhet orata [slash] guaruruq[ue]t⁴⁸ / orata [slash] Enxuete orata [slash] utahaque orata [slash] y anduque Orata / xueca orata [slash] qunaha orata [slash] vastu orata [slash] y juntamente / con ellos Joara mico cacique de la dha ciudad a los quales / El dho senor capitan por los mejor atraher al s[er]vicio de FOLIO 18 VERSO Su magestad y los tiener mas conthentos les dio a Joara / mico como a mas principal una cuña grande y unos pedaços / de paño colorado y a otros cinco caciques de los suso dhos / tambien principales A cada uno una cuña y a los demas / caciques y sus mandadores diez y seys Escoplos con que / los suso dhos fueron muy conthentos que de todo lo suso / dho yo El dho escrivano doy fee a lo qual que dho Es fueron pre/sentes por testigos El señor alverto Escudero de billamar / alferez de la dha conpania y pedro de hermosa y pedro de / olivares y luis ximenez soldados de la dha conpania Ju[an] / pardo guillermo Rufin [slash] Ju[an] de la vand[er]a escrivano /

Despues de lo suso dho En veynte y quatro dias del dho / mes de nobienbre del dho año de mill E quinien/tos E sesenta E siete años En presencia de mi Juan de / la vandera escrivano y de los testigos di yuso esc[ri]ptos estando En / la ciudad de quenca y fuerte de San Juan que En lengua / yndiana se dieze Joara El señor Capitan Juan pardo con / su conpania por su m[erce]d considerado y visto que se abia / de yr de la dha ciudad con la dha su compania y que en el / dho fuerte de san juan hera menester que quedasen numero / de gente para siguridad de la dha tierra y en su propio / lugar al señor alverto Escudero de billamar alferez para q[ue] / como tal se juzgase y tubiese quenta con conservar / El amistad de los caciques E yndios de toda la tierra / y para que bisitase y governase los fuertes y alcaydes / y soldados dellos que estan desde el lugar llamado / chiaha hasta la dha ciudad de quenca y desde la dha ciudad / de quenca hasta la punta de santa helena que son en el / lugar de chiaha el fuerte de san pedro que Esta cinquenta / leguas de la dha ciudad en la parte del puniente y en el / lugar de cauchi El fuerte de san pablo que Esta desde la FOLIO 19 dha ciudad veynte y ocho leguas Entre norte y puniente / y desta parte de la dha ciudad En el lugar llamado / guatari El fuerte de santiago que esta quarenta / leguas della Entre norte y lebante y desde guatari / al lugar llamado canos El fuerte de santo tomas que / Esta del dho guatari quarenta y cinco leguas al lebante / Entre el lebante y el sur y desde el dho lugar lla/mado canos hasta la punta de santa helena cinquenta / y cinco leguas al sur y con la consideracion suso dha En pre/sencia de mi El dho escrivano encargo y mando al dho señor / alverto Escudero de villamar alferez que tubiese / gran quenta y cuydado con lo suso dho pues sabia y en/tendia quanto conbenia al servicio de su magestad / y para ello le dio su poder Cumplido bastante aquel / que el dho señor Capitan de su magestad tiene y el / dho señor alferez prometio de lo ansi cumplir y lo fir/maron de sus nonbres a lo qual que dho Es fueron pre/sentes por testigos hernando moyano de morales sar/gento de la dha conpania y pedro de hermosa y pedro de / olibares y luis Ximenez soldados de la dha conpania [slash] Juan / pardo alverto Escudero de billamar Ju[an] de la vand[e]ra escriv[a]no /

otro si yo El dho escrivano doy fee como en mi presencia y de / los testigos di yuso esc[ri]ptos el dho senor capitan dexo al s[eñ]or / alverto Escudero de billamar alferez para las cosas de / que tenga nescesidad En nonbre de su magestad / las cosas siguientes

    primeramente sesenta y ocho libras de cuerda /
    mas cien libras de plomo en balas /
    mas ochenta y cinco libras de polbora /
    mas treynta y quatro libras de clabos /
    mas quarenta y dos azolejos[49] a manera de escoplos /
    mas tres palas de hierro /
    mas dos hachas de ojo / FOLIO 19 VERSO
    mas quatro Cuñas de hierro [slash] y dos azadones y dos picos /

otro si este dho dia veynte e quatro dias del dho mes / de nobienbre del dho año de mill E quinientos E / sesenta E siete años ante el dho señor Capitan / y en presencia de mi El dho escrivano binieron a la dha / ciudad dos caciques que se llamaban [slash] chara [slash] orata [slash] / y adini orata [slash] a traher bastimento para la con/pania los quales solian estar subjetos a guatari y a/gora querian benir a hazer el ya [sic] a Joara mando les el dho / señor capitan que fuesen a do solian y para Esto por los / contentar les dio tres Escoplos con que fueron muy con/tentos y prometieron de cumplir lo que El dho señor ca/pitan les mando fueron testigos hernando moyano de / morales sargento y pedro de hermosa y pedro de olivares / soldados de la dha compania de todo Esto yo El dho escrivano / doy fee Juan pardo Juan de la vandera escrivano /

Despues de lo suso dho en presencia de mi Juan de la Van/dera Scrivano El dho señor capitan Juan pardo prosigui/endo en el dar de la dha buelta en veynte E quatro dias del / mes de nobienbre del dho año de mill E quinientos E / sesenta E siete años salio con su conpania de la ciudad de / quenca que en lengua yndiana se disze Joara y este dia En / cierta parte despoblada hiço alto y durmio aquela noche / y otro dia siguiente llego mas adelante cinco leguas de / alli a un lugarejo que es de un mandador de yssa qe se / disze dudca [slash] y alli hico alto y durmio aquella noche y otro / dia siguiente que se contaron veynte E seys dias del dho / mes yendo marchando con la dha su conpania derecho a ysaa / cantidad de un quarto de legua mas adelante paso por / un lado del camino que traya a la mano yzquierda y o/tro poco mas adelante a la mano derecha por donde le guio / hernando moyano de morales sargento de su conpania / y andres xuarez platero que por mandado del dho señor FOLIO 20 *capitan en siete dias deste dho mes abian benido por / El dho camino A rreconozer una mina de christal que / abia En el y en presencia de mi El dho escrivano mostra/ron al dho señor capitan el principio de la dha mina / la qual yo El dho escrivano bi como Estaba començada a cabar / de que doy fee y vista como Esta dho El dho hernando / moyano de morales sargento dixo que El como per/sona que antes de agora abia visto la dha mina la rre/gistraba y rregistro y se obligaba y se obligo con licencia / de su mag. de beneficiar y labrar la dha mina y della a/cudir conforme las leyes que su mag. tiene puestas E yn/dias con la parte que es obligado y desde agora pide a / su mag. o a quien en su nonbre lo mandare que cada E quando / ubiere Repartimiento de la dha tierra le adquiera la / dha mina de christal porque el como tiene dho la beneficiara / y labrara y acudira a su mag. con lo que Es obligado juntam[en]te / con el yll[ust]re señor Esteban de las Alas governador en las / probincias de la florida y el dho señor Capitan Ju[an] pardo q[ue] / como tiene dho En siete dias deste dho mes Estando En / la dha ciudad de quenca aloxado con la dha su conpania a / le mando que biniese al dho lugar de dudca a Reconozer / la dha mina y en ocho dias del dho mes se bolbio a la dha ciudad / de quenca con algunas piedras del dho cristal que abia / sacado con un pico de azadon de la dha mina y las mostro y / dio al dho señor Capitan y este dia hiço El dho rregistro ante / mi El dho escrivano de que doy fee y [a] el dho andres xuarez pla/tero y a mi el dho escrivano y a el señor alverto Escudero de billa/mar alferez del dho señor Capitan juan pardo que todos / cinco Es su boluntad y quiere que sean* y bayan a la p[ar]te / de la dha mina ygualmente con el y que todos seys / cada uno de por si ansi El uno como El otro y el otro como El / otro lleben y ayan la parte que ubiere por yguales partes / porque con mas fuerca bigor y brebedad se labre y beneficie la   FOLIO 20 VERSO dha mina y para que cumplira asi lo uno como lo otro que / en esta esc[ri]ptura se contiene obligo su persona e bienes muebles / E Rayzes los que tiene y abia y para que no lo quebrantara / ni contradira agora ni en tiempo alguno y aunque lo quiera / contradeszir no le balga Renuncio todas E qualesquier / leyes fueros y derechos que en su fabor sean o ser puedan p[ar]a / que no le balgan ni aprobechen en cosa alguna y lo firmo de su / nonbre syendo presentes por testigos a todo lo que Es dho / pedro de hermosa y francisco corredor y francisco hernan/dez de tenbleque y luis ximenez soldados de la dha con/pania fernando moyano de morales Ju[an] de la bandera scriv[an]o /

despues de lo suso dho Este dho dia veynte E seis dias del / dho mes de nobienbre del dho año de mill E quinien/tos E sesenta E siete años yendo marchando El dho señor / capitan Juan pardo con la dha su conpania en presencia de / mi el dho Juan de la vandera escrivano ya que yria un quar/*to de legua mas adelante de la mina que arriba Esta* / *declarada a la mano derecha del dho camino hallo otras* / piedras de christal que hera sigun parescio El sytio donde / Estaban principio de otra mina y como la hallo parescio ante / mi El dho escrivano y me mostro las dhas piedras E sitio y en / mi presencia y de los testigos di yuso esc[ri]ptos dixo que la rre/gistraba E Registro y hera su boluntad de meter y me/tio a la parte de la dha mina al señor Esteban de las Alas / governador de las probincias de la florida y a alverto Escudero / de billamar su alferez y hernando moyano de morales su / sargento y a andres xuarez platero y a mi el dho Juan de / la vandera escrivano para que todos cinco como Estan non/brados y declarados juntamente con el ansi el uno como El / otro y de otro como El otro lleben y ayan tanta parte / como El y le ayuden a beneficiar y labrar la dha mina cada / y quando que por su magestad o por quien en su nonbre le   FOLIO 21 fuere mandado para que todos seys como Estan nonbrados y de/clarados acudan y den a su mag. la parte que conforme a las / leyes y hordenancas de las yndias le cupiere y para que cum/plira lo que En esta esc[ri]ptura se trata y contiene dixo / que obligaba y obligo su persona E bienes muebles E rayzes / avidos E por aver y renunciaba y Renuncio quales/quier leyes que sean en su fabor para que no le balgan ni a-/ {#} probechen en ningun tiempo y lo firmo de su nonbre / syendo testigos los dhos pedro de hermosa y fran[cis]co corredor / y fran[cis]co hernandez de tenbleque y luis ximenez sol/dados de la dha conpania Ju[an] pardo [slash] Ju[an] de la vandera escrivano /

despues de lo suso dho En presencia de mi Juan de la / vandera escrivano el dho señor capitan Juan pardo / prosiguiendo En el dar de la dha buelta en veynte y ocho / dias del dho mes de nobienbre del dho año de mill E qui/nientos y sesenta E siete años llego con la dha su con/pania a un lugar que se disze yssa En el qual fue bien / rrescibido y luego como llego se aloxo en una casa grande / nueba de madera y tierra que el cacique del dho lugar / abia hecho para el servicio de su mag. como antes de / agora por el dho señor Capitan se le abia mandado y de/mas desto otro dia siguiente binieron de su lugar / que se disze guatarimico dos yndios de p[ar]te del dho ca/cique del dho lugar a los quales El dho señor Capitan / Recibio bien y les dio dos hachas de ojo que llebasen al dho ca/cique para que hiziese cortar numero de madera para / un fuerte que El dho señor Capitan abia de hazer En el / dho lugar y estando en el dho lugar de yssa En cinco dias / del mes de dizienbre del dho año de mill E quinientos / y sesenta E siete años binieron al dho lugar [slash] subsaquibi [slash] / orata cacique y con el algunos yndios a traher bastimento / para la dha conpania y por el dho señor capitan visto se le   FOLIO 21 VERSO hico El parlamento acostumbrado por guillermo Rufin yn/terpetre y por lo tener mas grato y contento le dio un / cuchillo grande y un Espejo con que fue muy contento / y asimismo dio a un yndio que este dia bino al dho lugar / porque es lengua de guatari un Escoplo y un gargan/tilla y otro dia siguiente bino al dho lugar El man/dador de cauchi con ciertas cartas para El dho señor / capitan al qual su m[erce]d dio un cuchillo grande con que / fue muy conthento otro si En nuebe dias del dho / mes de dizienbre del dho ano El dho señor Capitan / dio a yssa cacique y a sus mandadores una hacha y dos / cuchillos grandes y dos Escoplos y quatro gargantillas / Respeto de que estubo en el dho lugar con la dha su con/pania doze dias y el dho cacique quedo muy conthento / de que yo El dho escrivano doy fee fueron testigos pedro / de hermosa sargento y lucas de canizares y luis perez [?] soldados de la dha con pania [slash] Juan pardo [slash] guillermo / Rufin Juan de la vandera escrivano /

despues de lo suso dho Este dho dia nuebe dias del mes de / dizienbre del dho año de mill E quinientos E sesenta / y siete años ante el muy magnifico señor Capitan Juan / pardo y en presencia de mi Juan de la vandera escrivano y de los / testigos di yuso esc[ri]pto parescio andres xuarez platero y dixo / que pedia E pidio a su m[erce]d como capitan que Es de su mag. / y lugar theniente del Ill[ustr]e señor Esteban de las alas governador / y theniente de capitan general En las probincias de la / florida mande que ninguna ni alguna persona sea osada / so graves penas que para ellos su m[erce]d ponga de cabar en / termino de media legua y mas en torno como bienen de / un lugar que se disze dudca un quarto de legua mas adelante / derecho a un lugar que se disze ysa a donde su m[erce]d biniendo / marchando con la dha su conpania hallo cierta señal y piedras / de mina de christal y hernando moyano de morales sar-⁵⁰   FOLIO 22 gento de su compania porque el dho andres xuarez fue a / la p[ar]te de las dhas minas con el dho señor Capitan y el dho sar/gento y al dho andres xuarez y sus companeros que / son p[ar]te En las dhas minas les conbiene lo ques a Su m[erce]d pide / hasta tanto que su mag. otra cosa mande y lo pidio / por testim[oni]o y a los presentes que dello El sean testigos / E ymploro El oficio del dho señor capitan andres / xuarez Juan de la vandera escriv[an]o /

despues de lo suso dho Este dho dia mes E año dho antel dho / señor Capitan parescio hernando moyano de morales E dixo / que hace y hiço el mismo Requerimiento aRiba contenido / E pidia E pidio lo mismo que El dho andres xuarez platero / pide y que no se llege a cabar ni labrar En termino de / la dha media legua de tierra Entorno de lo que Esta dho / syn consentimiento de todas las personas que son p[ar]te En / las dhas minas y pide lo y en lo nescesario ymploro El oficio del / dho señor capitan fernando moyano de morales juan de la / vandera escrivano /

 Luego yncontinente este dho dia mes E año suso dho por el dho s[eñ]or / capitan visto lo pedido por los dhos hernando moyano de / morales sargento y andres xuarez platero y que lo que pi/den es sin perJuicio de p[ar]te alguna syno antes cosa que / conbiene al servicio de Su mag. dixo que mandaba e mando / que ninguna ni alguna persona sea osado de cabar ni labrar / en las dhas minas ni en el termino de la dha media legua / de tierra Entorno que se le pide syn mandato E con-sen/timiento de todas las personas que son a la p[ar]te de las / dhas minas so pena de la bida fueron testigos a lo suso dho fran[cis]co / hernandez de tenbleque y gaspar Rodriguez y lucas de / canizares soldados de la dha conpania [slash] Juan pardo Juan de / la vandera escrivano /

 despues de lo suso dho diez dias del dho mes de dizienbre del / dho año de mill e quinientos E sesenta y siete años FOLIO 22 VERSO El señor Capitan Juan pardo estando en un lugar que se llama / yssa fue ynformado que una legua larga del dho lugar / {*ojo*} al rrio abaxo como salen del dho lugar de la otra parte del / dho Rio en un alto como quatrocientos o quinientos / pasos del dho rrio zerca de donde bibe mati yssa [slash] que es / mandador de yssa una quarto de legua como ban de yssa antes / {*ojo*} que se llegue al un Rio del dho mati abia ciertas piedras / que demostraban ser christal y para saver la verdad / de lo suso dho mando a mi El escriv[an]o di yuso esc[ri]pto que Junta/mente con hernando moyano de morales que fue sargento / de la dha conpania y con lucas de Canizares soldado della / fuese y por bista de ojos Entendiesen y biesen si hera / mina de christal y en Cumpli-mento de lo suso dho yo El dho / escrivano Juntamente con los suso dhos Este dho dia fue/*mos a ver y bimos El sitio y lugar donde Estaban las dhas / piedras y entendimos y bimos como hera christal* y se/ñalamos el dho lugar y otro dia siguiente bolbimos al / dho lugar de yssa y dimos quenta de lo suso dho al / dho señor Capitan y por su m[erce]d Entendido lo suso dho / dixo que Registraba E registro la dha mina y lo mismo / dixo que hazia E hiço El dho lucas de canizares y hernando / moyano de morales esto con licencia de su magestad / y el dho señor Capitan y los suso dhos juntamente con el / dixeron que metian y metieron a la p[ar]te de la dha mina / para q[ue] todos juntamente ayan yguales p[ar]tes tanto El / uno como El otro y el otro como El otro al señor Esteban de / las alas lugar theniente de capitan general en las p[ro]bincias / de la florida y a El señor alv[er]to Escudero de billamar al/ferez y a pedro de hermosa sargento y a Juan de la vandera / caporal para que todos siete como Estan nonbrados y de/claclarados beneficien y labren la dha mina o quien poder / dellos o de qualq[uier] dellos ubiere y acudan y den a su mag. / la p[ar]te que Justa E liquidamente le biniera conforme las / leyes y hordenancas que su mag. tiene puestas En yndias[51] FOLIO 23 y para que cun-pliran y abran por firme lo que En esta esc[ri]ptura / se contiene cada uno por lo que le toca y es obligado / a cumplir obligaron sus personas E bienes muebles E Ra/yzes avidos E por aver los que an E abran y dieron / poder a las Justicias de su mag. para que les Compelan E a/premien al cumplimiento de lo suso dho bien ansi como / por cosa sentenciada E pasada En cosa juzgada En tes/timonio de lo qual otorgaron Esta carta antel / escrivano y testigos di yuso esc[ri]ptos testigos que fueron pre/sentes a lo que dho Es gaspar R[odrigue]s caporal y fran[cisc]o her/nandez de tenbleque y luis ximenez y Juan garcia de / madrid soldados de la dha compania y firmaron lo El dho s[eñ]or / capitan y los dhos hernando moyano de morales y lucas / de canizares En el registro Juan pardo hernando moyano / de morales lucas de Canizares Juan de la bandera escriv[an]o /

despues de lo suso dho Este dho dia mes E año dho en / presencia de mi El dho escrivano ante el dho señor Ca/pitan parescieron los dhos hernando moyano de / morales y lucas de canizares y dieron que [a] su derecho / E Justicia y al de sus conpaneros conbiene que a la mina / que an Registrado no se llegue ni se cabe en ella ni en/trecho de un quarto de llegua Entorno porque la dha / mina sigun la demuestra della Es de piedras diferentes / que otras de christal porque de de-muestran ser de otra / hechura Respeto de que son puntas afiladas y entabladas / por tanto que piden E Requieren al dho señor Capitan / so grabes penas mande que ninguna ni alguna persona / sea osado de llegar a la dha mina ni un quarto de legua / En torno della a cabar ni hazer otro beneficio ninguno / syn licencia E consenti-miento de todas las p[ar]tes y pi/dieron lo por testimonio y a los presentes que dello / les sean testigos fernando moyano de morales lucas / de canizares [slash] Juan de la vandera escrivano /

despues de lo suso dho Este dho dia e mes E año dho por el FOLIO 23 VER-so dho señor Capitan visto oydo y entendido lo pedido por / los dhos hernando moyano de morales y lucas de canizares / dixo que mandaba E mando que ninguna persona sea osado / de llegar a cabar ni a hazer otro ningun beneficio de la dha / mina en un quarto de legua Entorno della syn consen/timiento de todas las personas que son parte En la dha / mina so pena de muerte hasta tanto que por su mag. sea / probeydo otra cosa y asi mando E firmo lo Juan pardo Juan / de la vandera escrivano /

despues de lo suso dho en honze dias del dho mes de di/zienbre del dho año de mill E quinientos y sesenta / y siete años El dho señor Capitan en presencia de mi El / dho escrivano prosiguiendo En el dar de la dha buelta salio / del dho lugar llamado yssa y este dia llego con la dha su com/pania tres leguas de alli a otro lugar que se llamaba tan/bien yssa porque es del mismo yssa a donde fue bien Re/civido y alli dio El dho senor Capitan al mandador del dho / pueblo un escoplo con que fue muy conthento de que yo / El dho escrivano doy fee y otro dia siguiente que fueron doze / dias del dho mes de dizienbre del dho ano de mill E qui[nient]os / y sesenta E siete anos salio con la dha su compania del dho lu/gar llamado yssa y este dia prosiguiendo En el dar de / la dha buelta llego cinco leguas mas adelante a un lugar / que se llama quinahaqui a donde fue bien Recibido y otro / dia siguiente treze dias del dho mes descanso En el dho / lugar y dio al cacique del dho lugar y a otro que se llama / otari yatiqui[52] orata y a sus mandadores dos Escoplos / y quatro gargantillas con que estubieron muy Con/tentos y demas desto por guillermo Rufin ynter/petre se declaro E dixo al dho quinaqui [sic] y guaquiri / y yssa caciques porque a entendido que estan desconten/tos Respeto de que entienden que los dexa debaxo del / dominio de Joada mico que su m[erce]d no los dexa ni quiere / que esten debaxo del dho dominio syno que bayan a / hazer El yaa a su alferez que lo dexa En su lugar En el FOLIO 24 dho Joada y a el acudan con lo que estan obligados y no al dho / Joada y por los dhos caciques entendido lo suso dho hi/zieron muchas vezes El yaa [slash] En que en dieron a entender que El / dho señor Capitan les hiço muy gran m[erce]d y que estaban y / heran muy contentos de hazer lo que su m[erce]d les manda / de que yo El dho escrivano doy fee fueron testigos a todo lo / suso dho pedro de hermosa sargento y lucas de canizares / y luis ximenez soldados de la dha compania [slash] Juan pardo / guillermo Rufin [slash] Juan de la vandera escrivano /

despues de lo suso dho Em presencia de mi Juan de la van/dera escriv[an]o y de los testigos di yuso esc[ri]ptos El dho señor / capitan con la dha su compania prosiguiendo En el dar / de la dha buelta En catorze dias del mes de dizienbre del / dho año de mill E quinientos y sesenta y siete años salio / del lugar llamado quinahaqui la bia de guatari y este / dia andubo cinco leguas y en fin dellas que hera en un des/poblado hico alto y aquella noche durmio y otro dia si/guiente salio de la dha canpaña y este dia llego con la / dha su compania al dho lugar llamado guatari mico que / Es seis leguas del dho lugar donde durmio a donde fue bien / Recivido de las cacicas del dho lugar y luego como llego trato / con las dhas cacicas por guillermo Rufin ynterpetre / que mandasen venir al dho lugar a todos los caciques / sus basallos porque le ayudasen a hazer un fuerte que / abia de hazer En el dho lugar para dexar dentro del ciertos / Españoles que los defendiesen de sus enemigos y las dhas / cacicas hizieron El yaa dando a entender que heran muy con/tentas de lo ansi hazer y otro dia siguiente que se con / taron diez y seis dias del dho mes de dizienbre del dho año / de mill E quinientos y sesenta E siete años començaron / A benir al dho lugar algunos caciques y asi como fueron / benidos que fuen harto de mañana por el dho guillermo Ru/fin ynterpetre [above the line: El dho señor capitan] hico llamar a las dhas cacicas y por las / contentar les dio para ellas y dos hijos suyos y mandadores[53] FOLIO 24 VERSO y otros cinco caciques ocho cuñas y nuebe cuchillos grandes / y tres Escoplos y tres espejos y quatro gargantillas / y unos pedaços de tafetan Roxo con que estubieron / muy conthentos y luego yncontinente bino otro ca/cique al qual y a dos mandadores suyos El dho señor Ca/pitan dio otra cuña y a otro cacique que bino juntamente / con el un Escoplo y un cuchillo con que fueron muy conten/tos y este dia se dio traça En el hazer del dho fuerte y se / començaron a hazer dos bastardos con muy gran praesa[54] / Respeto de que si por p[ar]te de su mag. le biniese al dho señor / capitan mandato que se partiese para otra parte / que fuese mas nesces[ari]a porque los soldados que quedasen / en el dho lugar tubiesen defensa para sus Enemigos / sy algunos ubiese y asi se hizieron y acabaron en cinco dias / en fin de los quales bisto por el dho señor Capitan que no / avia avido cosa de nuebo por donde dexase de acabar del / todo El dho fuerte mando hazer llamamiento de todos / los yndios subjetos al dho lugar los quales se

junta/ron En tres dias y como fueron juntos juntamente / con los soldados de la dha
compania se dio horden en el / acabar del dho fuerte En el qual se hizieron quatro
ca/valleros altos de madera gruesa y tierra y su muralla / de paliçada alta y tierra
con quedo muy fuerte El qual / con la buena horden y priesa que se dio En el hazer
se acabo / En seis dias del mes de henero del año de mill E quinien/tos y sesenta y
ocho años y porque quedase Con la fuerca E / bigor que conbenia dexo por alcayde
del dho fuerte / A lucas de canizares cabo de Esquadra de la dha compania que / Es
persona En quien concuRen las calidades de derecho / y con el diez y seis soldados
de los de la dha compania a los quales Juntamente Con el dho alcayde Encargo E /
mando que no saliesen ni quebrantasen los bandos / que su m[erce]d En nonbre de
su mag. tiene mandado que se guar/den y cumplan En el dho fuerte y demas desto
considerado   FOLIO 25 y visto por el dho señor capitan que conbenia que al dho
lucas / de Canizares se le diese el poder que su m[erce]d de su mag. / tiene Em
presencia de mi Juan de la vandera escrivano y de los / testigos di yuso escriptos
dixo que le daba E dio su poder cumplido / y bastante como lo tiene y de derecho se
Requiere al dho / lucas de Canizares cabo de Esquadra Especialmente p[ar]a que /
por el y en su nonbre y como El mismo Representando / su propia persona Este E
Resida En el dho fuerte puesto / por nonbre señor santiago y el lugar donde queda
sytuado / En nuestra lengua castellana la ciudad de salamanca / y rrixa[55] E
gobierne los dhos diez y seis soldados a los quales m[an]do / que hagan E cumplan
lo que en servicio de su magestad con/venga y no lo cumpliendo o En alguna
manera descrepando / En poco o en mucho de lo que les mandare los pueda
apre/miar y castigar como los apremiara E castigara y los puede / apremiar y
castigar El dho señor Capitan conforme al delito / o delitos que qualquiera de los
dhos soldados cometiere y demas desto para que gobierne y tenga En par los
caciques / y yndios del dho lugar de guatari y los demas que Estan / subjetos a el de
manera que siempre Esten En buena quie/tud y sosiego y porque al servicio de su
magestad conbiene / que El dho lucas de canizares alcayde no salga del dho fuerte /
hasta tanto que por su m[erce]d o por quien en nonbre de / su magestad se lo
mandare le sea mandado El dho señor / {ojo} capitan En el dho nonbre del dho
señor Capitan le mando // / de su magestad le mando que asi lo cumpla y que no
sea osado / de noche ninguna meter muger alguna En el dho fuerte / y no salega
[sic] del dho mandato so pena que sera muy bien / castigado y asimismo mando y
encargo a los dhos diez E / seis soldados que cumpliesen y guardasen lo que por el /
dho lucas de canizares les fuese mandado porque a El / dexa su m[erce]d por su
superior que quan cumplido E bas/tante {ojo} poder El dho señor [above the line:
capitan] de su mag. dixo que para lo suso / dho tenia tal lo dio y otorgo al dho lucas
de canizares con / sus yncidencias E dependencias anexidades E conex-
idades[56]   FOLIO 25 VERSO e con libre E general administracion en lo que dho Es
que para lo aver / por firme obligo sus bienes y dio poder a las justicias de su mag. /
para que le compelan e apremien a lo que dho es como por cosa / sentenciada E
pasada en cosa Juzgada en testimonio de lo / qual ortogo Esta carta antel escrivano
y testigos di yuso / escriptos y lo firmo de su nonbre y asimismo lo firmo El dho

The "Long" Bandera Relation

lu/cas de canizares al qual su m[erce]d sobre una señal de Cruz que / hiço con su mano derecha le tomo E recivio juramento por / dios E por santa maria En forma de derecho y por las palabras / de los santos quatro Ebangelios sobre que cumpliria lo que / su m[erce]d En nonbre de su mag. le tiene mandado y por el / dho lucas de canizares Entendido lo suso dho absolbiendo al / dho Juramento dixo que asi lo cumpliria so pena de / perjuro E ynfame y de caher En caso de menos valer a lo qual / que dho Es fueron presentes por testigos pedro gutierrez pa/checo cabo de esquadra y pedro de hermosa sargento y luis / ximenez soldado de la dha conpania y yo El dho escrivano di / yuso esc[ri]pto doy fee que conozco al dho señor capitan otor/gante El qual considerando y advertiendo que hera / menester dexar municiones las que fuesen nescesarias / para la defensa del dho fuerte de santiago y soldados del / en presencia de mi El escrivano di yuso esc[ri]pto dio y entrego al dho / lucas de canizares cinquenta y una libras del plomo y tre/ynta y quatro libras de cuerda y treynta y quatro libras / de polbora y algunas cosas de hierro y otras diferentes p[ar]a / que contentase a los caciques que biniesen a la ubidien/cia de su mag. y dos hachas de ojo para las cosas nescesarias En el / dho fuerte de que yo El dho escrivano doy fee testigos los / dhos Juan pardo lucas de canizares guillermo Rufin / Juan de la vandera escrivano /

despues de lo suso dho en siete dias del mes de henero / del año de mill E quinientos y sesenta y ocho años en pre/sencia de mi Juan de la vandera escrivano y los testigos / di yuso esc[ri]ptos El dho señor Capitan Juan pardo con FOLIO 26 {X} la dha su compania prosiguiendo en dar la dha buelta / Este dia salio de la ciudad de salamanca que en lengua / yndiana se disze guatari la buelta de aracuchi y an/dubo con la dha su conpania cinco leguas y en fin de ellas / hico alto y durmio en un despoblado y otro dia siguiente / salio deste lugar donde abia dormido prosiguiendo su / camino adelante y andubo otras cinco leguas y en fin / dellas hico alto y durmio en campana y por esta horden / Andubo otros tres dias durmiendo los dos dellos en cam/pana con algun trabaxo Respeto de los pocos bastimen/tos que ubo hasta que en fin de los cinco dias llego al dho lu/gar llamado Aracuchi En donde por el cacique del dho / lugar fue bien rrecivido y asi hiço alto y durmio aque/lla noche y por el dho senor Capitan visto la buena bo/luntad con que fue Recivido dio al dho cacique y a sus / mandadores y a otros caciques que binieron al dho lugar / una cuña y una gargantilla y un poco de tafetan Roxo / y cinco cuchillos grandes y tres Escoplos a manera de / cuñas con que los dhos caciques E yndios principales / quedaron muy contentos fueron testigos pedro de her/mosa sargento y pedro gutierrez caporal y hernando / moyano de morales que fue sargento de la dha compania / y firmo lo El dho señor capitan y guillermo Rufin / ynterpetre y yo El dho escrivano Juan pardo guillermo / Rufin juan de la vandera escrivano /

despues de lo suso dho Este dho dia honze dias del dho mes / de henero del dho año de mill E quinientos y sesenta y o/cho años ante my El escrivano E testigos di yuso esc[ri]ptos parescio / El dho señor Capitan y dixo que por quanto al serbicio de / Su mag. conbiene que su m[erce]d con la dha su conpania no os/tante El Camino que lleba derecho a la ciudad y punta / de santa helena baya dando buelta algun trecho de / tierra hacia oriente hasta llegar a un lugar que se llama / ylasi Respeto de que alli Es ynformado que an de benir FOLIO 26 VERSO Algunos caciques a la ubidiencia de su mag. que antes de / agora no an benido y por ebitar a la dha su conpania de algun / trabaxo de la carga quiere ynbiar algunos soldados con algunos / yndios derecho a un lugar que se llama canos que lleban / ciertas municiones porque Es el camino derecho a la dha ciudad / de santa helena que pide a mi el dho escrivano lo asiente / asi y por bista de ojos bea los dhos soldados y las municiones / que enbia y por mi el escrivano di yuso esc[ri]pto entendido lo suso dho / bi como El dho señor capitan ynbio treynta y quatro libras / de balas y diez y siete libras de cuerda y un barrile de polbora de / arcabuz En que abia veynte libras y dos hachas de ojo y con estas / municiones los soldados siguientes /

A fernan gomez portugues a pedro alonso natural de xerez de / la frontera [slash] a luis ximenez natural de quenca [slash] a pedro / de barcena natural del moral de la Reyna [slash] los quales dhos / soldados como Esta dho yo El escrivano di yuso esc[ri]pto bi como se / partieron con las municiones declaradas derecho al dho lugar / llamado Canos de que doy fee testigos El señor pedro de hermosa / sargento y pedro gutierrez pacheco y Fran[cis]co hernandez de ten/bleque soldados de la dha conpania Ju[an] de la vandera escriv[an]o /

despues de lo suso dho En doze dias del dho mes de henero / del dho año de mill E quinientos y sesenta y ocho años / en presencia de mi Juan de la vandera escrivano y de los / testigos di yuso esc[ri]ptos El dho señor Capitan Juan pardo pro/siguiendo En hazer lo que al servicio de su mag. conbiene Este dia / con la dha su compania salio del dho aracuchi la buelta de ylasi / en donde se detubo andando de hordinario quatro leguas / cada dia cinco dias que en ellos se paso trabaxo Respeto del poco / bastimento que de hordinario ubo y en fin de los dhos cinco dias / llego con la dha su conpania al dho lugar llamado ylasi En donde / fue bien Recivido por el cacique E yndios del dho lugar En el / qual estubo con la dha su conpania quatro dias Respeto de / muchas aguas que llobio Estando En el y tambien por aguadar / como aguardo algunos caciques que binieron a la ubidiencia / de su mag. En los [quales] dhos quatro dias dio al cacique del dho lugar FOLIO 27 y a sus mandadores porque hallo que tenia hecha una casa gran/de para el servicio de su magestad y dos camaras altas con / alguna cantidad de mayz una hacha de ojo grande y dos gar/gantillas y un pedaço de paño Roxo y quatro Escoplos con que / quedaron muy contentos otro si estando En el dho lugar / abiendo benido con el tagaya orata con algunos / yndios le hiço cacique porque antes de agora no lo a/bia sido y rrespeto de los muchos serbicios que a hecho / a Españoles y cada dia haze le dio un Escoplo y una Cuña / con que quedo muy contheto otro si Estando / en el dho ylasi bino a El un cacique de la marina q[ue] / se deszia uca orata a la ubidiencia de su magestad / y saruti orata a lo mismo porque antes de agora / no abian benido a los quales por guillermo Rufin / ynterpetre se les hiço El parlamento acostumbrado y / por ellos entendido hizieron El yaa [slash] dando a entender / que heran muy conthentos de hazer E cumplir lo que / se les mandaba y por el dho señor Capitan visto y en/tendido su ubidiencia les dio a Ellos y a sus mandadores / dos cuñas y siete escoplos con [que] quedaron muy conten/tos fueron testigos a todo lo que dho Es pedro de her/mosa sargento y pedro gutierrez pacheco y fran[cis]co her/nandez de tenbleque soldados de la dha compania Juan / pardo guillermo Rufin [slash] Juan de la vandera escrivano /

despues de lo suso dho Em presencia de mi Juan de la / vandera escrivano y de los testigos di yuso esc[ri]ptos El dho / señor Capitan Juan pardo con la dha su compania prosi/guiendo en el dar de la dha buelta En veynte y uno dias del mes de henero del dho año de mill E quinientos / E sesenta y ocho años salio del lugar de ylasi la buelta de / canos y al salir del dho lugar nuestro camino derecho fue/mos trecho de una legua por un pantano grande que / nos daba El agua casi a la Rodilla y algunos Ratos / mas alta bien llena de yelos[57] tantos que Respeto dellos FOLIO 27 VERSO Salieron muchos de los soldados de la dha conpania bien / mal heridos tanto que si no los ayudaran pongo En dudba la / salida al fin no ostante Este trabaxo y que todo El dia / no faltaron otros pantanos se camino cinco leguas En / fin de las quales se hiço alto y durmimos en despoblado con / {ojo} algun trabaxo Respeto del poco bastimento y que ubo / y otro dia de mañana salimos de aqui y andubimos seis / leguas en fin de las quales llegamos a un lugar que se disze / yca orata en donde dormimos aquella noche y otro / dia de mañana El dho señor Capitan con la dha su / compania no ostante que desde aqui enbio un cabo de / Esquadra con veynte honbres y veynte yndios derecho / a un lugar que se llama guiomaE en donde Estaba cierto nu/mero de mayz p[ar]a que en ciertos costales que les dio lle/bagen del lo que pudiesen hasta un lugar que se llama coçao[58] / en donde su m[erce]d mando a hernando moyano de morales / que fue sargento de su compania y que Estubiese alli y / rrecoxiese todo El mayz que fuese y lo guardase y estu/biese con ello hasta tanto que su m[erce]d alli llegase con la / dha su conpania porque desde alli se llebaria hasta la ciudad / de santa helena donde Estaba ya ynformado que hera me/nester Respeto del poco l [sic] bastimento que abia en la dha ciudad y fuertes della se partio desde alli y llego a canos hasta / donde abia dos leguas En donde Estaba hecha una casa grande / para su mag. y en ella mucho numero de mayz para que / estando alli daria como dio horden En llebar del dho / mayz por un rrio abaxo hasta guiomaE que es como Esta / dho desde donde con yndios y soldados se a de llebar hasta / la ciudad de santa helena En donde como llego no ostante / que El cacique del dho lugar no Estaba en el fue bien / Recivido El qual dho cacique bino a veynte E seis dias del / dho mes y como fue benido visto por el dhos señor Capitan[59] FOLIO 28 que traya carne y mayz para el sustento de su conpania / le dio una gargantilla y un poco de tafetan rroxa y unos / botones de atauxia con que estubo muy contento y / luego yncontinente su m[erce]d dio horden con el dho cacique / que le hiziese traher algunas canoas para hazer lle/bar p[ar]te del dho mayz por el Rio abaxo hasta guiomaE / que Es de donde la dha su compania tenia de llebar lo hasta / la ciudad de santa helena y asi El dho cacique junta/mente con otros caciques que le ayudaron hico traher / algunas canoas en las quales luego que binieron El dho / señor capitan hiço meter todo El mayz que fue posi/ble y metido mando a pedro de hermosa sargento de su / compania con tres soldados della que fuese en las dhas / canoas y llebase El dho mayz al dho lugar llamado guiomaE / y alli lo hiziese descargar y meter en una casa alta q[ue] / en el dho lugar Estaba hecha para el dho hefeto y des/cargado y puesto En la dha casa se estubiese En el dho / lugar con los dhos soldados y con otros que ya quando lle/gase Estarian en el dho lugar que heran los que antes / abian ydo a llebar El mayz que estaba en el dho lugar / hasta coçao y con esto se partio El dho sargento y fue / hasta El dho lugar de guiomaE En donde hallo los sol/dados que abian llebado El primera mayz y alli Estubo con ellos / {ojo} hasta honze dias del mes de hebrellro[60] del dho año que El dho / señor Capitan bino con la demas gente de su compania / y algunos yndios con el con los quales y con los sol/dados de la dha su compania hiço que se llebase todo El / mayz que estaba en el dho lugar en muchos sacos / de gamuças que el dho señor Capitan abia hecho hazer / para El dho hefeto y con esto otro dia siguiente que fue / doze dias del dho mes de hebrero El dho señor Capitan con / la dha su compania y todos los yndios que se pudieron Re-[61] FOLIO 28 VERSO coxer despidiendo como despidio casi todas las lenguas / que traya dando les como les dio paño y lienço para se / bestir y hachas y cuñas y escoplos y gargantillas y botones / de atauxia syn otros muchos rrescates que antes / de agora les abia dado de la misma suerte de que yo El / dho escrivano doy fee quedando como solamente que/do guillermo Rufin lengua castellana en la dha conpa/nia a quien demas de todas las cosas que el dho señor / capitan a dado a las dhas lenguas agora en fin que lle/gamos al dho lugar de guiomaE le dio un bestido de / pano azul del bestir del dho señor Capitan y un ca/pelo pardo y camisas y gamucas para hazer Jubon y me/dias calcas conque asi El dho guillermo Rufin como / las demas lenguas Estubieron y quedaron muy conten/tos de que asimismo yo El dho escrivano doy fee se partio la / buelta de la ciudad de santa helena con harto trabaxo / Respeto de los pantanos y cargas de mayz y el poco bas/timento que abia y prosiguiendo la dha buelta En / catorze dias del dho mes de febrero llego a unas casas / caydas donde solia morir un cacique que se solia lla/mar ahoyaca orata El qual salio a Recibir al dho señor / capitan con alguna carne y Rayzes de tierra que se / llaman batatas para comer de que toda la gente / de la dha compania Recivio muy gran contento por/que estaba muy nescesitada Respeto del poco basti/mento que abia al qual y a un mandador suyo El / dho señor capitan les dio dos mantas de yndias pin/tadas conqued [sic] quedaron muy contentos de que yo El dho / escrivano doy fee y con esto otro dia siguiente salimos / del dho lugar prosiguiendo en el dar de la dha buelta / y este dia pasamos harto trabaxo Respeto de los muchos FOLIO 29 pantanos que ubo que fueron mas que an sido nin/gun dia muy hondos E muy largos tanto que en tres / dellos quando los ubimos pasado tubimos por cierto que / nuestro señor demostraba alguna manera de milagro / En los aver pasado porque siendo tan trabaxosos / y la gente benir tan cargada no aver Caydo nadie ny a/verse le caydo carga ninguna y asi toda la dha compania / Juntamente con el dho señor capitan dimos mu/chas gracias a nuestro señor como hera Justo y con la / buena horden que trayamos andubimos siete leguas / En fin de las quales ya que hera tarde hezimos / alto y durmimos aquella noche En cierta p[ar]te de / despoblado y por esta horden otro dia siguiente / que llegamos a un lugar que se llama coçao En donde / hallamos al dho cacique y a un su mandador con / muchos yndios para yr con el dho señor Capitan / y con alguna Cantidad de Carne y batatas y villota / para que comese la dha gente y asimismo a orista / orata con treynta yndios y uscamacu orata / con otros yndios p todos para yr con el dho señor / capitan todos los quales como Estaban en el dho / coçao Recibieron al dho señor Capitan muy bien y el / dho señor capitan se holgo muy mucho porque abia en/tendido sigun las muchas nobelas que abia abido En / algunos caciques que los dhos coçao y orista y usca/macu se abian lebantado y huydo de nuestra amistad / y asi como llego les dio a cada uno dellos una manta pin/tada de los yndios de la tierra adentro con que / estubieron muy contentos y por el dho señor Capitan / visto y entendido la mucha cantidad que abia de / yndios hiço aderecer Espuertas y sacos de gamuça de / manera que se puediese llebar todo El mayz que tenia[62] FOLIO 29 VERSO junto en el dho coçao que serian como hasta sesenta / fanegas de mayz ya que abia Recaudo para lo traher / y puesto E metido El dho mayz En los dhos sacos / Este dho dia nos partimos dos leguas mas adelante / de Coçao donde hezimos alto y dormimos aquella / noche poniendo demas de la guardia acostumbrada / la guardia y centinelas nescesarias Respeto del / dho mayz y de los muchos yndios que benian / con el dho señor capitan y con esta horden y buen / Recaudo otro dia siguiente salimos del dho lugar / prosiguiendo En el dar de la dha buelta y an/dubimos cinco leguas en fin de las quales porque / toda la gente de la dha conpania y p[ar]te de los yndios / benian muy cansados hezimos alto y dormimos / aquella noche En donde El dho señor Capitan / mando dar Racion del dho mayz Entre dos honbres / una almoçada y con esto otro dia siguiente llegamos / a un lugar que se llama ahoya que esta siete leguas / de la dha ciudad de santa helena y es en donde El dho s[eñ]or / capitan tenia yntento de hazer una casa y fuerte / Respeto de que El cacique del dho lugar y sus yndios / antes de ahora abian muerto un cabo de Esquadra / de la dha conpania y tanbien porque desde El / dho lugar hasta la dha ciudad Es menester yr por / agua y hasta agora todas las vezes que an benido Es/pañoles o yndios de la tierra adentro con alguna / Recaudo p[ar]a la dha ciudad El dho cacique E yndios del / dho lugar de ahoya an sido muy Rebeldes en dar / El Recaudo que a sido menester para allegar / a la dha ciudad y asi por todo lo que Esta declarado / como por Thener al dho cacique E yndios del dho lugar FOLIO 30 debaxo del dominio Real El dho señor Capitan tenia / El dho ynthento de hazer El dho fuerte y agora por/que El dho lugar paresce que esta despoblado y que/mado y el dho cacique preso en la dha ciudad y orista / orata averse le ofrescido de que en su lugar Estara / la dha casa fuerte mejor porque El y sus yndios ser/biran muy de boluntad a los soldados que estubieren / En ella y que tendra Recaudo Cumplido de Canoas E yn/dios para que se haga lo que cumpla al serivçio de su mag. / determino de que se dexe de hazer el dho fuerte / en el dho ahoya y se baxa a hazer En el dho lugar del dho / orista orata y con este yntento por guillermo Ru/fin ynterpetre que estaba presente dixo y mando / a ciertos mandadores E yndios del dho ahoya que salieron / a Recibir al dho señor Capitan que luego yncontinente /

fuesen a los yndios que Estaban ausentes pues que he/ran sus hermanos y de su p[ar]te les mandase que biniesen / al dho lugar y lo poblasen que su m[erce]d solamente / tenia por enemigo a los que abian hecho deservicio a su / mag. y no a otros por eso que en todo caso biniesen y / poblasen su pueblo y si no lo hiziesen ansi que entenderia / que le abian sido traydores y como tales desde agora / en adelante les tenia de hazer guerra y por los / dhos mandadores y yndios Entendido la boluntad / del dho señor Capitan hizieron muchas vezes El yaa [slash] dan/do a entender la m[erce]d que se les hazia y que luego harian / que los yndios bolbiesen y poblasen El dho lugar y por el / dho señor Capitan visto la boluntad de los dhos man/dadores e yndios les dio a cada uno dellos un mandil a ma/nera de sobremesa pintado con que quedaron muy con/tentos de que yo El dho escrivano doy fee fueron testigos FOLIO 30 VERSO a todo lo que dho Es pedro de hermosa sargento y / pedro gutierrez pacheco y francisco hernandez de / {#} tenbleque soldados de la dha compania [slash] Juan pardo [slash] gui/llermo Rufin Juan de la vandera escrivano /

despues de lo suso dho en presencia de mi Juan de / la vandera escrivano y de los t[estig]os di yuso esc[ri]ptos / El dho señor Capitan Juan pardo con la dha su conpania / prosiguiendo En el dar de la dha buelta y hazer de la / dha casa fuerte en diez y nuebe dias del mes de fe/brero del dho año de mill E quinientos y sesenta / y ocho años salio del dho lugar nonbrado ahoya la buelta / de orista y con todo El trabaxo posible con los sol/dados de la dha conpania hiço llebar a cuestas toda / la cantidad de mayz que El traya de la tierra a/dentro para el sustento de la dha gente una legua / del dho ahoya para lo enbarcar y llebar donde hera / nescesario y asi como acabo de traher a la p[ar]te / y lugar donde abia de enbarcar en las canoas / donde se abia de traher lo enbarcamos en las dhas canoas / y toda la dha conpania con ello y desta manera / ya que seria a las dos despues de media dia llegamos / a nos desenbarcar un quarto de legua del dho orista / y desde alli llebamos todo el dho mayz y municiones / de su mag. lo mejor que pudimos al dho orista y lo / pusimos en una casa grande del dho cacique y con esto / Reposamos y dormimos aquella noche y otro dia si/guiente El dho señor capitan dio horden En el / hazer de la dha casa fuerte y se comenco a cortar / y a serrar la madera nescesaria y en dos dias del / mes de marco del dho año se acabo de la hazer al qual / por ser El primero que se a hecho para principio de la FOLIO 31 tierra adentro se le puso nonbre nuestra señora y al pueblo / {la villa} de buena Esperança y por el dho señor capitan visto / que conbenia dexar por alcayde del dho fuerte / un honbre que tubiese quenta con la guardia y de/fensa del y conserbase en buena Amistad los caciques / e yndios de la dha villa E de los contornos della / puso E dexo por alcayde a gaspar Rodriguez cabo de / Esquadra de su conpania y con el para que hiziesen / lo que El dho alcayde les mandase que cumpliese / Al servicio de su mag. doze soldados de la dha su con/pania al qual dho gaspar Rodriguez le dio El poder / que su m[erce]d de su mag. tiene y demas de lo suso dho / ante mi El dho escrivano le tomo E Recivio jura/mento por dios E por santa maria En forma de / derecho y por las palabras de los santos quatro / evangelios donde quiera que mas largam[en]te /

estan esc[ri]ptos sobre que guardaria y cumpliria / lo que esta dho y declarado y no lo quebrantaria / so pena de perjuro e ynfame e fementido y de caher / en caso de menos baler El qual dho gaspar Rodri/guez alcayde absolbiendo al dho juramento s di/xo si juro E amen y porque dixo que no sabia escrivir / Rogo a un testigo que firmase por el y por el dho / señor Capitan Entendido lo suso dho E bisto que para / defensa del dho fuerte conbenia dexar algunas mun/iciones y bastimento para la gente del ante mi / El dho escrivano Entrego al dho alcayde lo siguiente /

diez y ocho libras de polbora [slash] diez y ocho libras de cuerda [slash] / diez y ocho libras de plomo [slash] una hacha de ojo [slash] un azadon / grande [slash] un Escoplo [slash] una barrena [slash] bastimento de mayz para dos meses a Respeto de un albra de[63] FOLIO 31 VERSO mayz a cada uno cada dia testigos que fueron presentes / a todo lo que dho es hernando moyano de mo/rales y antonio de aguirre y pedro gutierrez / pacheco cabo de Esquadra de la dha compania Estantes / en la dha villa de buena esperanca Juan pardo / por testigos antonio de aguirre [slash] juan de la vandera / Escrivano /

despues de lo suso dho en la villa de buena Es/perança veynte y ocho dias del dho mes de / hebrero del año de mill E quinientos y sesenta / y ocho años em presencia de mi Juan de la vandera / escrivano y de los testigos di yuso esc[ri]ptos por el señor / capitan Juan pardo Entendido que En los fuer/tes y ciudad de santa helena ay muy poco bastim[en]to / para el numero de gente que trahee en su con/pania y que se haze mas servicio a dios y a su mag. / en que la dha gente se Reparta y este en p[ar]te don/de se pueda sustentar que no que baya a la dha / ciudad de santa helena a donde por ser much la g[en]te / {ojo} podria padeszer gran trabaxo y detribemento pro/beyo y mando a pedro de hermosa sargento de su / conpania que baya a la ciudad de toledo y fuerte / de santo tomas que en lengua yndiana se disze canos / con treynta honbres soldados de su conpania y alli este / y rresida en la dha ciudad y fuerte como Alcaydel / y rrixa E gobierna y administre los dhos treynta / soldados y guarde El dho fuerte y no salga del sin su / licencia o de quien por p[ar]te de su mag. se lo mande / y demas desto le encomendo y mando que tenga gran / quenta y cuydado con todo sienpre conserbar El amis/tad del cacique E yndios del dho canos y de los contornos del[64] FOLIO 32 y para todo lo que esta dho y declarado le dio El / poder que su m[erce]d de su mag. tiene y para que pueda / castigar a qualquiera de los dhos soldados que En qual/quier manera fueren contra El servicio de su mag. con/forme El delito que cada uno cometiere y para cun/plimiento de todo lo que esta dho le tomo juram[en]to /[65] En forma sobre que lo guardaria y no discreparia dello / so pena de perjuro E ynfame e fementido y el / dho pedro de hermosa sargento E alcayde absolviendo / al dho juramento dixo si juro E amen y prometio / de lo ansi cumplir y lo firmo de su nonbre juntam[en]te / con el dho señor Capitan syendo presentes por t[estig]os / hernando moyano de morales y antonio de aguirre / Estantes en la dha villa Juan pardo [slash] pedro de / hermosa Juan de la vandera escrivano /

despues de lo suso dho veynte y ocho dias del / dho mes de febrero del dho año de mill E / quinientos y sesenta y ocho años Em presencia / de mi Juan de la vandera escrivano y de los testigos / di yuso esc[ri]ptos parescio El muy magnifico señor / capitan Juan pardo y pidio y Requirio A mi El / dho escrivano le diese por fee E testimonio de como si / su m[erce]d dexo a su alferez y a algunos cabos de esquadra / de su compania Repartidos en ciertas p[ar]tes con los / soldados que an sido nescesario como Es notorio y / Agora enbia a su sargento con otros treynta honbres / de su compania a sido y es por lo que toca al ser/vicio de dios nuestro señor y de su mag. Respeto / de la sigüridad y fuerca que Es menester se tenga / en estas probincias de la florida y porque En la / ciudad y puerto de santa helena de presente⁶⁶ FOLIO 32 VERSO no ay bastimento bastante con que se puedan / sustentar dos meses cinquenta honbres quanto / mas El numero de gente que al presente ti[en]e / {ojo} en la dha su conpania que son mas de cien quenta / honbres de que yo El dho escrivano doy fee lo qual dixo q[ue] / me pedia E pidio para lo presentar en el descargo q[ue] / a su mag. diere si acaso se le tomare quenta / de que que asido la causa porque tiene Repartida / su compania y por mi El dho escrivano Entendido E / visto que El dho pedimiento Es justo como lo / an visto y entendido las personas de toda la dha / conpania le di al dho señor Capitan la dha fee E en / {ojo} testimonio como me lo pidio El dho dya mes E ano dho / syendo presentes por testigos El dho pedro de her/mosa sargento y hernando moyano de morales / y antonio de aguirre Estantes En el fuerte de / nuestra señora de buena Esperança que Es donde / se me pidio Juan pardo [slash] paso ante mi Juan de la / vandera escrivano /

otro si Este dho dia veynte y ocho dias del dho mes de / febrero del dho año de mill E quinientos y se/senta y ocho años ante mi El escrivano E testigos / di yuso esc[ri]ptos El dho señor capitan dio horden / y mando a pedro de hermosa sargento de su / compania que de la gente que lleba a su cargo de la / dha su conpania dexe en la villa de madrid que En / lengua yndiana se disze guiomaE a fernan gomez / portuguese y otros tres soldados con el al qual dho fer/nan gomez le mande que haga al dho guiomaE cacique / {ojo} y a los demas caciques que [above the line: son] sus companeros y amigos / que se den prisa En el hazer de quatro canoas que / su m[erce]d les dio horden y mando que hiziesen para⁶⁷ FOLIO 33 El servicio de su mag. pues que para El dho hefeto le / dexo al dho fernan gomez y a los dhos caciques una / hacha de ojo y una azuela y por la misma horden / encargo E mando a alonso bela soldado de la dha su / conpania que Es una de los personas que ban a cargo del / dho sargento que tubiese gran quenta y cuydado / con yr a ylasi orata que es donde algunos / dias a Residido El dho alonso bela y al dho ca/cique y a los demas que le son subjetos les mande / hazer y haga que hagan tres canoas que su m[erce]d les dio / horden y mando quando Estubo en el dho ylasi / que hiziesen pues para ello su m[erce]d dio al dho / cacique una hacha de ojo grande y agora de presente / otra al dho Alonso bela de que yo El escrivano di yuso / esc[ri]pto doy fee pues que sabia y entendia que com/benia hazerse asi para el servicio de su mag. o/tro si en presencia de mi El escrivano y testigos di yuso / esc[ri]ptos Este dho dia el dho señor capitan dio hor/den y mando antonio diaz pereyra portugues / y a domingo de leon y a Juan toribio soldados que / bayan a guando orata que es en donde esta una / casa de mayz de su mag. y del dho mayz procuren de / traher con los yndios del dho cacique todo lo que / pudieren hasta la villa de buena essperança que en / lengua yndiana se disze orista porque E para El dho / hefeto ante mi El dho escrivano les dio y entrego / cierto numero de costales y esto se lo encomendo E / mando que lo hagan con gran diligencia y cuydado E / por el consiguiente les encargo E mando a todos / los sobre dhos que procurasen de hazer con los dhos caciques   FOLIO 33 VERSO e yndios juntamente con ellos de senbrar y que / senbrasen mucha cantidad de mayz porque de todo / Esta serbirian a nuestro señor y a su mag. muy mu/cho a lo que dho es fueron presentes por testigos / hernando moyano de morales que fue sargento de / la dha su compania y pedro gutierrez pacheco cabo de / esquadra della Estantes En la dha villa de buena / esperança y firmo lo El dho señor Capitan Juan / pardo Juan de la vandera escrivano /

despues de lo suso dho En la dha villa de buena / Esperança que por otro nonbre se disze o/rista primero dia del mes de marco del dho año de / mill E quinientos y sesenta y ocho años Em pre/sencia de mi El dho Juan de la vandera escrivano y / de los testigos di yuso esc[ri]ptos Estando presente El dho / señor Capitan Juan pardo con la dha su compania / binieron a la dha billa de buena esperança qua/tro caciques que se llaman yetencunbe[68] orata [slash] / ahoyabe orata [slash] coçapoye [slash] orata [slash] uscamacu / orata y algunos yndios principales con ellos / a traher vastimento para la dha compania a los / quales El dho señor Capitan Recivio bien y por / su m[erce]d visto que con boluntad trayan de lo que / tenian a los dos mas principales dellos les dio dos / mantas grandes de gatos pintadas y quatro / mandiles pintados de los de los yndios de la / tierra adentro con que fueron muy contentos / y luego yncontinente su m[erce]d hiço llamar a gui/llermo Rufin ynterpetre por el qual su m[erce]d hico que los dhos caciques y orista orata y todos   FOLIO 34 sus yndios principales se juntasen en un buhio grande / donde se suelen juntar y ya que todos Estu/bieron juntos se les hiço El parlamento acos/tum brado y demas del dho parlamento por el / dho guillermo Rufin ynterpetre se les dixo que / ya sabian que su m[erce]d con la dha su conpania abia / hecho en el dho orista una casa fuerte En la / qual por les dar contheento dexaba treze hon/bres para que les Enseñasen y diesen a entender / como abia de hazer servicio a dios nuestro señor / y a su mag. y para que los defendiesen de su Ene/migos que les rrogaba que los tratasen bien y les / diesen de lo que tubiesen y que en todo caso les sen/brasen un gran pedaço de tierra de mayz porque / le harian mucho placer los quales dhos caciques / e yndios hizieron muchas vezes El yaa [slash] En que / dieron a entender que lo cumplirian como su m[erce]d / se lo mandaba y demas de los suso dho por el dho gui/llermo Rufin ynterpetre se les declaro y / dixo como hera menester pues que heran her/manos del dho señor capitan y basallos de su mag. / que desde oy en adelante truxesen y biniesen / al dho orista a traher de lo que tubiesen y a o/bedeszer todos al dho orista a lo qual que esta / dho El dho arista hiço muchas vezes El yaa [slash] dando / a entender la gran m[erce]d que se le hazia y los demas / por el consiguiente hizieron muchas vezes El yaa / dando a entender que heran muy contentos de / hazer lo que se les mandaba y benir a la dha ubi/diencia lo qual visto por el dho señor capitan / se lo agradescio muy mucho y demas de lo suso dho FOLIO 34 VERSO por el dho guillermo Rufin se les dixo y declaro / que ya sabian que ahoya orata y otros yndios / suyos le abian muerto uno de los Españoles de / su compania por lo qual El dho ahoya lo tiene preso / y sus yndios An despoblado El dho lugar y quema/do lo y se an huydo del syn thener Razon ninguna / por tanto que conbiene que tornen a poblar El / dho lugar y todos los que andubieren alterados / que se tornen a sus casas y Reposo y para que En/tiendan que los que son sus buenos hermanos lo a / de hazer bien con ellos y por el consiguiente / castigar a los que no lo son que desde oy en viente / dias Esten juntos En el dho orista y que Estan/do juntos que su m[erce]d les declarara y dira lo que / tiene de hazer castigando al que tubiere cul/pa y teniendo por sus hermanos a los que no la / tubiesen como antes lo heran a lo qual todos los / dhos caciques hizieron muchas vezes El yaa [slash] En / que dieron a entender que en ello les hazia El dho / señor Capitan muy gran m[erce]d y que asi Ellos / en fin de los dhos veynte dias procurarian de que / todos Estubiesen juntos y enbiarian a llamar / a su m[erce]d que le rrogaban que viniese a lo qual El dho / señor Capitan les rrespondio que hera muy con/thento y asi se despidio dando les a entender / que otro dia se queria yr juntamente con la dha / su compania que le mandasen thener juntas ciertas / canoas que heran menester y todos Respondieron / que lo harian a lo qual que dho Es fueron presentes / por testigos hernando moyano de morales que fue / sargento de la dha compania y pedro gutierrez FOLIO 35 pacheco cabo de Esquadra della y luis ximenez soldado de la dha / conpania Estantes en la dha villa Juan pardo guillermo Rufin / Juan de la vandera escrivano /

despues de lo suso dho en dos dias del dho mes de marco del / dho año de mill E quinientos y sesenta y ocho años en presen[ci]a / de mi El escrivano y testigos di yuso esc[ri]ptos El dho señor Capitan / Juan Pardo prosiguiendo En el dar de la dha buelta salio con / la dha su compania derecho a la punta y ciudad de santa helena / a do llegamos y de donde para el dho Efeto salimos El dia / que esta declarado y de buelta tornamos Este dho dia / y seria como las tres horas de la tarde con la buena bentura / no ostante los muchos trabajos que El dho señor cap[it]an / con la dha su conpania a pasado y con aver que a oy dia / seis meses y dos dias que salio con yntento de difinir y acabar / Esta jornada en el servicio de dios n[uest]ro señor y de su mag. a ver / que no se difinio no a thenido desgracia ninguna a nuestro / señor sean dados muchas gracias por tan gran m[erce]d y plega le El / por la su santa y sagrada pasion que nos de gracia que las / cosas que hizieremos En su santo nonbre que nos Encamine / de suerte que tengan tan buena principio y fin como / Esta jornada a thenido amen fecha usupra a pasado ante / mi Juan de vandera escrivano /

{aprobacion} despues de lo suso dho En presen[ci]a de mi El dho juan de la vandera / escrivano postr[er]o dia del mes de março año de mill E quinientos y sesenta / y nuebe años Estando En la ciudad de santa helena q[ue] Es En las pro/bincias de la florida visto por el muy Ill[ustr]e señor pedro menendez / de aviles governador y capitan general destas dhas pro/bincias lo hecho y efetuado por Juan pardo capitan por su mag. / En la Jornada y descubrimiento de tierra por las dhas / probincias de la florida derecho a la nueba España con hor/den del dho señor adelantado En nonbre de su mag. para / saver y entender si todo lo que Esta esc[ri]pto en Relacion / de lo que en la dha Jornada se hizo Es ansi quiso hazer / E hizo aprobacion dello con soldados honbres de fee E de / Credito que se hallaron En la dha Jornada a los quales / que son El dho capitan Juan pardo y fran[cis]co hernandez de / tenbleque cabo de Esquadra y francisco coRedor y francisco FOLIO 35 VERSO de cisneros y anton muñoz y Juan de Villela soldados de la dha con/pania del dho capitan Juan pardo a los quales su señoria En pre/sencia de mi El dho escrivano les tomo juramento por dios / E por santa maria En forma de derecho cerca de lo suso / dho para que debaxo del dho Juramento dixesen y decla/rasen si lo que en relacion Esta esc[ri]pto por mi el dho escri[van]o / Es ansi como Esta esc[ri]pto los quales todos 8 como son dichos E / declarados absolbiendo El dho Juramento a una boz di/xeron que todo lo que Esta esc[ri]pto zerca de lo que declara / que Es la dha tierra y el parlamento que se hizo a los / caciques de Cada lugar y como quedaron debaxo del dominio / de su mag. sigun lo que por p[ar]te de su mag. se les declaro Es / ansi como Esta esc[ri]pto En la dha rrelacion porq[ue] a todo / Ello se hallaron presentes lo qual Es verdad por El juram[en]to / que hizieron y los que sabian firmar lo firmaron de sus / nonbres y por los demas que no sabian firmar firmo / un testigo Juntamente con su señoria y comigo El dho escri[van]o / syendo presentes por testigos alonso garcia alferez de la / conpania del dho capitan Juan pardo que firmo por testigo / y gaspar Ro[drigue]z cabo de Esquadra y el alferez goncalo de la / Ribera y garcia martinez de Cos [?] Estantes En la dha ciudad / de santa helena pedro menendez [slash] Juan pardo [slash] Fran[cis]co de / cisneros anton muñoz por testigo alonso garcia Juan de la vandera escri[van]o /

T[esti]go del pedir y dar corregir y conzertar de lo que se Contiene / en esta escriptura con el original el señor alonso garcia / Alcayde del fuerte de san felipe y alferez del conpania / del dho señor capitan y fran[cis]co maldonado y Luis paz / soldados Estantes en la dha punta de santa helena / ba esc[ri]pto en el margin escri[van]o vala /

Juan de la vandera escrivano de mucha tierra destas probincias por el / muy Ill[ustr]e señor pedro menendez de aviles adelantado dellas / en nonbre de su mag. presente fuy en uno con los dhos t[estig]os y doy fee / de lo suso dho porq[ue] ante mi paso y en testim[on]io de verdad fize my signo /

Notes

1. Eugenio Ruidiaz y Caravia, *La Florida; su conquista y colonización por Pedro Menéndez de Avilés*, 2 vols. (Madrid: Hijos de J. A. Garcia, 1893), 2:48.
2. Herbert E. Ketcham, "Three Sixteenth Century Spanish Chronicles Relating to Georgia," *Georgia Historical Quarterly* 38 (March 1954): 66–67. Cited hereafter as Ketcham, "Three Chronicles."
3. Michael V. Gannon, "Sebastian Montero, Pioneer American Missionary,

1566-1572," *The Catholic Historical Review* 51 (October 1965): 343, note 24, with references on following pages.
4. Mary Ross, "With Pardo and Boyano on the Fringes of Georgia Land," *Georgia Historical Quarterly* 14 (December 1930): 267-85.
5. Woodbury Lowery, *Spanish Settlements within the Present Limits of the United States*, 2 vols. (New York, 1901-5), 2:294-6.
6. Pardo's Relation was published in Ruidiaz, *La Florida*, 2:465-473. English translations are in Ketcham, "Three Chronicles," 68-74, and by Gerald W. Wade (trans.), edited by Stanley J. Folmsbee and Madeline K. Lewis, "Journals of the Juan Pardo Expeditions, 1566-1567," *East Carolina Historical Society's Publications* 37 (1965): 112-116. The shorter Bandera relation was first published by Buckingham Smith in*Colección de varios documentos para la historia de la Florida y tierras adyacentes*, 2 vols. (London, 1857), 1:15, from a copy in the Colección Muñoz, Tomo 24, folios 42-4 (new numbers). French published another copy from the same source in 1875 in Benjamin F. French (ed.), *Historical Collections of Louisiana and Florida, 2nd Series, Historical Memoirs and Narratives, 1527-1702* (New York: A. Mason, 1875), 289-92. Ruidiaz, *La Florida*, 2:481-486, took his copy from one in the archive of the Condes de Revillagigedo, Legajo 2, No. 3, F. (This has apparently received a new signature since his time. It is now in Canelejas 46.) English translations have been made by Ketcham, "Three Chronicles," 78-82, and by Wade (trans.), "Journals of the Juan Pardo Expeditions," 118-21.
7. Old signature: 54-5-9.
8. Ketcham, "Three Chronicles," 66.
9. I.e., *Jesu Cristo*.
10. I.e., *de ayuso* (meaning below or infrascript).
11. I.e., *escribiese*.
12. Bandera normally uses this form of *dice*. I have transcribed it as it appears in the document. Curiously enough, in some of the marginal corrections he writes *dize* instead of the textual *disze*.
13. I.e., *cristianos*.
14. Previous transcribers have rendered this word as *Guiomas* or even *Guiomai*, but a careful study of the document shows that it is *GuiomaE*. Bandera's initial "e's" all have this form, with the tail straight to the right, as in the final letter of this word. His final "z's," on the other hand, have a similar form but with the tail dropping below the line.
15. An evident error for the name "Moyano" used elsewhere in the document.
16. I.e., *otro*.
17. I.e., *ataujía*. In the sixteenth century these were a type of enamel on copper button, not the "damaskeening" commonly given in modern dictionaries. See Martín Alonso, *Enciclopedia de la Idioma*, 3 vols. (Madrid: Aguilar, 1958), 1:552. A shipment of 2,880 of these buttons is recorded for 1566, and another of 7,200 was made in 1578. In both cases they seem to have been intended as

trade goods. Archivo General de Indias, Contaduría 442, no. 2, fol. 237, and Contaduría 312, no. 2, fol. 142 verso.

18. An alternative reading is *Sonapa*.
19. The spelling of this word is not clear. *Vora*, *Avora*, and *Aora* are all possible readings of the paleography.
20. The reading is not entirely certain. *Unharca* appears most likely and seems consistent with other names from this area, many of which begin with a "u." Another spelling would be *Unarca*.
21. I.e., *se* (?).
22. I.e., *cuñas*. Bandera frequently forgets the tilde or makes it in such a way that it is not obvious. I have taken the liberty to correct all subsequent occurrences of this word to *cuña* in order to avoid cluttering the text with a repetitive note. A *cuña* was a small wedge of iron, sometimes called a chisel.
23. In the bottom margin is written: *va enmendado do dize q y testada una l*. This is the first of these notes, which were added to the document by the notary wherever he had to make textual corrections. The notes serve to identify the notary's corrections and allow the reader to detect corrections by other, unauthorized hands.
24. Again, the final letter of this word has usually been read with an "s" or "z," but is an "E" in this document, as in *GuiomaE* and *EmaE*, above (note 14). See note 42 for the same name spelled *Tocahe*.
25. In the bottom margin is written: *va testada una q*.
26. The interior letters "au" might also be read as "av" giving *Cavchi*, but this is not very likely. On the next folio the "au" in *caudalosos* appears to be identical to the "au" in this word, supporting the "au" reading. Further support for the "au" is found in the fact that this vowel combination appears in other names in the document.
27. I.e., *ataujía*.
28. I.e., *ombligo*.
29. I have here separated the title *mico* from the name *ola* but elsewhere have left them together as *Olamico* and *olimeco*, both spellings referring to the same person.
30. I.e., *donde*; *do* is a common abbreviation for *donde*.
31. Alternate readings are *humio* and *junio*, neither of which makes any sense.
32. In the bottom margin is written: *va testado una s*.
33. I.e., *acoger*.
34. In the bottom margin is written: *va testado de dia con esta [slashl no vala*.
35. Could also be read as *Uluga*.
36. Could also be read as *a escundidos* which is probably the Spanish *a escondidos*.
37. The rendering of the last two letters of this word, which I have as *Aguacamu* has been read by others as *aguacanni*, a spellinq that seems to me inconsistent

with the norms of Spanish orthography and with the strokes Bandera used to write the letters. The difficulty is that his "u," "n," "i," and "m" are often not very distinct when written in combination, as is evidently the case here.
38. The division of this set of syllables is not clear. I have given the groups between slashes as *E ypeape*, but it is also possible to divide it as *E ype ape*.
39. In the bottom margin is written: *va escripto entre renglones do dize ante vala*.
40. This name appears below as *canasahaqui*. The present spelling is a variant using the capital "A" to begin the last syllable, which is written as a separate word, whereas in the subsequent use the two "a's" appear in a single word, necessitating the use of the silent "h" to separate them. The variant *Canos* here as against *canas* below cannot be explained, but the *Canos* may be more correct because that name appears as a separate name elsewhere in the document.
41. An alternative reading is *Guauuguaca*. Again the problem is in distinguishing Bandera's "u's" from his "n's".
42. I.e., *TocaE*? Here a silent "h" is used between the vowels to indicate that they are pronounced separately. Elsewhere the "e" is capitalized to signify this same fact.
43. See note 39.
44. I.e., *gamuzas*.
45. An alternative reading is *Guaquiri*.
46. An alternative reading is *Osugueu*.
47. An alternative reading is *Aubesau*.
48. An alternative reading is *Guaruruquete*. This name appears above on folio 16 (near note 41 in the text). A possible alternative reading is *Guaruruque*.
49. I.e., *azuelejas*.
50. In the bottom margin is written: *va enmendado do dize En*.
51. In the bottom margin is written: *va puesto entre re[nglon]es do dize parte y enm[enda]do si bala*.
52. Elsewhere this appears as *Otary y atiqui*.
53. In the bottom margin is written: *va testado en [slash] y escripto entre Renglones do dize El dho señor capitan*.
54. I.e., *prisa*.
55. I.e., *rija*.
56. In the bottom margin is written: *va testado do dezia dei dno señor capitan le mando [slash] y escripto entre / Renglones do dize capitan*.
57. I.e., *hielos*.
58. An alternative reading is *Cuçao*. However, *Coçao* appears elsewhere in the document.
59. In the bottom margin is written: *va testada una y*.
60. I.e., *Febrero*.
61. In the bottom margin is written: *va Emendado do dize hebr'o un la*.

62. In the bottom margin is written: *va testada una p y enmendada una t*.
63. In the bottom margin is written: *va enscriptos los margenes y al pueblo la villa y testada una s*.
64. In the bottom margin is written: *va testado do dize be [slash] no vala*.
65. There appears to be an *ojo* in the margin, but it is not very distinct.
66. In the bottom margin is written: *va enmendado do dize que vala*.
67. In the bottom margin is written: *va enmendado do dize cien [slash] dia mes [slash] vala y testado do dizia [slash] quenta [slash] no vala / y va escripto entre renglones do dize [slash] son*.
68. An alternative reading is *yentencumbe*.

Introduction

Herbert E. Ketcham began a translation of this document about 1950. He had become interested in it while completing his Ph.D. in French at the University of North Carolina, Chapel Hill, during the late forties.[1] After preparing a rough translation of some 90 pages in length, he filed his typescript with the photostats of the document and did not touch it again.

The Ketcham translation is a faithful and somewhat idiomatic translation that I found accurate in many respects but generally inaccurate in the transcription of the Indian names. My initial concern was to check that aspect of the translation, but I soon found other small discrepancies. Further, the Ketcham translation seemed unnecessarily formal. Thus in the end, I have completely revised Ketcham's work.

The translation that follows, while owing much to Ketcham's text, drops out almost all of the formula "the said" before proper names and provides interpolations where necessary to fully convey the meaning of the Spanish. Where the original Spanish is needed to clarify the meaning, I have provided it in parentheses. In addition, I have made revisions in sentence structure, generally splitting the endlessly linked phrases of the original by suppressing the conjunction "and" and applying appropriate punctuation and capitalization. I have also scrupulously followed the orthography of the Spanish text for all Indian names and words. Portions of the text that were underlined in the original have been italicized, as in the transcription. I believe that the result is an accurate rendering in idiomatic modern English.

Several conventions used in my retranslation of the text should be noted.

1. PLACE/village: The Spanish term translated with these words is *lugar*. In Spanish juridical terminology, a *lugar* is a collection of houses that is a satellite of a hamlet (*aldea*) or village (usually *pueblo*). Because the Spanish term conveys no exact idea of size or organization, I have used the English "place" in preference to "village," which implies size and organization. Where Bandera actually said *pueblo* in his text, I have used

the English word "village" and followed it with the Spanish in parentheses. Ketcham used "village" throughout his text.
2. TO DO: The Spanish term is *cumplir*. I have generally used the English "to do" instead of the verbose "to carry out," which is the literal meaning of this word. "To do" is more forceful and seems to better convey the meaning.
3. PERMISSION: The Spanish is *licencia*. The English cognate "license" suggests a formal document, which is not necessarily what the Spanish term suggests. It suggests both the formal documented act and the informal "OK" carried by a person obviously authorized to order something.
4. TO GO TO / TO RENDER TO: The Spanish is *acudir a*. The sense is to go up to with a load and suggests a very physical action as well as the fulfillment of an obligation.

Translation

COVER TEXT

1569
The Lieutenant Govenor of Florida

FOLIO 1
Santo Domingo, April 1, 1569
Juan Pardo 34 Folios

{*Proceedings for the account which Captain Juan Pardo gave of the entrance which he made into the land of the Floridas*} In the very noble and loyal city and the forts of the Point of Santa Elena which is in the Provinces of Florida, the first day of the month of April [in the] year of the birth of Our Savior Jesus Christ, one thousand five hundred and sixty-nine, in the presence of me, Juan de la Bandera, notary of this city and point and of its district and [in the presence] of the witnesses listed below there appeared the very excellent gentleman, Juan Pardo, captain of the Spanish infantry for His Majesty, *contino* of his house, lieutenant of the governor of the provinces of Florida, who said that in the year just past of 1567, His Grace with the greater part of his company left this city and forts with an order from the very illustrious gentleman Pedro Menéndez de Avilés, captain general of the provinces and adelantado of them in His Majesty's name, for the interior of Florida, in the direction of New Spain, in which journey he was commanded by the adelantado that with all possible care he should try to pacify (*allanar*) and calm (*quietar*) the caciques or Indians of all the land and to attract them to the service of God and of His Majesty and likewise to take possession of all the land in his royal name, for which effect the adelantado commanded me, the said notary, since I was notary of this point and port of Santa Elena and of its district, to go with the captain on this journey, and to write and set down that which might be done and happen in my presence in due form, with the solemnities that are required in order for it to be shown to His Majesty when necessary. All of which [record] it was suitable for the captain to have in his power in order to give an account to His Majesty or to whomever might request it in his name. Therefore he gave and asked me, the notary, to give him a clearly written copy signed and sealed in public form and in a manner which attests to all that which he effected in the service of God and of his Majesty that was done on the journey, which I, the notary, gave him as I was requested to do, together with an approval which the Adelantado made before me, the notary, of all that which was done on the journey [together] with a solemn oath which he received from the captain and from other persons who were found present at all of it and likewise with an order which the Adelantado gave to me, the notary, and another to the captain. The tenor of the above, one after the other, is as follows:

{*Instruction as to what Juan de la Bandera is to do on this journey that he is making with Captain Juan Pardo to the interior and the road to the mines of San Martín and Zacatecas.*}

Firstly, to any cacique who arrives, you are to remind the said captain, [that] he treat with the chiefs as to whether they would like to be Christians, and give obedience to the pope and to His Majesty. If they would like to have monks to teach them to be Christians, they will be sent to them. If they would like to give the possession of the land to His Majesty, it will be taken in his royal name with all formalities. Of all of these [matters] testimonies will be taken in due form so that at the proper time they can be given to His Majesty.

Having arrived at the land of Zacatecas and the mines of San Martín, at the first village of Christians the captain is to send you to the royal audiencia of New Galicia and to the Viceroy of New Spain for the purpose of informing them of your journey and of the other things which seem to you to be suitable to the service of His Majesty. You will make this trip with all the diligence and fidelity which is expected of you, and may Our Lord give you a prosperous trip. In this fort of San Felipe and point of Santa Elena, May 28, 1567. Pedro Menéndez

{*Instructions as to what you, Captain Juan Pardo, are to do on this present trip to the interior to procure the friendship of the caciques in order that they come to a knowledge of our Holy Catholic faith and to obedience to His Majesty.*}

In the first place, you will leave this port of Santa Elena and Fort of San Felipe at the beginning of September of this present year with as many as one hundred and twenty soldiers, harquebusiers and crossbowmen, and you will take the road which seems to you most convenient and direct to go to Zacatecas and the mines of San Martín trying to make all friendship with the caciques who may be on and along the way, delaying two or three or four days, as seems best to you, in suitable places in order that the caciques of the surrounding country may go to see you. FOLIO I VERSO Be very friendly with them, trying to persuade them to the obedience of His Majesty and to remain and give their word as to whether monks should go to tell them the things of God, Our Lord, and as they will be Christians they will treat them [the Indians] very well. Wherever you arrive where there is a principal cacique leave a cross and Christians who may teach them the Christian doctrine.

You will see that the persons whom you take live in a Christian manner and in very good discipline, as is right, and rest and take recreation. Do not work them on the road, making a halt in the places and parts which seem suitable to you in order that they may be refreshed and rested, in order that they may well conserve and have [their] health on the going as well as the return.

On arriving at any village of Christians of the mines of San Martín or Zacatecas you will make a halt and will inform the viceroy of New Spain and the audiencia of New Galicia of your arrival and of the journey which you have made. If it is necessary, let them give some monk, wise and of good spirit and life, in order that on the return to this fort you may leave him in the part which seems suitable to you with some people in order that, being there, he may try to attract the Indians to the service of God, Our Lord, and to the obedience of His Majesty. You will do [this].

You will try to be back in this fort of San Felipe in the course of the month of March if it is possible so that if by chance the French should come to it next summer, they can be resisted and [the fort] defended.

And on the return trip, if it seems suitable to you to return by the road on which you went you will do so and if not you will return by the way which seems best to you.

The persons whom you will take, and whom you will find and who will be given over to you [of those] who have left these provinces without my order, you will bring with you and if anyone goes away from you or absents himself or wishes to leave or absent himself you will punish him according to the crime. I give you full [legal] power to do so according to that [power] which I hold from His Majesty.

On the part of His Majesty, I ask and require and on my part I ask as a favor of all and whichever justices of His Majesty wherever you will arrive that they give you every favor and aid that you ask of them and which you may need. May Our Lord by his goodness give you a prosperous and successful trip. Done in San Felipe and the point of Santa Elena, May 25, 1567.

In all the places at which you arrive you will take possession of the lands and sites which the caciques may give to you, in His Majesty's name and before a notary who may give testimony of it, which testimonies you will bring signed and authorized in due form. Pedro Menéndez. By the command of His Lordship, Juan de Zuiniga.

{*Journey*} In the name of Our Lord Jesus Christ and of the glorious, always-stainless virgin, Holy Mary, his most blessed mother, in the honor and reverence of whom and in his holy name and for the service of His Holiness and of the King Don Philip, our lord, may all those who see this writing know how that, being in the city and point of Santa Elena which is in the provinces of Florida, Monday the first day to the month of September in the year of the birth of Our Savior Jesus Christ, one thousand five hundred and sixty-seven, in the presence of me, Juan de la Bandera notary for these provinces, and of the witnesses that are written below, the said captain appeared and said that inasmuch as today he is to depart from this city in order to make a certain trip through the interior of the land of Florida by order of the very illustrious gentleman Pedro Menéndez de Avilés, adelantado of these provinces in His Majesty's name, attracting and pacifying the Indians of Florida until arriving at some part of New Spain, [that therefore] he carries in his company Guillermo Rufín [Guillaume Rouffi], a Frenchman and interpreter for much [of the] land of Florida. From whom His Grace, in my presence and that of the said witnesses, took an oath on the sign of the cross, by God and by Holy Mary in due form de jure in order that the above mentioned, being a person who is an interpreter of the said Indians and ordinarily understands them all, since this journey is made in the service of God and His Majesty, should always perform and tell the truth-

FOLIO 2 in everything that will be treated and agreed upon with the Indians in order that of all of it a very extensive report may be given to His Majesty and if he does it thus, God, Our Lord, will aid him in this world in his body, and in the other in his soul where it will endure longer, and doing the contrary, it will be demanded of him and dearly as of one who swears and perjures himself swearing His Holy name in vain. Of which I, the notary, am a witness. To the oath Guillermo Rufín said, promising, as he promised, to tell the truth, "Yes, I swear" and "Amen," and he signed it with his name. Señor Alberto Escudero de Villamar, ensign for the captain, Juan Pardo, and Pedro de Hermosa and Pedro Gutierrez Pacheco and Pedro de Olivares, soldiers of the company and many other soldiers of it were present as witnesses at what is said. Juan Pardo, Guillermo Rufín, Juan de la Bandera, notary.

Likewise the captain commanded me, the notary, to write and set down how from the said city to the village that is called GuiomaE, which is forty leagues, His Grace does not wish to set down in detail the villages and their caciques which are in the forty leagues because the land is rough and full of swamps and in addition to this because the caciques and Indians {*Look*} of the forty leagues are very subject and obedient to the service of His Majesty and also because it appears to him, as it has appeared, that it is a thing and jurisdiction subject to the city of Santa Elena whence, whenever some command is sent to the caciques and Indians they do it very carefully and quickly, which I, the notary, attest. Present as witnesses at what is said where Alberto Escudero de Villamar, ensign, and Pedro de Hermosa and Pedro Guiterrez Pacheco and Pedro de Olivares, soldiers of the company. Juan Pardo, Juan de la Bandera, notary.

In the name of Our Lord Jesus Christ and of the glorious, always stainless Virgin His blessed mother, the *first* day of the month of December in the year of the birth to Our Savior Jesus Christ, one thousand five hundred and sixty-six, the very magnificent gentlemen Estebano de las Alas, governor in the said provinces, and Juan Pardo, *contino* and captain of Spanish infantry, lieutenant for the gentleman Estebano de las Alas, being in the city and point of Santa Elena, which is in the provinces of Florida, the captain with an order from the very illustrious gentleman Pedro Menéndez de Avilés, captain general and adelantado in the said provinces in the name of his Majesty, set out together with Alberto Escudero de Villamar, his ensign, and Hernando Moyano, his sergeant, [and] with the soldiers of his company into the interior of the land of Florida to subject (*subjetar*) and pacify (*allanar*) the caciques and Indians of the land in order that they may be under the dominion and obedience of His Holiness and of the king, Don Philip, our lord. With this order and motive he made the journey subjecting and attracting the caciques and Indians to the said obedience until arriving, as he arrived, at a place which is called *Joara*, where His Grace, with his company built a fort, *San Juan* by name, from which to the city and point of Santa Elena *it is one hundred and twenty* leagues. While he was building the fort an Indian arrived with a letter written and signed by Estebano de las Alas, governor, in which [letter] he informed him that in the city and point of Santa Elena there was need of help from his person and company because it was suitable to the service of His Majesty. The captain, having seen the letter and the grave necessity for returning with his company to the city and point of Santa Elena, left in the fort of San Juan, Hernando Moyano, sergeant, with thirty men of his company as the garrison of it. Then he returned with his ensign and the other people of his company to the city and point of Santa Elena where he arrived on March 7, 1567. He was in it with his company until September 1, 1567, whence on the said day he went out continuing the journey with an order which the Adelantado gave him in the month of May just past of the year of sixty-seven. {*EmaE Orata, a great lord.*} Continuing the journey as is said, he arrived at a place which is called *GuiomaE* where its cacique, who is called EmaE Orata, had built a large house for His Majesty which he had made as a result of what he had been commanded by the captain in His Majesty's name, and now today, Monday the eighth day of the month of September of the said year, which was when the captain together with his company arrived a second time at the place called GuiomaE, where, having seen the said house, His Grace summoned, by means of Guillermo Rufín, interpreter of that tongue, who he carried with him in his company, EmaE Orata, to whom through the interpreter, in the presence of me, Juan de la Bandera, notary, it was declared and said that in the service of God, Our Lord, and in the obedience which is owed to His Holiness and to His Majesty, it was fitting that he and the Indians who were subject to him should become Christians and, in addition to this, that he should gather a certain amount of maize and have a house built where it might be put, to which maize he should not come except with permission from His Majesty or of one who has the authority which the captain had. This said, EmaE Orata said and through the interpreter declared that as to what was said concerning [the fact] that he and the Indians who were his subjects should become Christians, that he was happy to do so whenever the captain wished it and that as for having the maize gathered and brought and having the house to hold it built, that he has already gathered the maize and that when the maize is cured he will make the house which is to hold it, and from it neither he nor any other for him will take out any amount except with the said permission. As a demonstration that he carries out that which he is commanded by the captain in His Majesty's name, he asked the captain's permission that he might take out of the maize that he has gathered for His Majesty a canoe [load] of it in order to give something to eat to the captain and to his people. FOLIO 2 VERSO

Moreover the speech which is declared [above] was made to *Pasque* Orata, cacique, who was present, who in the presence of me, the notary, said and declared through Guillermo, interpreter, that he was ready to do all that which he was commanded by the captain in His Majesty's name and that he also has gathered the maize which has been declared to him for which he will build a house in which it may be kept when it is cured, to which neither he nor any other for him will come except with the said permission. In demonstration of this he asked the captain's permission to take out a canoe [load] of the maize in order to feed the captain and his people. The captain, in view of the above, and that it was suitable to the service of His Majesty to give the permission to the caciques to take out the maize since, after all, it was for the sustenance of the people of his company, he gave it to them. In addition to this, in the presence of me, the notary, he gave to each one of the caciques an axe and certain enameled (*altajía*) buttons with which the caciques remained very content. At all of what is said there were present as witnesses Alberto Escudero de Villamar, ensign, and Pedro Hermosa and Pedro Gutierrez Pacheco and Pedro de Olivares, soldiers of the company. The captain signed it with his name [as did] Guillermo Rufín, interpreter and I, the notary. Juan Pardo, Guillermo Rufín, Juan de la Bandera, notary.

And after this in the presence of me, the notary, the captain, Juan Pardo, continuing the journey, on September 11, 1567, arrived at a place which is called *Canos* where he found a large house with a certain quantity of maize, which [house] by command of His Majesty had been made, as the captain had ordered, by Canos Orata, Ylasi Orata, Sanapa Orata, Unuguqua Orata, Vora Orata, Ysaa Orata, Catapa Orati, {*Here many lords gathered, as the letter declares.*} Vehidi Orata, Otari Orata, Uraca Orata, Achini Orata, Ayo Orata, [and] Canosaca Orata, very principal chiefs, without [counting] many others who are subjects and under the dominion of some of the above mentioned, which the above-mentioned captain found together in the place called Canos and thus in the presence of me, the notary, and of the witnesses noted below, he had them all summoned together by Guillermo Rufín, interpreter of their language, and through him it was declared to them how they already knew that before now it had been declared and said unto them by the captain how they were to make the house and maize which they had for His Majesty, and that it was suitable for them to turn Christian and be under his dominion and that from this they were not to vary a point and that now that His Grace saw that they actually fulfill the things stated above, it is necessary that in the presence of me, the notary, they ratify and approve it. They, in the presence of me, the notary, and through the interpreter declared and said that they will be very happy to do that which has been commanded and declared to them by the captain in His Majesty's name and as such, remaining as they remained under the said dominion, they made the "Yaa," which I, the notary, attest. {*Yaa means "I am content to do what you command me to do."*} They promised not to consume or take out any quantity of the maize for any reason except with permission of His Majesty or of one who might have his authority. Alberto Escudero de Villamar, ensign, and Pedro de Hermosa, and Pedro de Olivares, soldiers of the company were witnesses of what is said and of how the captain gave to each of the caciques a hatchet and a little colored taffeta and certain enameled [*altajía*] buttons as a sign of good friendship, with which [gifts] the above mentioned remained very content, which fact I, the notary, attest.

Moreover, this said day, month, and year, before me, the notary, and in the presence of the captain, there appeared Ylasi Orata, cacique named above, and through the interpreter he declared and said that notwithstanding [the fact] that he helped build the house of the place called Canos, that the maize which for his part he is to give and the house in which he keeps it, he has built, and gathered the said maize for His Majesty in his village, to which [store] he promised that neither he nor any other would come except with the permission of His Majesty or of one who might have his authority. Witnesses were the said ensign and the above-mentioned men. Juan Pardo, Guillermo Rufín, Juan de la Bandera, notary.

After this in the presence of me, Juan de la Bandera, notary, the captain, Juan Pardo, continuing the trip on September 13, 1567, arrived at a place which is called Tagaya. Its cacique is called *Tagaya*, who had built a good new house for His Majesty in that place. The captain, having seen the house, in my presence summoned Tagaya before him by {*Tagaya, chief, lord.*} means of Guillermo Rufín, interpreter, and through him he made the customary speech. He replied through the interpreter that he was ready to do that which he was commanded to do by His Grace, which I, the notary, attest. At which there were present as witnesses, Alberto [Escudero] de Villamar, ensign, and Pedro de Hermosa, and Pedro Gutierrez Pacheco, and Pedro de Olivares, soldiers of the company. Juan Pardo, Guillermo Ruffin, Juan de la Bandera, notary.

And after this in the presence of me, Juan de la Bandera, notary, the captain, Juan Pardo, continuing the journey on September 14, 1567, arrived with his company at a place which is called Gueca in which the cacique was (*estaba*) who is called Gueca Orata. Notwithstanding [the fact] that he was among the caciques who built the house of Canos, there was found built in his village a new wooden house rather large. [This] having been seen by the captain, in his presence and mine, the cacique said and declared through Guillermo Rufín, interpreter, who was present, that he had built the house for the service of His Majesty. This being said by the [cacique] through the interpreter there was declared and said to him the accustomed speech. The cacique replied that he was very happy to fulfill it thus. In addition to this there were in that place, Unharca Orata and Herape Orata and Suhere Orata and Suya Orata and Uniaca Orata and Sarati Orata and Ohebere Orata all caciques of other places different from the above mentioned, to whom it was said and declared through the interpreter that it was suitable for them as well as for all the others that they have a reckoning and [take] care that each one of them attend to His Majesty with the quantity of maize that each one could [give] according to his possibilities, which [quantity] they should put and carry to a house which is built for His Majesty in the village of Canos where there is a certain amount of maize that other caciques have given, and if they do not have maize, he who does not have any shall be obliged to go to the said house with some deer skins and if they do not have them, let them attend [to His Majesty], by the [same] order[2] and in the above-mentioned place with the amount of salt which each can [supply]. These [caciques], having heard and understood what was said above through Guillermo Rufín, interpreter, all replied with one voice, "Yaa," which means that they were ready to fulfill it as it had been declared and commanded, which I, the notary, attest. At which there were present as witnesses Alberto Escudero de Villamar, ensign, and Pedro Gutierrez Pacheco, and Pedro de Hermosa, and Pedro de Olivares, soldiers of the company. Juan Pardo, Guillermo Rufín, Juan de la Bandera, notary. FOLIO 3

After this in the presence of me, Juan de la Bandera, notary, the captain, Juan Pardo, continuing the journey on September 15, 1567, arrived with his company at a place which is called Aracuchi, in which the captain found built a good new wooden house and inside it an elevated room with a certain quantity of maize. The captain having seen it, by means of Guillermo Rufín summoned the cacique of the place, who was called Aracuchi Orata. Through the interpreter he said and declared to him how the house with the maize was for His Majesty. In addition to this there was made to him the customary speech. The cacique declared through the interpreter that for that very reason he had made the house and gathered the maize. As for the rest of the speech, in sign of obedience he made the "Yaa," which means that he is ready to do it and not to consume or take away any amount of the maize except with the permission of His Majesty or of one who has his power. To [all] of this I, the notary, attest. Witnesses present at what is said were Alberto Escudero de Villamar, ensign, and Pedro de Hermosa, and Pedro Gutierrez Pacheco and Pedro de Olivares, soldiers of the company. Juan Pardo, Guillermo Rufín, Juan de la Bandera, notary.

After this in the presence of me, Juan de la Bandera, notary, the captain, Juan Pardo, continuing the journey on September 17, 1567, arrived with his company at a village which is called Otari in which the captain found a new wooden house which the cacique of that place, who is called Otari Orata, had built for His Majesty. The captain having seen it, summoned the cacique, to whom he made the customary speech through Guillermo Rufín, interpreter. The cacique replied and said "Yaa," which means that he is ready to do it and he promised to come with what he had and could bring to His Majesty, which I, the notary, certify. Witnesses present at what was said were Alberto FOLIO 3 VERSO Escudero de Villamar, ensign, and Pedro de Hermosa, and Pedro Gutierrez Pacheco and Pedro de Olivares, soldiers of the company. Juan Pardo, Guillermo Rufín, Juan de la Bandera, notary.

Likewise on the aforesaid day, month and year Captain Juan Pardo being with his company in the village called Otari, there came Guatari Meco and Orata Chiquini, cacicas from a place which is {*Meco is a great lord. Orata Chiquini [is] a lesser lord.*} called Guatari where there are a cleric and two Spanish boys teaching the doctrine to the Indians of the country. These [cacicas], through Guillermo Rufín, interpreter, said that they came to say that they had a house built in the said village for His Majesty according to that which earlier they were commanded by the captain and to learn what henceforth they were to do in the service of His Majesty. The captain, seeing the above, made them the customary speech through the interpreter. Having understood it, they said "Yaa," which means that they are ready to do it, which I, the notary, attest. Witness were Alberto Escudero de Villamar, ensign, and Pedro de Hermosa, and Pedro Gutierrez Pacheco and Pedro de Olivares, soldiers of the company.

Likewise, through Guillermo Rufín, interpreter, the cacicas declared that they have authority over thirty-nine caciques, who helped them to build the house, and they, as their leaders, promised to fulfill that which the thirty-nine caciques ought to do in His Majesty's service and thus they promised together with the caciques from today, the declared day, to collect and have collected two rooms (*camaras*) of maize in performance of what is due to His Majesty, which I, the notary, attest. Likewise, for more satisfaction that this house is built for that end, because we, the captain and I, the notary, could not see it, being on the road continuing the journey, I, the notary, in the presence of the captain, took and received an oath in legal form from Antón Múñoz and Francisco de Apalategui, FOLIO 4 soldiers of the company, who for the last three months, at the command of the captain, have been in the village called Guatari Mico until today, when they came to the company, in order that they might say and declare whether it was true that the cacicas had the house built. Which [men] under charge of the oath said and declared that the cacicas, together with their caciques, in the presence of the above mentioned [men] had built the said large house of new wood and [had] completely covered [it] within with matting, which is the truth because of the oath which they have taken and they signed it with their names. I, the notary, attest how the captain gave to the cacicas, to each one of them an axe and likewise to the other caciques who before now are declared, to those understood to be principals, to each one an axe and to the others, subject to them, to some a chisel and to others enameled buttons and some red taffeta. Witnesses were the [afore] said. Juan Pardo, Antón Múñoz, Francisco de Apalategui, Guillermo Rufín, Juan de la Bandera, notary

After this in the presence of me, Juan de la Bandera, notary, the captain, continuing the journey, on September 20, 1567, arrived with his company at a village which is called Quinahaqui in which the captain found the cacique, who is called Quinahaqui Orata who had built a house of new wood for His Majesty. The captain, having seen the house, lodged in it with his company. He summoned Guillermo Rufín, interpreter, through whom, in His Majesty's name, he made the customary speech to the cacique. The cacique, having understood that which had been declared to him through Guillermo Rufín, replied "Yaa," which means that he is ready to do it. The captain, having seen the obedience of the cacique, gave him two chisels and a knife, before me, the notary, and witnesses, with which the cacique remained very happy   FOLIO 4 VERSO and promised not to make any changes in the house or the maize without permission of His Majesty or of one who would have his power, rather, through the interpreter, he promised and obligated himself to hasten with the maize to the place called Otari, which I, the notary, attest. Present as witnesses were Alberto Escudero de Villamar, ensign, and Pedro de Hermosa, and Pedro Gutierrez Pacheco and Pedro de Olivares, soldiers of the company. Juan Pardo, Juan de la Bandera, notary.

Likewise in the village of Quinahaqui, Ytaa Orata, a very important cacique, and Cataba Orata, also a very important cacique, and Uchiri Orata, cacique, came to see and visit the captain. They, in the presence of me, the notary, said and declared through Guillermo Rufín, interpreter, who was present, that they came to learn and understand from the captain what it was that he commanded them to do in His Majesty's service. The captain, in view of the obedience of the caciques, said and declared to them through Guillermo Rufín, interpreter, the customary speech. Having learned, heard, and understood the speech, they replied and said "Yaa," which means that they are ready to do it. Likewise Ysaa [*sic*] and Cataba Oratas promised and obliged themselves, each one as concerns what he is to [do] to carry and go with the said maize to the village called Otari, which [maize] they will not consume without permission from His Majesty or from one who might have his power. Likewise Uchiri Orata promised and obligated himself that he will put the maize which he is to bring within a village called Canos [and that] he will not consume it without permission from His Majesty or from one who might have his authority, which I, the notary, attest. There were present as witnesses Alberto [Escudero] de Villamar, ensign, and Pedro de Hermosa and Pedro Gutierrez Pacheco and Pedro de Olivares, soldiers of the company. Juan Pardo, Guillermo Rufín, Juan de la Bandera, notary.

After this in the presence of me, Juan de la Bandera, notary, the captain continuing the journey, on September 21,   FOLIO 5 1567, arrived with his company at a place which is called Guaquiri in which the captain found a house built of new wood which the cacique had built for the service of His Majesty. Having seen it, the captain, by means of Guillermo Rufín, interpreter, summoned the cacique, who is called Guaquiri Orata, to whom he made and said through the interpreter the customary speech. The cacique, having heard and understood it, said "Yaa," which means that he is ready to do it and that the maize which the captain orders him to gather, he is ready to gather in the house and not to come to it without permission from His Majesty or of one who might have his authority. The captain, in view of the obedience of the cacique, gave him two wedges (*cuñas*), with which the cacique remained content, which I, the notary, attest. Witnesses [were] the ensign and the others listed above. Guillermo Rufín, Juan Pardo, Juan de la Bandera, notary.

In the noble and loyal city of Cuenca and fort of San Juan which is in a place which is called Joara which is at the foot of the range of mountains which the very magnificent gentleman, Juan Pardo, *contino* and captain of Spanish infantry for His

Majesty, discovered in the present year 1567, on September 24, 1567, in the presence of me, Juan de la Bandera, notary, on the said day, the captain, Juan Pardo, continuing the journey, arrived at the place called Joara to which he gave the name of city of Cuenca because His Grace is a native of the city of Cuenca that is in the kingdom of Spain and at the foot of a range of mountains, surrounded by rivers, as is the place called Joara. Likewise His Grace gave to the fort called San Juan, which previously His Grace had had built FOLIO 5 VERSO by his company in His Majesty's name, the name of San Juan because the captain is named Juan and also because the year before this present one he arrived with his company at the place called Joara on the day of San Juan, Apostle and evangelist. [Joara] is one hundred and twenty leagues from the city and point of Santa Elena which is in the provinces of Florida. Also [he made the fort] because His Grace saw that the mountain range was full of snow which could not be passed, and because for the service of His Majesty it was convenient for His Grace to return to the city of Santa Elena, [so that] what they had done and worked until the place called Joara should not remain a wilderness, he had built by his company the fort called San Juan in which, in order that it should remain with the strength which was suitable, he left Hernando Moyano, his sergeant, with thirty men of his company. Thus he returned to the city of Santa Elena with all the other people, where he remained until the first day of the month of September when he began a second time to pursue the journey. Pursuing these daily travels, with all [of them] he arrived on the said day at the place called Joara where he found built a new house of wood with a large elevated room full of maize, which the cacique of the village, who is called Joada [*sic*] Mico, had built by the command of the captain for the service of His Majesty. The captain summoned the cacique by means of Guillermo Rufín, interpreter. When he had come, in the presence of me, the notary, there was declared to him through Guillermo Rufín how it was suitable to His Majesty's service that ordinarily he should have and go with an amount of maize to the house for the support of a certain number of soldiers whom His Grace, for the present, leaves in the fort called San Juan. In addition to the above, he made him the customary speech. FOLIO 6 The cacique having heard and understood what was said by Guillermo Rufín, the interpreter, replied saying "Yaa," which means that he is ready to do it. The captain, in view of the obedience of the cacique, in addition to a little battle axe on a handle which previously he had given him in my presence, now he gave him an axe and for him and other caciques, his subjects, eight small long wedges like chisels and eight large knives, and a piece of satin and another of red taffeta, with which the cacique and the others were very content, which I, the notary, attest. Likewise, the captain having seen that it was suitable for His Majesty's service that the fort remain with the force of people that was necessary and also some munitions and the necessary foods, left in the fort, as governor of it, Lucas de Caniçares, corporal of his company, a person in whom are gathered the qualities [demanded by the] law, to whom he commanded and charged that which is suitable to His Majesty's service, and together with him for the fortress certain soldiers and a certain quantity of maize with certain munitions,[3] those which were enough for the soldiers, which I, the notary, likewise attest. At all of what is said there were present as witnesses Alberto Escudero de Villamar, ensign, and Pedro de Hermosa, and Pedro Gutierrez Pacheco and Pedro Olivares, soldiers of the company. Juan Pardo, Guillermo Rufín, Juan de la Bandera, notary.

After this in the presence of me, Juan de la Bandera, notary, the captain, Juan Pardo, continuing the journey on October 1, 1567, arrived with his company at a place which is over the top (*desecabo*) of the ridge [of mountains] and FOLIO 6 VERSO which is called TocaE in which the captain stopped for four hours. While he was there, he summoned the cacique of the place, who is called TocaE Orata, by means of Guillermo Rufín, interpreter and by other interpreters, to whom by command of the captain, was declared and said the customary speech by Guillermo Rufín and the other interpreters. The cacique having heard and understood, replied "Yaa," which means that he is ready to do what he is commanded now and for all time. The captain, in view of the obedience of the cacique, gave him a small wedge and a large knife and a little green taffeta, with which the above mentioned remained very content, which I, the notary, attest. Likewise there came to the place a cacique who is called Uastique Orata, from a distance of seventeen days journey to the west of the place. In the presence of me, the notary, the captain went and commanded Guillermo Rufín, interpreter, to declare to the cacique what he was to do in His Majesty's service. The cacique having understood it replied through the interpreters that he was very happy to do what he was commanded because for that reason he had come seventeen days journey in order to carry out and do that which was suitable to His Majesty's service now and in all time, and to everything he made the "Yaa." The captain, having seen this obedience, gave him a small wedge and a large knife and a large conch shell[4] and a little green taffeta, with which the cacique was very content, which I, the notary, attest. Likewise, there came to that place four other caciques from the neighborhood of the place, who were called Enuque Orata, and Enxuete Orata, and Xenaca Orata, and Atuqui Orata. They, in the presence of me, the notary, appeared before FOLIO 7 the captain and through Guillermo Rufín and the other interpreters, they said that they came to learn what they were commanded by the captain. The captain, in view of what has been said above, commanded Guillermo Rufín and the other interpreters to declare to the caciques what it was suitable for them to do in His Majesty's service, which is the customary speech. All four together made the "Yaa," letting it be understood that they are ready to fulfill it. The captain, in view of the obedience of the above mentioned, gave them seven small wedges and a piece of green taffeta with which the caciques were very content, which I, the notary, attest. At all of which there were present as witnesses Alberto Escudero de Villamar, ensign, and Pedro de Hermosa, and Pedro de Olivares, and Luís Ximenez, soldiers of the company. Juan Pardo, Guillermo Rufín, Juan de la Bandera, notary.

After this in the presence of me, Juan de la Bandera, notary, the captain, Juan Pardo, continuing the journey, on October 2, 1567, arrived at the place which is called Cauchi in which he found built a new house of wood and earth, which the cacique of the place, who is called Cauchi Orata, had built for His Majesty because previously he had been ordered [to do so] by the captain when the captain was in the place which is called Joara where the said cacique went to reconnoiter. The captain, having seen how the cacique had built and completed that which His Grace had previously commanded him, had him summoned by Guillermo Rufín, interpreter, FOLIO 7 VERSO and through him and the other interpreters there was declared and said to him the customary speech and the cacique replied saying "Yaa." The captain in view of this obedience gave him a small wedge and a large knife and a little green and red taffeta with which the above mentioned was very content, which I, the notary, attest. Likewise, this said day, in the presence of me, the notary, five other caciques {*Look*} who were called Neguase Orata, and Estate Orata, and Tacoru Orata, and Utaca Orata, and Quetua Orata, came to subjection and submission before the captain. To them was declared and said through Guillermo Rufín, interpreter, and the other interpreters, what they were to do in the service of God and of His Majesty and they, all five together, said "Yaa," whichmeans that they were ready to do it now and in all time. The captain, in view of this obedience, gave them certain small wedges and a little green and red taffeta and some enameled [*altajía*] buttons with which the above said were very content, which I, the notary, attest. Witnesses of what is said were Alberto Escudero de Villamar, ensign, and Pedro de Hermosa, and Pedro de Olivares and Luís Ximenez, soldiers of the company. Juan Pardo, Guillero Rufín, Juan de la Bandera, notary.

After what has been said above, on October 3, 1567, the captain, being in the place called Cauchi, saw an Indian walking among the Indian women with an apron before him as [the women] wear it and he did what they did. The captain, having seen this, summoned Guillermo Rufín, interpreter, and the other interpreters and when FOLIO 8 they were thus called, the captain, before many soldiers of his company, told them to ask why that Indian went among the Indian women, wearing an apron as they did. The interpreters asked the above mentioned of the cacique of the place and the cacique replied through the interpreters that the Indian was his brother and that because he was not a man for war nor carrying on the business of a man, {*Look*} he went about in that manner like a woman and he did all that is given to a woman to do. The captain, having learned the above, commanded me, Juan de la Bandera, notary, to write it in the above form in order that it may be known and understood how warlike are the Indians of these provinces of Florida in order to give it as truth and testimony whenever I am asked for it. I attest all that has been said, because all that has been said happened in my presence, and [in that] of many soldiers of the company. Juan de La Bandera, notary.

After this in the presence of me, Juan de la Bandera, FOLIO 9 VERSO notary, the captain, Juan Pardo, continuing the journey left the place called Olameco (*sic*) on called Tanasqui, which was situated on a certain piece of solid ground, like an island surrounded by water because the place was like that, *surrounded by two copious rivers, which* join one with the other, at a tip of the said island, which is where the site of the village (*lugar*) is and by the road on which the captain came, which is that which leads to the said place, by which, in order to enter into it the captain with his company, crossed on foot one of the rivers which was FOLIO 8 VERSO a great labor because it was navel-deep, and rather more than less. There was a good distance [from] the spot where the captain with his company crossed the river to get to the village (*lugar*) and on that side the cacique and Indians of *the place had built a wall with three towers for its* defense. As the captain with his company arrived in it [the village], he summoned the cacique of the place, who is called Tanasqui Orata, by means of Guillermo Rufín, interpreter, and other interpreters. His Grace asked him through the interpreters why he had built the wall in that part where his Grace entered rather than in another place. To which question the *cacique replied that [he did it] for defense from his enemies, who, if they came to do him harm, had no place by which to enter his* town (*pueblo*) except by that place. The captain, in view of the good reason which the cacique gave him, commanded Guillermo Rufín and the other interpreters to tell the cacique what he should do in the service of God, Our Lord, and of His Majesty. The interpreters made the customary speech to the cacique and the cacique replied making the "Yaa," by which he gave [one] to understand that he was very content to do and carry out what he was commanded. The captain, in view of the obedience of the cacique, gave him a large wedge and a half yard of London cloth and a yard of linen and to the three principal Indians of the cacique, to each a small wedge, with which the cacique and the Indians were very content, which I, the notary, attest. Alberto Escudero de Villamar, ensign, and Pedro de Hermosa, and Pedro de Olivares, and Luís Ximenez, soldiers of the company were witnesses to what is said. Juan Pardo, Guillermo Rufín, Juan de la Bandera, notary. FOLIO 9

After this, in the presence of me, Juan de la Bandera, notary, the captain, Juan Pardo, continuing the journey on October 7, 1567, arrived at *a place which is called Chiaha. In order to enter it, the captain with his company crossed three copious rivers* and entered into the place. It is on an excellent, strong site because it is an island surrounded by copious rivers. He was in the place with his company for eight days because it was a large place and in it [were] many Indians and because he was well received by the cacique and the Indians. When the captain was in the place then immediately, by means of Guillermo Rufín, interpreter, he summoned the cacique who is called Ola Mico to whom through Guillermo Rufín and other interpreters, by the command of the captain, there was said at great length what he ought to do in the service of God and of His Majesty. The cacique, having understood the above, made the "Yaa," letting it be understood that he was very content to do what he was commanded. The captain, in view of the obedience of the cacique, gave him an axe and a little piece of London cloth and a vara of linen and a little red taffeta with which the cacique remained very content. In addition to this he gave to three principal Indians of the place, to each one a small wedge and certain enameled (*altajía*) buttons, which I, the notary, attest. Alberto Escudero de Villmar, ensign, and Hernando Moyano de Morales, sergeant, and Pedro de Hermosa, and Pedro de Olivares, soldiers of the company were present as witnesses at what is said. Juan Pardo, Guillermo Rufín, Juan de la Bandera, notary.

After this in the presence of me, Juan de la Bandera, FOLIO 9 VERSO notary, the captain, Juan Pardo, continuing the journey left the place called Olameco (*sic*) on October 13, 1567, and on this day went with his company five leagues from the said place [to a spot] which was in the countryside (*campaña*), where he stopped for the night and slept. Being there, three caciques who were called Otape Orata, Jasire Orata, Fumica Orata, and a principal Indian with them came to bring food for the company. They gave him to understand, through Guillermo Rufín, interpreter, that they came to learn what the captain in His Majesty's name wished to command them to do. The captain, Juan Pardo, in view of the above, commanded that the customary speech be made and said to them; which [speech] was said to them. The caciques having understood it, made the "Yaa," by which they gave one to understand that they were very content to do what they were commanded. The captain, in view of the obedience of the above mentioned, gave to each of the three caciques a small wedge and a piece of green taffeta and some passements[5] of white and red silk and to the principal Indian who came with them a large wedge (*cuñuela*) with which the above mentioned were very content, which I, the notary, attest. Alberto Escudero de Villarmar, ensign, and Hernando Moyano de Morales, sergeant, and Pedro de Hermosa, and Pedro de Olivares, soldiers of the company were witnesses to the above. Juan Pardo, Guillermo Rufín, Juan de la Bandera, notary.

After this, in the presence of me, Juan de la Bandera, notary, the captain, Juan Pardo, continuing the journey on October 14, 1567, departed from the spot where he had slept and on this day he went with his company five leagues further by a very rough way, FOLIO 10 where, climbing a very high mountain he slept on the other side. Being at the top of it he *found a small reddish stone and having found it, he summoned* before him Andrés Suarez, a smelter of gold and silver, to whom he showed the said stone. He saw it and having seen it, he said that the appearance which the stone had gave one to understand that it might be silver, which I, the notary, attest. Alberto Escudero de Villamar, ensign, and Hernando Moyano, sergeant, and Pedro de Hermosa, and Pedro de Olivares, soldiers of the company were present as witnesses at this. Juan Pardo, Guillermo Rufín, Juan de la Bandera, notary.

After this in the presence of me, Juan de la Bandera, notary, the captain, Juan Pardo, continuing the trip on October 15, 1567, arrived with his company at a place called Chalahume, where he made a halt and slept that night. Immediately he called the cacique of the place, who is called Chalahume Orata. Through Guillermo Rufín he was given the customary speech. The cacique made the "Yaa," in which he let it be understood that he was ready to do it. The captain, having seen the obedience of the above mentioned, gave him an axe and a necklace, with which the cacique was very content, which I the notary, attest. Alberto Escudero de Villamar, ensign, and Hernando Moyano sergeant of the company, were present as witnesses to it. Juan Pardo, Guillermo Rufín, Juan de la Bandera, notary. FOLIO 10 VERSO

After this in the presence of me, Juan de la Bandera, notary, the captain, Juan Pardo, continuing the trip on October 16, 1567, arrived at a place which is called Satapo where he was well received. This day he made a halt and stayed in the place. Then immediately he summoned the cacique of the place, who is called Satapo Orata, to whom, through Guillermo Rufín, interpreter, the customary speech was made. The cacique, having understood it, made the "Yaa," letting it be understood that he was ready to do as he was commanded. The captain, in view of the obedience *which the cacique showed, gave him an axe and a mirror and a* necklace in order to content him because there *the captain was informed that many other Spaniards who before now have come through these parts* on foot as well as on horseback, have been killed by the cacique and the Indians who are subject to him. In addition to this, the captain gave to a mandador[6] and two principal Indians of the village to each a small wedge and a necklace, with which, thus, the cacique and Indians remained very content, which I, the notary, attest. Witnesses present at what is said were Alberto Escudero de Villamar, ensign, and Hernando Moyano, sergeant, Pedro de Hermosa, and Pedro de Olivares, soldiers of the company. Juan Pardo, Guillermo Rufín, Juan de la Bandera, notary.

After this, this said day, the sixteenth day of this said month and year, in the afternoon, when it was beginning to grow dark, the sentinels having been posted which they were accustomed   FOLIO 11 to put as guard and protection (*amparo*) of the company, two of them who were posted at two high parts of the place reported how they had heard a very great noise of Indians outside of the place. The captain having understood this, commanded and advised his company to be alert because he distrusted a certain gathering and number of Indians who were outside of the place, close to it, at such an hour. Thus the company was very alert all night, which I, the notary, attest.

Likewise that night, when it was about midnight, more or less, Guillermo Rufín, interpreter of the company, being asleep in a hut, there came to him an Indian of that place who knew him because for two or three days he had come with the company. He said to Guillermo Rufín, interpreter, that if he would arrange for the captain to give him an axe he would discover and tell a certain treachery that the Indians and caciques of the place and the Indians and caciques of Cosa, and of Uchi, and of Casque and of Olameco, who until now had gone with the company, had prepared. The said Guillermo Rufín, together with the Indian went to where the captain was and told him what the Indian had said to him. The captain having understood the above, gave an axe to the Indian. Then the Indian gave him to understand and said that Cosa, who had arrived there that night with many Indians, his vassals, and Olameco, and Uchi, and Casque, caciques, with all the Indians whom they had there, and the cacique of that place, with all his Indians, were accustomed to kill many Spaniards who before now had come through those parts and that they would have discussed no less and [had] agreed among themselves to make three ambushes and put them in   FOLIO 11 VERSO a spot where the captain and his company had to pass on their way to Cosa and in addition to this that the said Cosa and the other caciques had agreed earlier not to give [anything] to eat nor to bring anything to the captain with his company in any place of theirs unless it was paid for. The captain, having understood the above, thanked the Indian very much for what he had told him. As his company was alert, for that night he dissembled. The following day when the sun had come up, he summoned the cacique of the place and asked him to give him certain Indians who were needed to carry certain burdens. The cacique, dissimulating, made as if to go to seek the Indians and after a while he came and he did not bring a one, giving excuses which occurred to him, by which the captain understood and saw that what the Indian had told him was true because neither the Indians who until now had gone with His Grace nor those of the place appeared, but only the women and children of the town (*pueblo*). Thus he made appear and called before him Alberto Escudero de Villamar, ensign, and Hernando Moyano, sergeant, and Pedro de Hermosa and Juan de la Vandera and Marcos Ximenez and Pedro Flores, and Juan de Salazar, and Miguel de Haro, and Gaspar Rodriguez, corporals of the company. Thus all being together, His Grace told them how he had learned that the cacique of that place and Olameco, who until then had gone with him, and Cosa and other caciques with their Indians had agreed to kill him and his company and for that [purpose] they had placed certain ambushes on the road by which he was to pass, that the said ensign, and sergeant, and corporals   FOLIO 12 might consider if it would be good for what was suitable to the service of His Majesty to go ahead or to return from there. With this he withdrew. The ensign and sergeant and corporals having heard what is said, each one in turn said what he thought. They agreed on neither the one nor the other, but rather returned to meet with the captain and each one gave his opinion as to what seemed good to him. After all of this the captain, together with the ensign, sergeant, and corporals agreed that for the service of God and of His Majesty that from there they should return directly to Olamico which was three days' journey from there, by another different road from that which they had followed until there and that in Olamico there should be built a fort and that in it there should remain the number of persons and munitions which seemed suitable to the captain. Thus with this opinion the captain together with his company returned directly to Olamico, by very rough roads and mountains in order to take his company safely, on which road he slept two nights in the countryside (*campaña*) which was crossed with very hard work. On the nineteenth day of the said month he arrived at a place which is called Chiaha where he summoned the cacique of the place, who is called Chiaha Orata. The customary speech was made to him and the cacique made the "Yaa," letting it be understood that he was ready to do what he was commanded. The captain gave the cacique an axe, with which he was very content. Pedro de Olivares, and Luís Ximenez, and Juan Garcia de Madrid, soldiers, and many other soldiers of the company were present as witnesses at all that is said. It was signed by the captain and ensign and sergeant and the corporals   FOLIO 12 VERSO who knew how to write. Juan Pardo, Alberto Escudero de Villamar, Fernando Moyano de Morales, Pedro de Hermosa, Juan de Salazar, Guillermo Rufín, Juan de la Bandera, notary.

After the above on October 17, 1567, in the presence of me, Juan de la Bandera, notary, and of the witnesses noted below, the very magnificent gentleman, Captain Juan Pardo, appeared and asked me and required that I give him certification and testimony how yesterday, which was the sixteenth day of the present month, he arrived with his company at a place which is called Satapo with the intention and will of continuing the journey and of finishing it as he had been ordered to do by the very illustrious gentleman, Pedro Menéndez de Avilés, adelantado of the provinces of Florida in the name of His Majesty; and how because of learning and understanding that the caciques and Indians of the places through which he was to pass from there on were irritated, and wished to kill him and the people of his company, and that if this should not be that in none of the villages (*lugares*) would they let him enter or give him food unless he paid for it and that in addition to this on the said day a great number of Indians were waiting for him in three places on the road by which he was to pass in order to kill him and destroy his company; and how for those reasons and for many others which were notorious in his company, His Grace returned on this day from [that] place, directly toward the city of Santa Elena with the intention of making and building in a place which is called Olamico, which by another name is called Chiaha, {*Look*} a fort where he intended to leave a number of people in order to pacify and keep safe that which until now, in the service of His Majesty, he has conquered and the rest that from here he might [conquer] because thus it was suitable to the service of His Majesty. I, the notary, FOLIO 13 having understood what was asked me by the captain in order better to verify the above and to give him as [written] testimony what His Grace asks of me, with the express authority which he gave me for it and enough legal authority, took and received an oath from some soldiers of the company who had talked with some Indians and understood what was asked of me. Their names are as follows. Juan Pardo, Juan de la Bandera, notary.

Then immediately I, Juan de la Bandera, notary, by virtue of the legal authority and license which is written above, took and received an oath in due form de jure from Alonso Velas, a soldier of the company, who, having sworn and being questioned concerning the above, said that what this witness knows is that being in a place which is called Satapo, which is where the company had been, he began to talk and he talked with an Indian who came with the company as an interpreter, who is a native of Oluga. This witness asked him what he felt or understood that the Indians of Satapo were treating or discussing, and as the Indian interpreter understood this witness, he replied to him that the Indians said that a grand cacique who was called Cosa, [who lived] on the way the captain had to pass, would not consent to the company entering his town (*pueblo*) nor would he give them food unless they paid for it and that before the captain would arrive at Cosa, many Indians would go out to meet the captain in war and not in peace and that it would trouble the Indian very much to go because in Cosa and in other places round about [it] this Indian had five brothers, captives, whom they captured while his brothers where in another company whose captain was called Soto and thus his brothers remained as slaves because that captain and the people FOLIO 13 VERSO were lost because the Indians of Cosa and the other places round about [it] killed them and this is what this witness knows and the Indian said, under the oath which he has taken. He did not sign it because he said that he did not know how to write. He declared that he was twenty-six years of age more or less. Juan de la Bandera, notary.

{*Witness*} Likewise, I, Juan de la Bandera, notary, for greater certification of the above took and received an oath in due form de jure from Diego de Morales, a soldier of the company, who, having sworn and being questioned concerning the above, said that what he knows is that, when this witness was talking with an Indian in the place of Satapo, because the Indian understood some things of those that he was saying, he said to this witness that if the captain was to go through Cosa with his company, they were not to receive them in the place, nor give him food for his people if they did not pay very well for it and that he heard this and nothing else, which is true and is what he knows, by the oath which he took. He did not sign it because he said that he did not know how to write and that he is twenty-four years of age, more or less. Juan de la Bandera, notary.

{*Witness*} Likewise, on this day, October 17, 1567, I, Juan de la Bandera, notary, for better certification of what is said took and received an oath in due form de jure from Juan Perez de Ponte de Lima, a soldier of the company, who having sworn in due form and being questioned, said that what he knows is that this witness on the night of this past Sunday, which was the nineteenth [*sic*] of the present month, being in a place which is called Chiaha where the captain arrived on that day with his company, [and] being on guard making the first [posting], an Indian came to him of those from the place and began to talk with this witness FOLIO 14 so that he understood him [to say] that Satapo, who had two sons, "a scundidos Españoles que Chafahane," which means that he is deceitful (*bellaco*) and that he and Cosa "aguacamu E ypeape," which means that they were going to kill him. This is what he knows and understood and nothing else, by the oath which he took. He did not sign because he said that he did not know how. He said that he was twenty-eight years of age, more or less. Juan de la Bandera, notary.

Likewise, I, the notary, attest that in agreement with what the witnesses have said, the reason that the captain returned from Satapo is certain. That night, in which the captain was in the place with his company, I, the notary, was on guard with my squad as corporal of a squad of the company, and I saw the gathering of Indians which was outside of the place [while] making the rounds of it. The day after the gathering not a single Indian appeared in the village (*lugar*). [Therefore] I say and testify that what the captain says is a positive fact and it happened thus as I was asked and thus, I, the notary, gave it to him as testimony, Alberto Escudero de Villamar, ensign, and Hernando Moyano de Morales, sergeant and Pedro de Hermosa, and Pedro de Olivares, soldiers of the company, being witnesses of the request and giving of it. Juan de la Bandera, notary.

Likewise on this day, October 17, 1567, the captain, Juan Pardo, asked and commanded me, the notary {*Look*} to set down how Guillermo Rufín, interpreter, came before His Grace and with him an Indian, a native of Satapo, [and that while] they [were] speaking of several things touching the journey which the captain was going to make, the Indian through the interpreter let it be understood that the journey would be much better going via a river which passed beside Olameco (*sic*), which by another FOLIO 14 VERSO name is called Chiaha, rather than going by that road directly to Cosa. Because by virtue of this and the other things declared above His Grace returned, as it was agreed that he should, I, the notary, wrote it as he commanded me and I signed it. Likewise it was signed by the captain and Guillermo Rufín, interpreter. Juan Pardo, Guillermo Rufín, Juan de la Bandera, notary.

After the above, on October 20, 1567, in the presence of me, Juan de la Bandera, notary, the captain, continuing his return with his company, arrived at the place called Olamico which by another name is called Chiaha. As soon as he arrived, on this day, he laid out the fort which it was agreed should be built in Olamico. Thus it was begun to be built then and after four days it was finished. When it was finished the captain left in the fort Marcos Jimenez, a corporal in his company, with twenty-five men, soldiers of the company, in order that they should be there and guard the fort. He entrusted and commanded Marcos Jimenez to concern himself with what ought [to be done] in the service of His Majesty and that he should not go out of nor depart from the fort except with license from His Majesty or of one who in his name might command it. As more security for the above, in the presence of me, the notary, he took from him a solemn oath that he would guard the above. The said Marcos Jimenez, acquitting [himself] of the oath, promised to do it in [this manner] under pain of perjury and of infamy and of falling into less value. The captain, in view of the above, understanding as he understood that it was necessary that in the fort there should be left some munitions for the defense of the people, it being necessary, he delivered and gave to Marcos Jimenez a Biscayan socketed axe and a FOLIO 15 hoe and a shovel and sixty pounds of powder and fifty pounds of matchcord and eighty-five pounds of [lead] balls, which I, the notary, attest. Likewise I attest of how in my presence the captain, in order to leave the cacique and mandador of the place grateful and content, in addition to the other tools and things which he had given them, on this day gave to the cacique a large Biscayan axe and to the mandador and a principal Indian of the place a large wedge apiece, and to Marcos Jimenez, corporal, two dozen chisels and knives in order that he might have [something] with which to content some Indians that it would be necessary [to please]. At all of what is said Alberto Escudero de Villamar, ensign, and Hernando Moyano de Morales, sergeant, and Pedro de Hermosa and Pedro de Olivares, soldiers of the company were present as witnesses. Juan Pardo, Guillermo Rufín, Juan de la Bandera, notary.

{#} After the above, on October 22, 1567, the captain, Juan Pardo, continuing the return on this day, left the place called Olamico with his company and on the twenty-seventh day of the month arrived at a place which is called *Cauchi*. As soon as he arrived in it, he laid out a fort, which he named San Pablo. He gave an order and commanded that it should be built in order better to preserve the friendship of the Indians of the land, which was done and [it was] finished on October 30. The captain, seeing that [and] understanding as he understood, that it was fitting to His Majesty's service that in the fort he should leave a dozen men of the soldiers of his company and with them a person who might rule, govern, and administer in the things FOLIO 15 VERSO which were suitable to the service of God and of His Majesty, and also so that the friendship of the Indians might be preserved, left Pedro Flores, a corporal of his company, and with him ten soldiers of it, whom His Grace commanded to obey and consider Pedro Flores as their superior henceforth. And he recommended and commanded that Pedro Flores take great account and care to guard the fort and not to leave it without the license of His Majesty or of one who might command it in his name and also [to take great care] of the government of the soldiers and the preservation of the friendship of the Indians and for all of this he gave him his complete legal authority, as His Grace has it from His Majesty. After it had been given, in the presence of me, the notary, on a sign of the cross which he made with his right hand, he took and received an oath by God and by Holy Mary, according to law, from the said Pedro Flores, corporal, that faithfully and loyally he would guard the fort and would do the things which he was commanded and would not break them nor would he leave the fort without the license of His Majesty or of one who might command it in his name. Pedro Flores, acquitting himself of the said oath said "Yes, I swear" and "Amen" and he promised to fulfill it thus under pain of perjury and the infamy of falling into less worth. The captain, in view of the above and understanding as he understood that Pedro Flores had a need for some munitions for the service of the fort and for what might be necessary, left him thirty-six pounds of lead balls and twenty-two pounds of powder and twenty-four pounds of matchcord and a socketed Biscayan axe, which, I, the notary, attest. Likewise he left him two dozen chisels and knives to give to some Indians as might be necessary. Present as witnesses of what is said were Alberto Escudero de FOLIO 16 Villamar, ensign, and Hernando Moyano de Morales, sergeant, and Pedro de Hermosa, and Pedro de Olivares, soldiers of the company. Juan Pardo, Guillermo Rufín, Juan de la Bandera, notary.

Likewise, on October 27, 1567, when the captain arrived, returning, to the place called Cauchi, he gave to the mandador of the village and to a brother of his, because the cacique was not there, a large wedge and a necklace and a chisel with which they were very content and to a cacique who was from another place which was called Canosaqui Orata, a wedge and a chisel and a necklace with which he was very content. Then, on the twenty-ninth day of the month, the captain being with his company in Cauchi, five [sic, seven] caciques came to see and visit him, who were called Utahaque Orata, Anduque Orata, Enjuete Orata, Guanuguaca Orata, Tucahe Orata, Guaruruquete Orata, [and] Anxuete Orata, to whom, His Grace, seeing the obedience with which they came, gave five chisels and two wedges and two necklaces and a piece of green taffeta. The customary speech was said and declared to them through Guillermo Rufín, interpreter. All said the "Yaa," by which they gave [him] to understand that they were ready to do it, which I, the notary, attest. Present as witnesses at what is said were Alberto Escudero de Villamar, ensign, and Hernando Moyano, sergeant, and Pedro de Hermosa, and Pedro de Olivares, soldiers of the company. Guillermo Rufín, Juan de la Bandera, notary. FOLIO 16 VERSO

After the above, on November 1, 1567, the captain, Juan Pardo, continuing the return, arrived at a place which is called Tocahe (*sic*) where the captain was well received by the mandador and principal men of the said town (*pueblo*) to whom the captain gave two chisels and two necklaces and a piece of red taffeta with which the above mentioned were very content. On the following day [while] the captain was resting with his company in Tocahe, Cauchi Orata came there. He had taken certain captive Indian [women] to Joara by the command of the captain. His Grace gave him a large wedge, with which the cacique was very content. He returned to his village. Of which I, the notary, am witness. Present as witnesses at what is said were Alberto Escudero de Villamar, ensign, and Hernando Moyano, sergeant, and Pedro de Hermosa, and Pedro de Olivares, soldiers of the company. Juan Pardo, Guillermo Rufín, Juan de la Bandera, notary.

After the above on October 29, 1567, which is when the captain was in the place called Cauchi, in the presence of me, Juan de la Bandera, notary, and of the witnesses noted below, there came before the captain Cauchi Orata, Canasahaqui Orata, Arasue Orata, Guarero Orata, Joara Chiquito {*YnihaEs are like what we might call justices or Jurados[7] who command the people.*} Orata, Arande Orata, and two ynahaes oratas and through Guillermo Rufín, interpreter, they said and declared that the captain [should] tell them FOLIO 17 where he wished them to go with the tribute that they were obliged [to give]. The captain, having understood through Guillermo Rufín, declared to them and said that they already knew how His Grace left there a principal man of his company with ten other soldiers of it [and] that they should take care that as for the food which they were obliged to bring to His Grace, that from this day onward they should bring it to the said principal man and soldiers that he left in Cauchi and that as concerns deer skins that they should go to Joara to another Guançamu, who was his brother, who was there, because by doing so His Majesty would be served. The caciques having understood the above, all made the "Yaa," which I, the notary, attest. Present as witnesses at what is said were Alberto Escudero de Villamar, ensign, and Hernando Moyano, sergeant, and Pedro de Hermosa, and Pedro de Olivares, soldiers of the company.

After the above on November 3, 1567, in the presence of me, Juan de la Bandera, notary, and of the witnesses noted below, the captain, Juan Pardo, continuing his return, on this day departed from TocaE and arrived with his company five leagues from the said place in the countryside (*campaña*) where he made a halt and slept. While he was there, five caciques from thereabouts came to bring food to the company. The captain gave them five small wedges with which they were very content and he did not say anything to them about the submission because before this day it was said to them and they are in the same submission as the others.

Witnesses were Alberto Escudero de Villamar, ensign, and Hernando Moyano, sergeant, and Pedro de Hermosa, and Pedro de Olivares, soldiers FOLIO 17 VERSO of the company. Juan Pardo, Juan de la Bandera, notary.

After the above, in the presence of me, Juan de la Bandera, notary, and of witnesses noted below, the captain, continuing his return, on November 4, 1567, departed with his company from the spot where he had slept the night before. On this day he arrived five leagues further, near a stream in a ravine, where he made a halt and slept. While he was there, a cacique came who was called Atuqui, who brought food for the company. The captain, having seen this, gave him a piece of green taffeta, with which he was very content. Present as witnesses of what is said were Messers Alberto Escudero de Villamar, ensign, and Hernando Moyano, sergeant, and Pedro de Hermosa, and Pedro de Olivares, soldiers of the company. Juan Pardo, Juan de la Bandera, notary.

After this in the presence of me, Juan de la Bandera, notary, and of the witnesses noted below, the captain, Juan Pardo, continuing the return, on November 5, 1567, departed with his company from the site and place where he had slept the previous night. On this day he arrived with his company four leagues further, where he halted and that night the company slept. While he was there, Joara Mico, cacique, came to bring food. Witnesses were Alberto Escudero de Villamar, ensign, and Hernando Moyano, sergeant, and Pedro de Hermosa, and Pedro de Olivares, soldiers of the company. Juan Pardo, Juan de la Bandera, notary.

After the above in the presence of me, Juan de la Bandera, notary, and of the witnesses noted below, the FOLIO 18 captain, Juan Pardo, continuing the return, on November 6, 1567, arrived with his company at the city of Cuenca and fort of San Juan which in the Indian tongue is called Joara, where he made a halt and remained twenty days because the people of his company were tired and poorly provided, that they might have a place to rest and to provide themselves. On this day the captain gave to a principal Indian of the said city, a son of the old cacica of it, a large wedge and to the cacica a chisel and a mirror, with which the above-mentioned were very content. While he was in the city on November 16, there came to it to make the "Yaa," which is the obedience, five caciques who were called Quinahaqui Orata, the two Catapes Orata, Guaquiri Orata, and Ysa Chiquito Orata, with some principal Indians. They brought food to the men. The captain, having seen this, gave them seven small wedges and three necklaces and some pieces of red cloth with which the above mentioned were very content. Being, as is said, in the city, on November 20, there came to it eighteen caciques to make the "Yaa," which is obedience, as before now they were accustomed to do, and to bring maize and other food for the men. [These] caciques where called Atuqui Orata, Osuguen Orata and his mandadores, Aubesan Orata, Guenpuret Orata, Ustehuque Orata, three Pundahaques Oratas, TocaE Orata, Guanbuca Orata, Ansuhet Orata, Guaruruquet Orata, Enxuete Orata, Utahaque Orata, Anduque Orata, Jueca Orata, Qunaha Orata, [and] Vastu Orata and together with them Joara Mico, cacique of the said city. The captain, in order better to attract them to the service of FOLIO 18

The "Long" Bandera Relation 277

VERSO His Majesty and to keep them more content, gave to Joara Mico, as the most important one, a large wedge and some pieces of red cloth and to five other caciques of the above mentioned, also principal [men], to each a wedge, and to the other caciques and to their mandadores sixteen chisels, with which the above mentioned were very content. I, the notary, attest all of the above. Alberto Escudero de Villamar, ensign of the company, and Pedro de Hermosa, and Pedro de Olivares, and Luís Ximenez soldiers of the company were present as witnesses at what is said. Juan Pardo, Guillermo Rufín, Juan de la Bandera, notary.

After the above on November 24, 1567, in the presence of me, Juan de la Bandera, notary, and of the witnesses noted below, the captain, Juan Pardo, being in the city of Cuenca and fort of San Juan, which in Indian language is called Joara, with his company, having considered it and seen that he had to go from the city with his company and that in fort San Juan it was necessary for a number of persons to remain for the security of the land and in his own place Alberto Escudero de Villamar, ensign, in order, that, as such, he should judge and have a care of the conservation of the friendship of the caciques and Indians of all the land and in order that he might visit and govern the forts and governors and soldiers of them which are from the place called Chiaha to the city of Cuenca and from the city of Cuenca to the Point of Santa Elena, which [forts and garrisons] are: in the place of Chiaha, the fort of San Pedro, which is fifty leagues from the city on the west; and in the place of Cauchi, the fort of San Pablo, which FOLIO 19 is twenty-eight leagues, between north and west; and from this part of the said city in the place called Guatari, the fort of Santiago, which is forty leagues from it between north and east; and from Guatari to the place called Canos, the fort of Santo Tomas, which is forty-five leagues from Guatari to the east between east and south; and from the place called Canos to the Point of Santa Elena, [which is] fifty-five leagues to the south. With the above-mentioned consideration, in the presence of me, the notary, he charged and commanded Alberto Escudero de Villamar, ensign, to take great account and care of the above mentioned, since he knew and understood how much it was convenient to the service of His Majesty and for it he gave him complete [legal] authority that [legal authority] which the captain had from His Majesty. The ensign promised to do it thusly and they signed it with their names. Present as witnesses at what is said were Hernando Moyano de Morales, sergeant of the company, and Pedro de Hermosa and Pedro de Olivares, and Luís Ximenez, soldiers of the company. Juan Pardo, Alberto Escudero de Villamar, Juan de la Bandera, notary.

Likewise, I, the notary, attest to how in my presence and in that of the witnesses noted below, the captain left Alberto Escudero de Villamar, ensign, the following things in the name of His Majesty for the needs which might arise:

First, sixty-eight pounds of matchcord
Further, one hundred pounds of lead balls
Further, eighty-five pounds of powder

Further, thirty-four pounds of nails
Further, forty-two azolejos [i.e., *azuelejos*], like chisels
Further, three iron shovels
Further two socketed axes
FOLIO 19 VERSO
Further, four iron wedges and two hoes, and two pickaxes.

Likewise, this day, November 24, 1567, before the captain and in the presence of me, the notary, two caciques who were called Chara Orata and Adini Orata came to the city to bring food for the company. They were accustomed to be subject to Guatari and now wished to come to make the "Yaa" at Joara. The captain ordered them to go where they were accustomed to. In order to content them in this, he gave them three chisels, with which they were very content and they promised to do what the captain commanded them. Witnesses were Hernando Moyano de Morales, sergeant, and Pedro de Hermosa, and Pedro de Olivares, soldiers of the company. I, the notary, attest all of this. Juan Pardo, Juan de la Bandera, notary.

After the above, in the presence of me, Juan de la Bandera, notary, the captain, Juan Pardo, continuing his return on November 24, 1567, departed with his company from the city of Cuenca, which, in the Indian language is called Joara. On this day he made a halt in a certain unpopulated place and slept that night. On the following day he reached a small place (*lugarejo*) five leagues beyond there which belongs to a mandador of Yssa, who is called Dudca. There he made a halt and slept that night. The following day, which was the twenty-sixth day of the month, marching with his company directly to Yssa, a distance of a quarter of a league further on, he passed along a left-hand bend of the road and another [bend] to the right a little further on, as he was guided by Hernando Moyano de Morales, sergeant of his company, and Andrés Suarez, silversmith, who by command of the FOLIO 20 *captain, on the seventh day of the said month had come by that road to examine a crystal mine which was on it. In the presence of me, the notary, they showed* the captain the beginning of the mine which I, the notary, saw, [and] how it had begun to be dug, which I attest. In view of what has been said, Hernando Moyano de Morales, sergeant, said that he, as a person who before now had seen the mine, was registering and registered it and was obligating and obligated himself with license from His Majesty to exploit and work the said mine and in conformity with the laws which His Majesty has placed on the Indies to render up the part for which he is obligated and henceforth he asks His Majesty or whoever will command in his name that every time there should be a division of land he should get the said crystal mine because he, as he has said, will exploit and work it and will take to His Majesty that to which he is obligated and [will do so] jointly with the illustrious gentleman Esteban de las Alas, governor in the provinces of Florida and the captain, Juan Pardo, [(]who, as he has said, on the seventh day of this month being lodged in the city of Cuenca with his company, commanded him to come to the place of Dudca to examine the mine and on the eighth day of the month he returned to the city of Cuenca with some rocks of the crystal which he had obtained with the point of a mattock from the said mine and he showed and gave them to the captain, and this day, he made the claim before me, the notary, which I attest[)] and Andrés Suarez, silversmith, and me, the notary, and Alberto Escudero de Villamar, ensign of the captain, Juan Pardo. [He further said that] it is his will and he wished that all five be and go to the division of the mine equally with him and that all six, each one for himself, one as well as the other and the other as he, carry and have the equal part that would be equal parts so that with stronger energy and speed the   FOLIO 20 VERSO said mine could be worked and exploited. In order to carry this out, one as well as the other, who is contained in this writing, he pledged his person, and movable goods and foundations, those which he has and may have, in order not to break or contradict it now or in any time and although he may wish to contradict it he may not do so. He renounced all and every law, privilege (*fuero*) and right which in his favor may or can be in order that they might not be valid or benefit him in anything. He signed it with his name, there being present as witnesses at all of what is said Pedro de Hermosa, Franciso Corredor, and Francisco Hernández de Tenbleque, and Luís Ximenez, soldiers of the company. Hernando Moyano de Morales. Juan de la Bandera, notary.

After the above on this said day, November 26, 1567, while the captain, Juan Pardo, was marching with his company, in the presence of me, Juan de la Bandera, notary, having gone *a quarter of a league beyond the mine which is declared above, on the right side of the road he found other* crystal stones which were, it seemed, the site of the beginning of another mine. As soon as he found it, he appeared before me, the notary, and showed me the stones and site and in my presence and in that of the witnesses noted below he said that he was registering and registered it and it was his wish to put and he did put in the shares of the mine Señor Esteban de las Alas, Governor of the provinces of Florida, and Alberto Escudero de Villmar, his ensign, and Hernando Moyano de Morales, his sergeant, and Andrés Suarez, silversmith, and me, Juan de la Bandera, notary, in order that all five, as they are named and declared, together with him, one as well as the other and the other as he may take and have as much part as he [has] and may aid him to exploit and work the mine whenever by His Majesty or by someone in his name he   FOLIO 21 may be commanded to do so, in order that all six, as they are named and declared, may render to and give to His Majesty that which pertains to him according to the laws and ordinances of the Indies. So that he will carry out what is discussed and contained in this writing he said that he was binding and bound his person and movable goods and foundations [those that he] had or was to have and [that] he was renouncing and renounced any laws which might be in his favor so that they might not avail or {#} profit him anything at any time. He signed it with his name, witnesses being Pedro de Hermosa and Francisco Hernández de Tenbleque and Luís Ximenez, soldiers of the company. Juan Pardo, Juan de la Bandera, notary.

After the above, in the presence of me, Juan de la Bandera, notary, the captain, Juan Pardo, continuing his return on November 28, 1567, arrived with his company at a place which is called Yssa, in which he was well received. As soon as he arrived he lodged in a large new house of wood and earth which the cacique of the place had built for the service of His Majesty as he had previously been commanded by the captain. In addition to this, on the following day two Indians came from their place, which is called Guatari Mico, on behalf of the cacique of the place. The captain received them well and gave them two socketed axes to take to the said cacique in order that he might have cut a quantity of wood for a fort that the captain was to build in the village (*lugar*). Being in Yssa, on December 5, 1567, there came to the village Subsaquibi Orata, cacique, and with him some Indians to bring food for the company. The captain, having seen this, FOLIO 21 VERSO had the customary speech made to him through Guillermo Rufín, interpreter. In order to keep him more grateful and content, he gave him a large knife and a mirror, with which he was very content. Likewise he gave to an Indian who on this day came to the village, because he is an interpreter of Guatari, a chisel and a necklace. On the following day the mandador of Cauchi came to the place with certain letters for the captain, to whom His Grace gave a large knife, with which he was very content. Moreover, on the ninth day of the said month of December of the said year, the captain gave to Yssa, cacique, and to his mandadores an axe and two large knives and two chisels and four necklaces, because he was in the place with his company twelve days. The cacique remained very content, which fact I, the notary, attest. Witnesses were Pedro de Hermosa, sergeant, Lucas de Canezares and Luís Ximenez, soldiers of the company. Juan Pardo, Guillermo Rufín, Juan de la Bandera, notary.

After the above, on this day, December 9, 1567, Andrés Suarez, silversmith, appeared before the very magnificent captain, Juan Pardo, and in the presence of me, Juan de la Bandera, notary, and of the witnesses noted below, said that he was asking and asked His Grace as the captain that he is for His Majesty and lieutenant for the very illustrious gentleman Esteban de las Alas, governor and lieutenant for the captain general in the province of Florida, to command that no one nor any person dare, under grave penalties which His Grace may put for it, to dig for the distance of a half league and more around about [a spot which is] a quarter of a league beyond [Dudca] directly toward a place which is called Yssa, as one comes from a place which is called Dudca, where His Grace coming marching with his company found a certain sign and stones of a crystal mine and Hernando Moyano de Morales sergeant FOLIO 22 of his company [also] because the said Andrés Suarez has shares in the said mine with the captain and the sergeant and what he asked His Grace was in the interest of the said Andrés Suarez and his companions who are partners in the said mines until His Majesty might command something else. He asked it by [written] testimony and of those present that they be witnesses of it and he implored the office of the said captain. Andrés Suarez, Juan de la Bandera, notary.

After the above on this said day, month, and year before the captain there appeared Hernando Moyano de Morales and said that he was making and made the same demand [as that] contained above and was asking and asked the same as the said Andrés Suarez, silversmith, asks and that no one should come to dig nor work for the space of the said half of a league of land around [the place] that is declared without the assent of all the persons who have a part in the said mines. He asked it and as far as was necessary he begged the office of the captain. Fernando Moyano de Morales, Juan de la Bandera, notary.

Then immediately this said day, month, and year, the captain, in view of what was requested by Hernando Moyano de Morales, sergeant, and Andrés Suarez, silversmith, and that what they ask is without prejudice to any party but rather is a thing which is suitable to the service of His Majesty, said that he was commanding and commanded that no one nor any person dare to dig or work the said mines nor the distance of the said half league around [them,] which is [what is] requested of him, without the command and consent of all the persons who have shares in the said mines, on pain of death. Witnesses of the above where Francisco Hernández de Tenbleque and Gaspar Rodriguez and Lucas de Canizares, soldiers of the company. Juan Pardo, Juan de la Bandera, notary.

After the above, on December 10, 1567, FOLIO 22 VERSO Captain Juan Pardo, being in a place which is called Yssa, was informed that a long {look} league from the place, down the river as you go from the said place, on the other side of the river on a hill some four or five hundred paces from the river near where Mati Yssa lives, who is the mandador of Yssa, a quarter of a league as you go from Yssa, before arriving at a {look} river of the said Mati, there were certain stones which showed that they were crystal. In order to know the truth of it, he commanded me, the infrascript notary, together with Hernando Moyano de Morales who was sergeant of the company and with Lucas de Canizares, a soldier of it, to go and with our own eyes to understand and see if it was a crystal mine. In fulfillment of the above I, the notary, together with the above mentioned on this said day, *went to see and we saw the site and place where the stones were and we understood and saw that they were crystal* and we marked the said place. The following day we returned to Yssa and we reported the above to the captain. His Grace, having heard the above, said that he was registering and registered the mine and Lucas de Canizares and Hernando Moyano de Morales said they were doing and did the same with the license of His Majesty and of the captain. The above-mentioned men together with him said that they were putting and put in the division of the mine, so that all together might have equal parts, one as much as the other and the other as the other, Señor Esteban de las Alas, lieutenant of the captain general in the provinces of Florida, and Alberto Escudero de Villamar, ensign, and Pedro de Hermosa, sergeant, and Juan de la Bandera, corporal, in order that all seven as they are named and declared may exploit and work the said mine, or one who has power from them or from any of them, and render and give to His Majesty the part which justly and as a net balance (*liquidamente*) comes to him in conformity with the laws and ordinances which His Majesty has put on the Indies. FOLIO 23 So that they will carry out and do for certain what is contained in this paper, each as concerns himself and as he is obliged to fulfill, they obligated their persons and movable and stationary goods which they had and will have and which they have and may have and they give legal power to His Majesty's justices to compel and oblige them to the fulfillment of the above just as in a case sentenced and passed as a judgment. In testimony of which they authorized this letter before the notary and witnesses noted below. Witnesses who were present at what is been said were Gaspar Rodriguez, corporal, and Francisco Hernández de Tenbleque, and Luís Ximenez, and Juan Gracia de Madrid, soldiers of the company. The captain and Hernando Moyano de Morales and Lucas de Canizares signed it in the register. Juan de la Bandera, notary.

After the above, on this said day, month, and year, in the presence of me, the notary, there appeared before the captain Hernando Moyano de Morales and Lucas de Canizares and said that for their right and justice and for that of their companions it is suitable that no one approach the mine which they have registered, nor dig in it, nor for a distance of a quarter of a league around because the mine, according to the appearance of it, is of rocks different from others of crystal because they show that they are of another form, since they have sharp and faceted points. Therefore, they ask and require of the captain to command under grave penalties that no one nor any person dare to come to the mine nor within a quarter of a league around it to dig or make any other exploitation without the license and consent of all the partners. They asked for [a written] testimony and of those present that they be witnesses of it. Fernando Moyano de Morales, Lucas de Canizares, Juan de la Bandera, notary.

After the above on this said day, month, and year   FOLIO 23 VERSO the captain, having seen heard and understood what was asked by Hernando Moyano de Morales and Lucas de Canizeres, said that he was commanding and commanded that no person dare to approach to dig or to make any other exploitation of the mine for a quarter of a league around it, without the consent of all the persons who have a part in the mine, under pain of death, until such time as something else may be provided by His Majesty. Thus he commanded it and signed it. Juan Pardo, Juan de la Bandera, notary.

After the above, on December 11, 1567, the captain, in the presence of me, the notary, continuing his return, left the place called Yssa and on this day arrived with his company three leagues from there at another place which is also called Yssa because it belongs to the same [chief] Yssa. He was well received there. There the captain gave to the mandador of the place a chisel, with which he was very pleased, which I, the notary, attest. On the following day, which was December 12, 1567, he departed with his company from the place called Yssa and on this day continuing his return, he arrived five leagues farther at a place called Quinahaqui where he was well received. The following day, the thirteenth day of the said month, he rested in the place and he gave to the cacique of the place and to another who is called Otari yatiqui Orata and to his mandadores two chisels and four necklaces, with which they were very content. In addition to this, through Guillermo Rufín, interpreter, there was declared and said to Quinahaqui and Guaquiri and Yssa, caciques, because he had heard that they are discontented because they understand that he leaves them under the dominion of Joara Mico, that His Grace does not leave them nor does he wish that they be under the said dominion but rather that they go to make the "Yaa" to his ensign, whom he leaves in his place in FOLIO 24 Joara and that they render to him what they are obliged [to bring] and not to Joara. The caciques having understood the above made the "Yaa" many times, by which they gave it to be understood that the captain did them a very great favor and that they were very content to do what His Grace commanded them, which I, the notary, attest. Witnesses of all the above said were Pedro de Hermosa, sergeant, and Lucas de Canizares, and Luís Ximenez, soldiers of the company. Juan Pardo, Guillermo Rufín, Juan de la Bandera, notary.

After the above, in the presence of me, Juan de la Bandera, notary, and of the witnesses noted below, the captain with his company continuing the return on December 14, 1567, left the place called Quinahaqui on the way to Guatari and on this day he went five leagues and at the and of them, which was in an uninhabited place, he made a halt and slept that night. On the following day, he left the countryside (*campaña*) and on this day he arrived with his company at the place called Guatari Mico which is six leagues from the place where he slept. He was well received by the cacicas of the place. As soon as he arrived, he treated with the cacicas through Guillermo Rufín, interpreter, that they should command to come to the village (*lugar*), all the caciques, their vassals, so that they could help him build a fort which he had to build in that place in order to leave in it certain Spaniards who might defend them from their enemies. The cacicas made the "Yaa," letting it be understood that they were very content to do it thus. On the following day, which was December 16, 1567, some caciques began to come to the village. When they had come, which was very late in the morning, through Guillermo Rufín, interpreter, the captain summoned the cacicas and in order to content them he gave to them and to two sons of theirs and to mandadores FOLIO 24 VERSO and five other caciques eight wedges and nine large knives and three chisels and three mirrors and four necklaces and some pieces of red taffeta, with which they were very content. Then immediately another cacique came to whom and to two of his mandadores the captain gave another wedge and to another cacique who came with him, a chisel and a knife, with which they were very content. This day he laid out the fort and they began to build two "bastards" [i.e., makeshift fortifications?] with very great haste so that if on the part of His Majesty there should come to the captain a command to leave for another place, which might be more necessary, the soldiers who remained in the place might have a defense from their enemies, if they have any. Thus it was made and finished in five days, after which the captain, having seen that there had been nothing unusual for which he should fail to finish all of the fort, ordered summoned all of the Indians subject to the place. They gathered in three days. When they were gathered together with the soldiers of the company, an order was given to finish the fort in which there were built four tall cavaliers [i.e., raised corner bastions] of thick wood and dirt and its wall of high poling and dirt, with which it remained very strong. With the good order and haste which was made in the building, it was finished on January 6, 1568. In order that it remain with the strength and vigor which was suitable, he left as governor of the said fort Lucas de Canizares, corporal of the company, who is a person in whom are united all the qualities [required by] law and with him sixteen soldiers of those of the company. He charged and commanded them together with the said governor that they should not leave nor break the edicts which His Grace, in His Majesty's name, has commanded that they should guard and do in the fort. In addition to this, the captain, having considered FOLIO 25 and seen that it was suitable that Lucas de Canezares be given the legal power which His Grace had from His Majesty, in the presence of me, Juan de la Bandera, notary, and of the witnesses noted below, he said that he was giving and gave his complete and sufficient authority as he has it and as according to the law was required to Lucas de Canizares, corporal, especially, so that for him and in his name as he himself, representing his own person, should be and should reside in the fort which is given the name of Santiago and [in] the place where it is situated, in our Castilian tongue [called] the city of Salamanca, and [that] he should rule and govern the sixteen soldiers. He ordered them to do and carry out that which might be suitable to the service of His Majesty and if they do not do it, or in any manner differ in much or little from what they might be commanded he [Canizares] can compel and punish them, as he will compel and punish them and as the captain would compel and punish them, according to the crime or crimes which any of the said soldiers might commit. In addition to this, in order that he govern and maintain at peace the caciques and Indians of Guatari and the others who are subject to it, so that they may always be in good quiet and peace and because for the service of His Majesty it is suitable that Lucas de Canizares, warder, should not leave the fort until His Grace, or someone in His Majesty's name should command it, the captain {Look} commanded him in the name of His Majesty, to do it thusly, and that no one should dare to bring any woman into the fort at night and that he should not depart from the command under pain of being severely punished. Likewise he commanded and charged the sixteen soldiers to do and guard what they were commanded by Lucas de Canizares because the captain left him as their superior, with complete and sufficient legal power as the said captain has it from His Majesty such that he gave and authorized it to Lucas de Canizares with its incidental rights, accessories, appurtenances, and annexed rights FOLIO 25 VERSO with a free and general administration in what has been said. So that it might be certain, he obligated his goods and gave power to His Majesty's justices to compel and oblige him to [do] what is said as a thing decided and passed in judgment. In testimony of this, he authorized this letter before the notary and witnesses noted below and he signed it with his name. Likewise Lucas de Canizares signed it. His Grace, on a sign of the cross which he made with his right hand, took and received from him an oath by God and by Holy Mary, in due form and by the words of the four Holy Gospels, that he would carry out what His Grace in His Majesty's name has commanded him. Lucas de Canizares, having understood the above, acquitting the oath, said that he would carry it out thusly, under pain of perjury and infamy and of falling onto a condition of little value. Present at what is said, as witnesses were Pedro Guiterrez Pacheco, corporal, and Pedro de Hermosa, sergeant, and Luís Ximenez, soldiers of the company, and I, the notary noted below, attest the fact that I know the authorizing captain. He, considering and observing that it was necessary to leave munitions which might be necessary for the defense of the fort of Santiago and the soldiers of it, in the presence of me, the infrascript notary, gave and delivered to Lucas de Canizares fifty-one pounds of lead, and thirty-four pounds of matchcord and thirty-four pounds of powder, and some iron articles and various other things in order that he might please the caciques who might come to give obedience to His Majesty and two socketed axes for the necessary matters of the fort, which I, the notary, attest. Witnesses [were] Juan Pardo, Lucas de Canizares, Guillermo Rufín, Juan de la Bandera, notary.

After the above, on January 7, 1568, in the presence of me, Juan de la Bandera, notary, and of the witnesses noted below, the captain, Juan Pardo, with FOLIO 26 {X} his company continuing his return departed on this day from the city of Salamanca which in Indian language is called Guatari, returning toward Aracuchi. He went with his company five leagues and at the end of them he made a halt and slept in an uninhabited place. On the following day he left this place where he had slept, continuing his road onward, and he went five more leagues and at the end of them he made a halt and slept in the countryside (*campaña*). In this way he went three more days, sleeping for two of them in the countryside, with some labor because of the scant food that there was until at the end of the five days he arrived at the place called Aracuchi, where he was well received by the cacique. Thus he made a halt and slept that night. The captain, having seen the good will with which he was received, gave to the cacique and to his mandadores and to other caciques who came to that place a wedge, and a necklace, and a little red taffeta and five large knives and three chisels like wedges, with which the caciques and principal Indians remained very content. Witnesses were Pedro de Hermosa, sergeant, and Pedro Gutierrez, corporal, and Hernando Moyano de Morales, former sergeant of the company. The captain signed it, and Guillermo Rufín, interpreter, and I, the notary. Juan Pardo, Guillermo Rufín, Juan de la Bandera, notary.

After the above, on this day, January 11, 1568, before me, the notary, and the witnesses noted below, the captain appeared and said that as concerns the service of His Majesty, it is suitable that His Grace with his company, in spite of the fact that the road leads straight to the city and Point of Santa Elena, return [to them] taking a detour a distance toward the east, until arriving at a place which is called Ylasi, because he is informed that FOLIO 26 VERSO some caciques are to come to [promise] obedience to His Majesty, who before now have not come. But in order to spare his company some of the burden of the load, he wishes to send some soldiers with some Indians directly to a place which is called Canos, who will carry certain munitions because it is on the direct road to the city of Santa Elena. He asks me, the notary, to set it down thus and by sight to see the soldiers and munitions which he sends. I, the infrascript notary, having understood the above, saw how the captain sent thirty-four pounds of [lead] balls, and seventeen pounds of matchcord and a barrel of harquebus powder, in which there were twenty pounds, and two socketed axes and with these munitions, the following soldiers: Fernán Gomez, a Portugese, Pedro Alonso, a native of Xerez de la Frontera, Luís Ximenez, a native of Cuenca, Pedro de Barcena, a native of the Moral de la Reina. They departed with the stated munitions directly to the place called Canos, which [fact] I attest. Witnesses [were] Pedro de Hermosa, sergeant, and Pedro Gutierrez Pacheco and Francisco Hernández de Tenbleque, soldiers of the company. Juan de la Bandera, notary.

After the above, on January 12, 1568, in the presence of me, Juan de la Bandera, notary, and of the witnesses noted below, the captain, Juan Pardo, continuing to do what is suitable to His Majesty's service, on this day left Aracuchi with his company, returning toward Ylasi, where he stopped, usually going four leagues each day for five days, and in them there was much work because of the usual scarcity of food. After the said five days, he arrived with his company at the place called Ylasi, where he was well received by the cacique and Indians. He was there with his company four days because of the heavy rains while he was there and also to await, as he awaited, some caciques, who came to [make] obedience to His Majesty. In the said four days he gave to the cacique of the place   FOLIO 27 and to his mandadores, because he found that he had built a large house for His Majesty's service and two elevated rooms with some quantity of maize, a large socketed axe and two necklaces and a piece of red cloth and four chisels, with which they remained very content. Moreover, being in the same place, and Tagaya Orata with some Indians having come with him, he made him cacique because before now he had not been one and because of the many services which he has done and daily does for the Spaniards. He gave him a chisel and a wedge with which he remained very content. Moreover while he was in Ylasi, there came to him a cacique from the coast who is called Uca Orata, to [give] obedience to His Majesty and Saruti Orata, for the same [purpose,] because before now they had not come. Through Guillermo Rufín, interpreter, the customary speech was made to them and having understood it, they made the "Yaa," letting it be understood that they were very content to do and carry out what they were commanded. The captain, having seen and understood their obedience, gave to them and to their mandadores two wedges and seven chisels, with which they remained very content. Witnesses of all that is said were Pedro de Hermosa, sergeant, and Pedro Gutierrez Pacheco, and Francisco Hernández de Tenbleque, soldiers of the company. Juan Pardo, Guillermo Rufín, Juan de la Bandera, notary.

After the above, in the presence of me, Juan de la Bandera, notary, and of the witnesses noted below, the captain, Juan Pardo, with his company continuing the return [trip] on January 21, 1568, departed from Ylasi, returning toward Canos. On leaving the place following our direct road, we went the distance of about a league through a large swamp whose water was up almost to the knee and sometimes higher, very full of so much ice that as a result of it FOLIO 27 VERSO many of the soldiers of the company came out very dangerously wounded, so much so that if they were not helped, I put the outcome in doubt. In the end, notwithstanding this work and the fact that all day other swamps were not lacking, they went five leagues, at the end of which a halt was made. We slept in an uninhabited place, with {*Look*} some difficulty, because of a scarcity of food. On the morning of the following day we departed from here and marched six leagues at the end of which we arrived at a place which is called Yca Orata, where we slept that night. On the following morning the captain with his company—notwithstanding that from here he sent a corporal with twenty men and twenty Indians directly to a place which is called GuiomaE where there was a certain quantity of maize, in order that in certain sacks which he gave them they might carry what they could of it to a place which is called Coçao where His Grace commanded Hernando Moyano de Morales, former sergeant of his company [to go], and be there and gather all the maize which there might be, and keep it, and be with it until such time as His Grace arrived there with his company, because from there it would be carried to the city of Santa Elena where, he was now informed, it was needed because of the scarcity of foods in that city and its forts—went out [and] arrived at Canos, about two leagues away, where a large house had been built for His Majesty and in it a large quantity of maize, in order that being there he might give, as he gave, an order concerning carrying the said maize down a river to GuiomaE, which is, as is said, whence, with Indians and soldiers, it is to be carried to the city of Santa Elena. As he arrived, notwithstanding the [fact] that the cacique of the place was not in it, he was well received. This cacique came on the twenty-sixth day of the said month. When he had come, the captain, having seen FOLIO 28 that he brought meat and maize for the sustenance of his company, gave him a necklace and a bit of red taffeta and some enameled buttons, with which he was very content. Then immediately His Grace gave an order to the cacique that he have some canoes brought to him, in order to transport part of the maize down the river to GuiomaE, which is whence his company was to carry it to the city of Santa Elena. Thus the cacique, together with other caciques who aided him, had some canoes brought in which, as soon as they came, the captain had put all the maize that was possible. Having put it there he commanded Pedro de Hermosa, sergeant of his company, to go with three soldiers in the canoes and carry the maize to the place called GuiomaE and there have it unloaded and put in an elevated house which was built in the place for that purpose and having unloaded and put it in the elevated house to remain in the place with the soldiers and with others who, when he arrived, would already be in the place, who were those who had previously gone to carry the maize which was in the village to Coçao. With this, the sergeant departed and went to the place of GuiomaE where he found the soldiers who had carried the first maize. He was there with them {*look*} until February 11, when the captain came with the rest of his company and some Indians with him by whom and by the soldiers of his company he had transported all the maize that was in the place, in many deerskin sacks which the captain had had made for that purpose. With this, on the following day, which was February 12, the captain with his company and all the Indians who could FOLIO 28 VERSO be gathered—dismissing, as he dismissed, almost all the interpreters whom he brought, giving them as he did give them cloth and linen to dress themselves and axes and wedges and chisels and necklaces and enameled buttons without [counting] many other exchanges (*rescates*) which before now he had given them in the same manner, which I, the notary, attest, remaining as he remained with only Guillermo Rufín, Castilian interpreter in the company, to whom, in addition to all the things which the captain has given to the interpreters, now that at last we arrived at the place of GuiomaE, he gave him a suit of blue cloth from the wardrobe of the said captain and a brown hat (*capelo*) and shirts and deer skins to make doublets and stockings, with which Guillermo Rufín, as well as the other interpreters, was and remained very content, which likewise, I, the notary, attest— departed, on the route to the city of Santa Elena with great labor, because of the swamps and loads of maize and the scarcity of supplies. Continuing the return on February 14, he arrived at some fallen houses where a cacique, who was called Aboyaca Orata, used to reside. He came out to receive the captain with some meat and local roots which are called sweet potatoes (*batatas*) to eat, with which all the people of the company were very content, because they were in great need because of the scarcity of food there had been. To him, and to a mandador of his, the captain gave two painted Indian blankets,[8] with which they remained very content, which I, the notary, attest. With this, on the following day, we left the place, continuing the return and on this day we suffered hard work because of the many FOLIO 29 swamps which there were, which were more than there had been on any day, very deep and very wide, so much so that in three of them, when we had passed them, we considered it certain that Our Lord had showed some manner of miracle in our having crossed them because, they being so difficult, and the people coming so loaded, no one had fallen, nor had the load fallen from anyone. Thus all the company together with the captain, gave many thanks to Our Lord, as was right. With the good order which we maintained we walked seven leagues, at the end of which, because it was late, we made a halt and slept that night in a certain uninhabited place. In this order on the following day we arrived at a place which is called Coçao where we found the cacique and one [who is] his mandador with many Indians to go with the captain and with some quantity of meat and sweet potatoes (*batatas*) and acorns in order that the people might eat, and likewise Orista Orata with thirty Indians, and Uscamacu Orata with other Indians, all in order to go with the captain. All of whom, as they were in Coçao, received the captain very well. The captain rejoiced greatly because he had understood, according to the plentiful fables which he had had from some caciques, that Coçao and Orista and Uscamacu had rebelled and fled from our friendship. As soon as he arrived, he gave to each of them a painted blanket from the inland Indians, with which they were very content. The captain, having seen and understood the great number of Indians, had baskets and deerskin sacks prepared so that they might carry all the maize which he had gathered   FOLIO 29 VERSO in the said Coçao, which would be about sixty bushels (*fanegas*) of maize. When the maize had been gathered and placed and put in the sacks, in order to carry it on this said day we left [and went] two leagues beyond Coçao where we made a halt and slept that night, placing in addition to the customary guard, the guard and sentinels necessary because of the maize and the many Indians who came with the captain. With this order and good collection (*recaudo*) on the following day we left that place, continuing the return [trip] and we went five leagues, at the end of which, because all the people of the company and part of the Indians were very tired, we halted and slept that night. The captain commanded that a ration of one handful (*almoçada*) of maize between two men be given and with this on the following day we arrived at a place which is called Ahoya which is seven leagues from the city of Santa Elena and it is where the captain had the intention of building a house and fort because the cacique of the place and his Indians had previously killed a corporal of the company and also because from that place to the city it is necessary to go by water and until now every time Spaniards or inland Indians have come with some collection for the city the cacique and Indians of Ahoya have been very rebellious about giving the aid which was necessary in order to approach the city. Thus because of all that has been declared as well as to have the cacique and Indians of the place   FOLIO 30 under the royal dominion, the captain had the intention of building the fort. Now because the place appears to be uninhabited and burned and the cacique a prisoner in the said city and [because] Orista Orata having offered that the fort would be better in his village (*lugar*) because he and his Indians would serve the soldiers who were in it very willingly and that he would have a complete supply of canoes and Indians in order that whatever fulfills the service of His Majesty might be done, he [the captain] decided to abandon building the fort in Ahoya and to go down to build it in the village of Orista Orata. With this purpose, through Guillermo Rufín, interpreter, who was present, he told and commanded certain mandadores and Indians of Ahoya who came out to receive the captain that they immediately go to the Indians who were absent, since they were his brothers, and on his part command them to come to the place and live in it, since His Grace only considered as enemies those who have done a disservice to His Majesty, and not others, and thus, in any case they should come to live in their town (*pueblo*). If they did not do it thus, he would understand that they had been traitors to him and as such henceforward he would have to make war upon them. The said mandadores and Indians, having understood the will of the captain, made the "Yaa" many times, letting it be understood the favor which was done to them and that they would immediately make the Indians return and settle the place. The captain, having seen the goodwill of the mandadores and Indians, gave to each one of them a painted breechclout[9] like a doily, with which they remained very content, which I, the notary, attest. Witnesses   FOLIO 30 VERSO of all that is said were Pedro de Hermosa, sergeant, and Pedro Gutierrez Pacheco, and Francisco Hernández {#} de Tenbleque, soldiers of the company. Juan Pardo, Guillermo Rufín, Juan de la Bandera, notary.

After the above in the presence of me, Juan de la Bandera, notary, and of the witnesses noted below, the captain, Juan Pardo, with his company, continuing the return and the building of the stronghouse on February 19, 1568, departed from the place named Ahoya, toward Orista, and, by means of all possible work from the soldiers of the company, had carried on their shoulders all the quantity of maize, which he brought from the interior for the support of the said people, for a league from Ahoya in order to embark it and carry it where it was necessary. As soon as he had finished bringing it to the port and place where it was to be loaded on the canoes in which it was to be brought, we embarked in the canoes and all the company with it. In this manner at about two o'clock in the afternoon we came to disembark a quarter of a league from Orista and from there we carried all of His Majesty's maize and munitions as best we could to Orista and we put it in a large house belonging to the cacique. With this we rested and slept that night. On the following day the captain gave an order to build the stronghouse and they began to cut and saw the necessary wood. On March 2, it was finished, to which, because it was the first that was built at the beginning of the FOLIO 31 interior, there was given the name "Nuestra Señora" and to the town (*villa*) {*The town*} "de Buena Esperança." The captain, having seen that it was suitable to leave as warder of the fort a man who will be concerned with the guard and defense of it and who will preserve in good friendship the caciques and Indians of the town and of its neighborhood, put and left as warder Gaspar Rodriguez, corporal of his company and with him, to do what the warder commanded them to do in the service of His Majesty, twelve soldiers of his company. To Gaspar Rodriguez be gave the legal power which His Grace has from His Majesty. In addition to the above, before me, the notary, he took and received from him an oath by God and Holy Mary in due form and in the words of the four Holy Gospels where they are most fully written, that he would keep and carry out what is said and declared and he would not break it under pain of perjury and infamy, and falsity and of falling into the case of being of little value. The said Gaspar Rodriguez, warder, acquitting himself of the oath said "Yes, I swear" and "Amen." Because he said that he did not know how to write he asked a witness to sign for him. The captain, having understood the above and having seen that for the defense of the fort it was fitting to leave some munitions and food for the people in it, before me, the notary, gave to the warder the following: eighteen pounds of powder, eighteen pounds of matchcord, eighteen pounds of lead, a socketed axe, a large hoe, a chisel, a drill, [and] provisions of corn for two months, allowing one handful (*albra*) FOLIO 31 VERSO of corn a day to each man. Witnesses who were present at all that is said were Hernando Moyano de Morales, and Antonio de Aguirre, and Pedro Gutierrez Pacheco, corporal of the company, being in the town of Buena Esperança. Juan Pardo. For witness, Antonio de Aguirre. Juan de la Vandera, notary.

After the above in the town of Buena Esperança, on February 28, 1568, in the presence of me, Juan de la Bandera, notary, and of the witnesses noted below, the captain, Juan Pardo, having understood that in the forts and city of Santa Elena there is very little food for the number of people whom he brings in his company and that greater service would be done to God and to His Majesty if his people separated and were in a place where they might support themselves and did not go to the city of Santa Elena where {look} because of the many people they might suffer great harm and detriment, he provided and commanded Pedro de Hermosa, sergeant of his company, to go to the city of Toledo and fort of Santo Tomás, which in the Indian language is called Canos, with thirty soldiers of his company and to be and remain there in the said city and fort as warder of it and to rule and govern and administer the said thirty soldiers and guard the fort and not to leave it without his permission or that of one who might order it on His Majesty's behalf. In addition to this, he enjoined and commanded him to take great care always to preserve the friendship of the cacique and Indians of Canos and of those around it FOLIO 32 and for all that is said and declared he gave him the legal authority which His Grace has from His Majesty, in order that he might punish any one of the soldiers who in any manner acts against the service of His Majesty, according to the crime which each one might commit. For the fulfillment of all that which is said he took from him an oath in due form that he would guard it and would not change it under pain of perjury, and infamy, and falsity. Pedro de Hermosa, sergeant and warder, acquitting himself of the oath said "Yes, I swear" and "Amen" and he promised to fulfill it thus. He signed it with his name, together with the said captain, there being present as witnesses Hernando Moyano de Morales, and Antonio de Aguirre, transients in the town. Juan Pardo, Pedro de Hermosa, Juan de la Bandera, notary.

After the above, on February 28, 1568, in the presence of me, Juan de la Bandera, notary, and of the witnesses noted below, the very magnificent captain Juan Pardo appeared and asked and required that I, the notary, attest and give him [written] testimony that if His Grace left his ensign and some corporals of his company distributed in certain places with the soldiers who were necessary, as is well known and that if now he sends his sergeant with thirty other men of his company, it was and is because it was for the service of God, Our Lord, and of His Majesty, because of the security and strength which it is necessary to have in these provinces of Florida and because in the city and port of Santa Elena at present FOLIO 32 VERSO there is not enough food with which to support fifty men for two months, much less the number of persons that he has at present {Look} in his company, who are more than one hundred men. I, the notary, attest it. He said that he asked (*pedía*) and asked (*pidió*) me this in order to present it in the discharge which he will give to His Majesty if, perhaps, he is asked to give account of the reason he divides his company. I, the notary, having understood and seen that the said request is just, as has been seen and understood by the persons of all the company, gave the attestation {look} to the captain in [written] testimony as he requested it of me, on the said day, month, and year, there being present as witnesses, Pedro de Hermosa, sergeant, and Hernando Moyano de Morales, and Antonio de Aguirre, transients in the fort of Nuestra Señora de Buena Esperança, which is where it was requested of me. Juan Pardo. Passed before me, Juan de la Bandera, notary.

Moreover on this said day February 28, 1568, before me the notary and the witnesses given below the captain gave an order and commanded Pedro de Hermosa, sergeant of his company, that from the people from the company whom he carries in his charge, he leave in the place of Madrid, which in the Indian language is called GuiomaE, Fernán Gomez, a Portuguese, and three other soldiers with him. He commanded this said Fernán Gomez to {look} have GuiomaE, cacique, and the other caciques who are his companions and friends make haste in building four canoes for which His Grace gave them an order and commanded them to build for FOLIO 33 the service of His Majesty. For the said purpose he gave to Fernán Gomez and to the caciques a socketed axe and an adze. By the same order he charged and commanded Alonso Bela, a soldier of his company, who is one of the persons who go in the charge of the said sergeant, to be greatly concerned and very careful in going to Ylasi Orata, which is where Alonso Bela resided for some days, and to command that cacique and the rest who are his subjects to build and to have built three canoes for which His Grace gave an order and [which] he commanded when he was in Ylasi that they might be built since for that [Purpose] His Grace gave to that cacique a large socketed axe, and now at present [he gives] another to Alonso Bela, which I, the infrascript notary, attest, since I knew and understood that it was suitable to do it thusly for the service of His Majesty. Moreover, in the presence of me, the notary, and of the witnesses noted below, on this said day the captain gave an order and commanded Antonio Diaz Pereyra, Portuguese, and Domingo de León, and Juan Toribio, soldiers, to go to Guando Orata which is where there is a house [full] of His Majesty's maize, and of that maize to try to bring, with [the aid of] the Indians of the said cacique, all that they could to the town of Buena Esperança, which in the Indian language is called Orista. For the said purpose, before me, the notary, he gave and delivered to them a certain number of sacks and he charged and commanded that it be done and that they do it with great diligence and care. Consequently he charged and commanded all of the above to try to have the said caciques FOLIO 33 VERSO and Indians with them sow a large quantity of maize because by all of this they would serve Our Lord and His Majesty very much. Present as witnesses at what is said were Hernando Moyano de Morales, former sergeant of his company, and Pedro Gutierrez Pacheco, corporal of it, transients in the town of Buena Esperança, and the captain signed it. Juan Pardo, Juan de la Bandera, notary.

And after the above in the town of Buena Esperança, which by another name is called Orista, on March 1, 1568, in the presence of me, Juan de la Bandera, notary, and of the witnesses noted below, the captain, Juan Pardo, being present with his company, there came to the town of Buena Esperança four caciques, who were called Yetencunbe Orata, Ahoyabe Orata, Coçapoye Orata, [and] Uscamacu Orata and some principal Indians with them to bring food for the company, [all of] whom the captain received well. His Grace, having seen that willingly they brought from what they had, gave to the two most important ones two large painted blankets of catskin,[10] and four painted breechclouts from the Indians of the interior, with which they were very content. Then immediately His Grace summoned Guillermo Rufín, interpreter, through whom His Grace had the caciques and Orista Orata and all FOLIO 34 his principal Indians gather in a large hut where they were accustomed to meet. When all were together, he made them the customary speech and in addition to the speech, through Guillermo Rufín, interpreter, he said to them that they already knew that His Grace with his said company had built in Orista a stronghouse in which, to make them content, he left thirty men in order to teach them and let them know how they are to do service to God, Our Lord, and to His Majesty and so that these soldiers might defend them from their enemies. He asked them to treat them well and give them of what they had and in any case to sow for them a large piece of land with maize because it would give him great pleasure. The caciques and Indians made the "Yaa" many times by which they let it be known that they would fulfill it as His Grace commanded it. In addition to the above, through Guillermo Rufín, interpreter, there was declared and said to them how it was necessary, since they were brothers of the captain and vassals of His Majesty, henceforth to come to Orista and to bring what they might have and for all to obey Orista. On hearing what is said, Orista made the "Yaa" many times, letting it be understood the great favor which was done him, and the others, consequently, made the "Yaa" many times, letting it be understood that they were very content to do what they were commanded and to come to the said obedience. The captain, having seen this, thanked them very much. In addition to the above FOLIO 34 VERSO through Guillermo Rufín there was said and declared to them that they already knew that Ahoya Orata and other Indians of his had killed one of the Spaniards of his company and for that [reason] he has Ahoya in prison and his Indians have depopulated the place and burned it and have fled from it without having any reason, for which reason it is fitting that they again settle the said place and [that] all those who are in rebellion return to their houses and repose. In order that they understand that he will do well to those who are his good brothers, and consequently punish those who are not [his good brothers], that on the twentieth day from today they shall gather in Orista and that when they are gathered His Grace will declare to them and will tell them what they have to do, punishing those who are at fault and considering those who are not as his brothers, as they were before. To which all the caciques made the "Yaa" many times by which they let it be understood that the captain did them a very great favor and that thus they would see that all were present at the end of the twenty days and they would send to call His Grace, whom they asked to come. To which the captain replied to them that he was very content and thus he took his leave giving them to understand that on the following day he would like to leave together with his company and that he wished them to gather certain canoes which were necessary. All replied that they would do it. Present as witnesses at what is said were Hernando Moyano de Morales, former sergeant of the company, and Pedro Gutierrez FOLIO 35 Pacheco, corporal of it, and Luís Jimenez, soldiers of the company, transients in the town. Juan Pardo, Guillermo Rufín. Juan de la Bandera, notary.

After the above, on March 2, 1568, in the presence of me, the notary, and of the witnesses noted below, the captain, Juan Pardo, continuing the return [trip], departed with his company directly to the point and city of Santa Elena where we arrived on the declared day and from which for the said purpose we departed. And returning, we arrived on this said day at about three o'clock in the afternoon with good fortune, notwithstanding the many difficulties which the captain with his company has passed and the fact that in six months and two days after he left with the intention of finishing and ending this journey in the service of God, Our Lord, and of His Majesty, although it is not finished, he has had no misfortune at all. Many thanks be given to Our Lord for such great mercy and may He by His holy and sacred passion give us grace that the things which we may do in His holy name, that He may lead us to, may have as good a beginning and end as this journey has had, amen. Dated as above and passed before me. Juan de la Bandera, notary.

{*Verification*} After the above, in the presence of me, Juan de la Bandera, notary, on March 31, 1569, being in the city of Santa Elena which is in the provinces of Florida, the very illustrious gentleman, Pedro Menéndez de Avilés, governor and captain general of these said provinces, having seen what was done and accomplished by Juan Pardo, captain for His Majesty, on the journey and discovery of land through the provinces of Florida on the way to New Spain with the order of the said Adelantado in His Majesty's name, in order to learn and understand whether all that is written in the report of what was done on the said trip was done thus, wished to make and made a verification of it with soldiers, men of good faith and of credit, who were on the said journey, who are the captain, Juan Pardo, and Francisco Hernández de Tenbleque, corporal and Francisco Corredor, and Francisco de FOLIO 35 VERSO Cisneros and Antón Múñoz and Juan de Billola, soldiers of thecompany of the captain, Juan Pardo, from whom, His Lordship, in the presence of me, the notary, took an oath by God and by Holy Mary in due form de jure, concerning the above in order that under that oath they may say and declare whether that which is written in the report by me, the notary, occurred just as it is written. All [of them,] as they are said and declared, acquitting themselves of the said oath, with one voice said that all that which is written concerning the land and the speech which is made to the caciques of each place and how they remained under the dominion of His Majesty according to what on His Majesty's part was declared to them, is just as it is written in the report because at all of it they were present, which is the truth by the oath which they made. Those who knew how to sign signed it with their names and for the rest, who did not know how to sign, a witness signed together with His Lordship and with me, the notary, there being present as witnesses Alonso Garcia, ensign of the company of the said captain, Juan Pardo, who signed as witness, and Gaspar Rodriguez, corporal and the Ensign Gonçalo del Ribero and Garcia Martinez de Bolo, transients in the city of Santa Elena. Pedro Menéndez, Juan Pardo, Francisco de Cisneros, Antón Múñoz, as witnesses, Alonso Garcia, Juan de la Bandera, notary.

Because of the request and in order to correct and compare what is contained in this writing with the original, Alonso Garcia, warder of the fort of San Felipe and ensign of the company of the captain, and Francisco Maldonado, and Luís Ximenez, soldiers, transients in the said Point of Santa Elena, have signed in the margin. The notary being I, Juan de la Bandera, notary for a large area of these provinces for the very illustrious gentleman Pedro Menéndez de Avilés, adelantado of them in His Majesty's name, was present together with the said witnesses and I attest the above, because it occurred before me. In testimony of truth I affix my sign [Rubric].

Notes

1. Ketcham, "Three Relations," 66–67.
2. That is, in accordance with the amount of tribute levied, which seems to have been proportionate to some base in each village, probably houses or possibly even a rough count of population supplied by the chief.
3. The Spanish is *municiones las que abastaban para los dichos soldados*.
4. The Spanish is *un caracol grande de la mar*, which is literally a large marine shell of the periwinkle type, i.e., a mollusk shell of the genus *Littorina* or genus *Thais*. Archaeological evidence from throughout the Southeast suggests that it was probably a large conch shell.
5. "Passaments": decorative braids, cords, tassels, fringes, etc. used on clothing.
6. Literally, a *mandador* is a "driver," possibly a war captain, or else a functionary with coercive powers.
7. A Jurado was a member of a special panel of officials elected and sometimes appointed to represent the public interest in various matters of city government. At Toledo they were the sworn defenders of the laws and privileges (*fueros*) of the city, had oversight of the judicial system, and saw to it that the city's patrimony was conserved and expanded and well administered (Gamero 1862:825–29). The term has nothing in common with the English jury, whose name in Spanish is the same.
8. These "painted blankets" could have been mulberry fiber matchcoats woven by the Indians (amply documented by the Soto chroniclers) or else skin matchcoats with designs painted on them.
9. Presumably these painted breechclouts were of the same stuff as the painted matchcoats. The reference to a "doily" is an attempt to describe what they looked like in terms of an object familiar to Spaniards: the often elaborate doilys placed under lamps or used as center decorations on tables.
10. These may have been matchcoats of cougar skins with designs painted on the inside.

The "Short" Bandera Relation

AGI, Patronato 19, R. 20.

Transcription

Florida   1566   Sta Elena

Memoria de los lugares y que tierra es cada Lugar de los de las provi[nci]as de la florida / por donde El capitan Ju[an] pardo Entro a descubrir camino para nueva españa desde la pun/ta de sancta elena de las dichas provi[nci]as los a[ño]s de qui[nient]os y sesenta y seis y qui[nient]os y sesenta y siete / q[ue] todo es como se sigue: ———

v. Primeramente salio de Sancta elena con su compania prosiguiendo El d[ic]ho / efecto y el dia que salio fue a durmir a un lugar que se dize uscamacu / aqui es ysla cercada de rrios tierra arenisca y de muy buen barro para ollas / y teja y otras cosas que sean nescesarias ay en esta ysla buenos pedaços de t[ie]rra / para mayz y mucha çepa de viña ———

v. Desde uscamacu salio derecho a otro lugar que se llama ahoya a do[nde] hizo auto / y durmio este ahoya es ysla algunos rrincones della cercados de rrios y los / demas como tierra firme y rrazonable tierra para mayzes y tan bien mu/chas çepas de viñas con muchos sarmi[ent]os ———

v. Desde ahoya salio derecho a otro lugar q[ue] se llama ahoyabe puebo pequeño / subjeto a ahoya y la misma tierra q[ue] es ahoya:

v. Desde ahoyabe salio derecho a otro lugar que se llama coçao que [e]s un caciq[ue] / algo grande y tiene mucha tierra buena como las demas dichas y muchos / pedaços de tierra pedrisca donde se puede cultivar el mayz el trigo la ceva/da la biña todo genero de frutas y huertas porque ay rrios y arroyos dulces / y rrazonable tierra para todo

v. Desde coçao salio derecho a otro lugar pequeño que [e]s de un mandador del mis/mo coçao la tierra deste lugar es buena pero poca

v. Desde este lugar salio derecho a otro que se dize el Enfrenado tierra es misera / aunque ay muchos rrincones de muy buena tierra como las demas d[ic]has

v. Desde el enfrenado salio derecho a otro lugar que se llama guiomaE[1] desde don/de hasta la punta de sancta elena ay quare[n]ta leguas el camyno por / donde se fue algo trabajoso pero tierra que se puede cultivar todo lo que / en coçao y aun mejor ay algunos pantanos grandes y hondables pero / causalo la mucha llanura de la tierra:

v. Desde guiomaE salio derecho a canos que los Indios llaman canosi y por otro non/bre cofetazque ay en el ter[mi]no de [e]sta tierra tres o quatro rrios rrazonables y el / uno muy caudaloso y aun los dos ay algu[n]os pantanos pequeños q[ue] qualquier / persona aunque sea muchacho los puede pasar por su pie ay en este trecho / valles altos de mucha piedra y peña y baxos es tierra bermeja muy buena / en efeto muy mejor que todas las dichas

v. Canos es tierra q[ue] Pasa uno de los dos Rios caudalosos cabe [a] el y otros arroyos / tiene muy grandes vegas y muy buenas y aqui y desde aqui adelante / Se coje mucho mayz y ay mucha uva gruesa y muy buena y mala / gruesa y menuda y de otras muchas maneras al fin es tierra a que / Se puede situar pueblo principal [slash] ay hasta Sancta elena cinq[uen]ta legu[a]s / y hasta la mar como veynte leguas puede se yr hasta el por el rrio dicho / cursando la tierra y mucho mas adelante por el mismo rrio y asy mysmo por / El otro que Pasa junto a guiomaE ———

v. Desde canos Salio derecho a otro lugar que se llama tagaya muy principal FOLIO I VERSO sin pantanos tierra rasa de poca arboleda prieta y bermeja muy buena y de mucha buen agua fuentes y arroyos ———

v. Desde tagaya salio derecho a otro lugar que se llama gueça[2] tierra ny menos / ny mas que la de arriba muy abundante de buena ———

v. Desde tagaya digo de gueça salio derecho a otro lugar que se llama aracu/chy tanbien t[ie]rra muy buena ———

v. Desde aracuchi salio derecho a otro lugar que se llama otari y atiqui que [e]s ca/cique y lengua de mucha tierra adelante tierra muy abundante de / buena desde este otari a otro lugar que se llama guatari ay como quinze / o diez y seis leguas a la mano derecha mas debaxo del norte que / este otro en este a avido y ay dos caçicas q[ue] son señoras y no poco en conpara/cion de los demas caciques porque en su traje se siruen Con pajes y damas / es tierra rrica ay en todos los lugares muy buenas casas y buhios terre/ros rredondos y muy grandes y muy buenas es tierra de sierra y Campiña / buena todo lo del mundo este lugar le vimos y estovimos[3] veynte dias / de buelta Junto a este lugar pasa un rrio muy caudaloso que viene a / dar a sauapa y usi donde se haze sal[4] Junto con la mar Sesenta leguas / de sancta elena desde sancta a este guatari ay ochenta leguas / y por este mismo rrio se puede entrar mas de veynte sigun dizen qual/quier navio: ———

v. Desde otari y atiqui salio derecho a otro lugar que se llama quinahaqui don/de pasa otro rrio muy caudaloso es t[ie]rra muy buena y muy buena:

v. Desde el lugar atras declarado a la mano yzquierda doze leguas del ay otro / lugar que se llama yssa [slash] que tiene muy lindas begas y toda la tierra / muy linda ay muchos rrios y fuentes en la Jurisdicion de [e]ste yssa hallamos / tres minas de cristal muy bueno estas estan rregistradas en feto como / si luego se oviera de Sacar provecho dellas todo esto vimos y entendimos / a la buelta que volvi[mo]s a Santa elena:

v. Desde quinahaqui salio derecho a otro lugar que se llama aguaquiri que es t[ie]rra / muy acabada y fertil de buena y fertil:

v. Desde aguacari [sic] salio derecho a otro lugar que se llama Joara que esta Junto a la / sierra y es donde Juan pardo a la primera Jornada que hizo llego y que/do su sargento se dezir que es tan linda tierra como la ay en la mejor de toda / españa para todos quantos generos de Cosas los onbres en ella quierea[n] / cultivar ay hasta S[an]ta Elena cien leguas:

v. Desde Joara Salio por la sierra adelante derecho a otro lugar que se llama / tocar [sic] donde en la pasar tardamos tres dias en esta sierra ay mucha / uva mucha castaña mucha nuez mucha cantidad de otras frutas es / mejor que sierra morena porque ay en ella muchas vegas y la tierra / muy poco fragosa en tocar es muy buena tierra donde se puede hazer / grandes labranças de qu[a]lqui[e]r suerte:

v. Desde tocar salio derecho a otro Lugar que Se llama Cauchi muy principal t[ie]rra / desde aqui adelante conpare esta tierra con el andaluzia porque es muy / rrica tierra toda ella:

v. Desde cauchi Salio derecho a tanasqui que tardamos en llegar a el tres dias por / despoblado es una t[ie]rra tan rrica q[ue] no se como me lo encarezca FOLIO 2

v. Desde tanasqui Salio derecho a otro lugar que se llama olameco y por otro nonbre / chiaha es tierra muy rrica y anchurosa lugar grande cercado de rrios muy / lindos ay en derredor deste lugar a legua y a dos leguas y a tres leguas y me/nos y mas muchos lugares pequeños todos çercados de rrios ay unas ve/gas de bendiçion mucha uva y muy buena mucho nispero en efeto es tierra / de angeles:

v. Desde olameco salio derecho al poniente a un lugar que se dize ohalahume [sic] / a donde tardamos en llegar tres dias por despoblado y a donde hallamos / sierras mas asperas q[ue] la sierra q[ue] nonbramos en estos fuertes por donde pasa/mos es tierra muy Rica y agradable y fresca al subir de una Sierra des/tas hallamos humo de metal y preguntando a los alquimystas dixe/ron con Juramento que hera de plata llegamos a chalahume que tie/ne tan buen sitio de tierra en conparacion como tiene la ciudad de cordoua / muy grandes vegas y muy buenas alli hallamos uvas tan buenas como las / ay en españa se dezir que [e]s tierra que paresce que [e]spanoles la [h]an cultivado /sigun es buena:

v. Desde chalahume salio derecho a otro lugar que esta dos leguas de alli y se / dize satapo desde donde nos bolvimos es pueblo rrazonable de buenas / casas y mucho mayz y muchas frutas silvestres pero la tierra rrica / y muy agradable y todos estos lugares y los de atras situados Cabe / muy lindos rrios :

v. Desde satapo aviamos de yr derechos a cosaque: creo yo sigun me informe / de yndios y de un soldado que llego alla de [e]sta conpania y bolvio y dio qu[en]ta / de lo que vido ay cinco Jornadas o seis hasta cossa tierra muy poco pobla/da porque no ay mas de tres lugares pequeños el primero que [e]sta dos Jor/nadas de Satapo q[ue]

The "Short" Bandera Relation

se dize tasqui en estas dos Jornadas ay buena tierra / y tres rrios grandes y un poco mas adelante otro lugar que se dize tasquiqui / y desde alli otra Jornada mas adelante otro pueblo destruydo q[ue] se dize olitifar / todo buena tierra llana y desde alli a otras dos Jornadas del despoblado / mas adelante esta un lugar pequeno y mas adelante de [e]ste Como una / legua cosa [colon] es pueblo grande el mayor que ay desde S[an]ta helena por don/de fuymos hasta llegar a el tendra como hasta 150 vezi[n]os esto sigun / el grandor del pueblo es lugar mas rrico que ninguno de los dichos ay en el / de hordinario gran cantidad de yndios Esta situado en tierra baxa a la hal/da[5] de una sierra ay en derredor del a m[edi]a legua y a quatro de legua y a le/gua muy muchos lugares grandes es tierra muy abundante esta su sitio / al sol de m[edi]o dia y aun a menos de a m[edi]o dia [colon, slash] desde cossa aviamos de yr derecho / a trascaluça que es el fin de lo poblado de la florida ay desde cossa a tras/caluça siete jornadas y Creo que ay en todas ellas dos lugares o tres todo / lo demas es despoblado trascaluça se dize que esta al sol de m[edi]o dia y que des/de aqui a tierra de nueva espana ay unos dizen que nueve Jornadas otros / que onze otros que treze y lo mas comun nueve Jornadas todo de despoblado y / en el m[edi]o de todo este camyno ay un lugar de 4 o 5 casas y despues pro-siguie[n]/do en el d[ic]ho efeto la primera poblacion q[ue] ay es de nueva espana sigun dizen / y rruego a n[uest]ro s[eño]r lo provea como se le haga servi[ci]o ame[n] f[ec]ha en la punta de S[an]ta Elena / veynitres dias del mes de henero ano de myl y qui[nient]os y ses[en]ta y nueve a[ñ]os Juan de la Vandera [rubric]

Notes

1. The text appears to read *hujomaE* but has been corrected at the "a" and perhaps had the "E" added.
2. Or *gueça*.
3. I.e., *estuvimos*.
4. An "A" is crossed out here with double lines.
5. I.e., *falda* or skirt; I have translated it as "lap."

Translation

Florida 1566 Santa Elena

A memorial of the places and the sort of land each place is of those [places] of the provinces of Florida that Captain Juan Pardo entered [while] discovering a road to New Spain from the Point of Santa Elena of the said provinces in the years 566 and 567. They are as follows:

First, he departed from Santa Elena with his company pursuing the said purpose. The day he left he went and slept in a place that is called Uscamacu. Here is an island surrounded by rivers. The ground is sandy and has very good clay for pots and roof tiles and other things that may be needed. On this island there are good parcels of soil for maize and [there are] many vinestocks.

From Uscamacu he went directly to another place that is named Ahoya, where he stopped and slept. This Ahoya is an island some of whose nooks and corners[1] are surrounded by rivers and the rest [are] like a mainland. The soil [is] reasonable for maize. [There are] also many vinestocks with many runners.

From Ahoya he went directly to another place that is named Ahoyabe, a small town [pueblo] [that is] subject to Ahoya. The soil is the same as Ahoya's.

From Ahoyabe he went directly to another place named Coçao, who is a chief of some importance. It has much good land, like the other [soils] already described, and many patches of stoney soil where maize, wheat, barley, grapevines, all sorts of fruits and vegetables may be grown because there are sweet rivers and creeks and fair soil for all [of them].

From Coçao he went directly to another small place that is [the village of] a *mandador* of the same Coçao. The soil of this place is good but [there is] little [of it].[2]

From this place he went directly to another place that is called *El Enfrenado* ["the Bridled"]. The ground is miserable although there are pockets of very good soils like those already noted.

From El Enfrenado he went directly to another place that is called GuiomaE.[3] From there to the Point of Santa Elena is [some] forty leagues. The road by which he went was somewhat difficult but ground that can be cultivated [is] all that [is] in Coçao and even better.[4] There are some large, deep swamps but the cause of them [is] the great flatness of the land.

From GuiomaE he went directly to Canos, which the Indians call Canosi and, for another name, Cofetazque. In the district of this land there are three or four fair rivers. One, or maybe two [of them] have a high volume of water. There are a few small swamps which anyone, even a boy, can pass on foot. In this stretch [of the countryside] there are high and low valleys with much rock and [many] boulders.[5] The soil is bright red [and] very good, in effect, very much better than all those noted so far.

One of the two rivers of high volume passes by Canos, [as do] other creeks. It has extensive, very good flatlands [i.e. valley floors]. Here and from here on much maize is harvested and there are many large, very good grapes and large and small *mala* [*mailla*?, i.e., crabapples?] and many sorts of other [wild fruits?]. In short, it is a land where a principal town could be situated. It is fifty leagues to Santa Elena and about twenty leagues to the sea. It [the sea] can be reached by [means of] the said river, crusing through the land, [as can areas] much further on [i.e., inland], by means of the said river. This [can] also [be done] by means of the other [river] that passes next to GuiomaE.

From Canos he went directly to another place that is named Tagaya, [which is] very important. [It is] without swamps. The land is plateaus with little tree cover. [The soils are] blackish and bright red, very good. [There is] much good water [from] fountains and creeks.

From Tagaya he went directly to another place that is called Gueça, a land just like that above and abundant in good. . . . [6]

From Tagaya, I mean Gueça, he went directly to another place that is named Aracuchy, [which is] also a very good land.

From Aracuchi he went directly to another place that is named Otari and Atiqui, which is the head and spokesman for a large area further on. This is a very abundant land of good [soil?]. From this Otari to another place that is named Guatari there are about 15 or 16 leagues, to the right, further below the north than this other one. In this [place] [Guatari], there have been and are two chieftanesses who are the lords and not unimportant in comparison to the other chiefs because in their going about[7] they are served by pages and ladies. It is a rich land. Good houses and humble, round huts as well as very large and very good [huts] are [to be found] in all the settlements. It is a land of mountain ridges and flat tracks of arable land, good [for] all [the crops] of the world.[8] We saw and were in this place for twenty days on the return. Next to this place passes a very full river that gives a way to Sauapa and Usi where salt is made next to the sea 60 leagues from Santa Elena. From Santa Elena to this Guatari is 80 leagues. They say that any [sort of] ship could sail more than 20 leagues up this river.

From Otari and Atiqui he went directly to another place that is named Quinahaqui, where another very full river passes by. It is a very good land and very good.

From the place declared above [i.e., Otari and Atiqui?], on the left hand [and] 12 leagues from it is another place that is called Yssa, which has very beautiful flat lowlands [i.e., valley floors]. All the land is very beautiful. There are many rivers and fountains of water. In the jurisdiction of this Yssa we found three mines of very good crystal. These are recorded so that later they might be worked. We saw and understood all this on the return when we returned to Santa Elena.

From Quinahaqui he went directly to another place that is named Aguaquiri, which is a well-finished land and fertile of good [omission?] and fertile.

From Aguacari he went directly to another place that is called Joara, which is next to the mountain ridges and is where Juan Pardo arrived on his first trip and left his sergant. It is said that it is as beautiful a land as there is in the best [that there is] in all of Spain for all sorts of things that men might wish to cultivate in it. It is about 100 leagues from Santa Elena.

From Joara he went over the mountain range before it directly to another place that is named Tocar [sic]. We took three days to pass [over] this [mountain range]. On this mountain range there are many grapes, many chestnut trees, many nuts, [and] quantities of other fruits. It is better than the Sierra Morena because there are many flat lowlands and the soil is not very rough to the touch. It is a very good land where great harvests of all sorts can be made.

From Tocar [sic] he went directly to another place that is called Cauchi, a very important land. From here onward I will compare this land to Andalucia because it is all a very rich land.

From Cauchi he went directly to Tanasqui. It took us three days to reach it [going] through an uninhabited area. It is such a rich land I don't know how to extoll it.

From Tanasqui he went directly to another place that is named Olameco and, for another name, Chiaha. It is a very rich and wide land. [It is] a large place [settlement] surrounded by very pretty rivers. Around about this place are many small settlements at [distances of] a league or two or three or more or less, all surrounded by rivers. There are some flat lowlands *de bendición*. [There are many] very good vines and many medlar trees. In sum, it is the land of the angels.

From Olameco he went directly to the west to a place that is called Chalahume. We were three days getting there [passing through] an uninhabited area. Here we found the mountain ranges rougher than the mountain ranges that we name in these forts where we passed [*sic*!]. It is a very rich, agreeable and fresh land. On climbing one of the ridges of these [mountain ranges] we found a trace of metals. When questioned, the alchemists swore that it was silver. We arrived at Chalahume which has as good a site of ground, in comparison, as the city of Cordoba has. [It has] very large and very good flat lowlands. There we found grapes as good as those in Spain. Is said that it is a land that looks like Spaniards have cultivated it because it is so good.

From Chalahume he went directly to another place that is two leagues from it and is called Satapo, from which we returned. It is a fair town with good houses and much maize and many forest fruits but the land is rich and very pleasing. All of these settlements and those behind them [i.e., already passed through] are situated next to very lovely rivers.

From Satapo we should have gone directly to Cosaque. I believe, according to what I was told by the Indians and a soldier of the company who went there and returned and gave an acount of what he had seen, [that it] is five or six days' journey to Coosa. The land [on the way] is very lightly inhabited because there are no more than three small settlements. The first is two days' journey from Satapo and is called Tasqui. During these two days' journey there is good land and three large rivers. A bit further on [from Tasqui] [is] another place that is called Tasquiqui. A day's journey further [is] a destroyed town that is called Olitifar. All [the ground covered during this part of the journey] is good, flat land. From there two days' journey further on through uninhabited lands is a small settlement and beyond that, about a league, [is] Cosa. It is a large town, the largest there is [in the area] where we went from Santa Elena until you arrive at it. It must have about 150 householders, judging from the size of the town. It is a place richer than any of those noted. Ordinarily there is a large number of Indians in it. It is situated in a low land in the lap of a mountain range. Around it at a half and a quarter of a league and at a league are many large settlements. It is a very abundant land. Its site is at the midday sun or less than at the midday sun.[9] From Cossa we should have gone directly to Trascaluça which is the last of the settled area of La Florida. From Cosa to Trascaluça is seven days' journey. I believe that in all of them there are two or three settlements, the rest [of the area] is unpopulated. Trascaluça is said to be at the midday sun. Some say that it is nine days' journey, others eleven, others thirteen, but the most common [estimate is] nine days' journey from there to the land of New Spain, all of them through uninhabited areas. In the middle of this road is a settlement of 4 or 5 houses. After that the next settlement is in New Spain, so they say. I pray to Our Lord that he will provide it as may be his service. Amen. Dated at the Point of Santa Elena, January 23, 1569.

Juan de la Vandera.

Notes

1. The Spanish is *rincón,* which can be "corner, angle, nook"; that is, the expression is "nooks and corners"—a way of designating various parts of the island. Bandera seems to suggest it is a very large island, parts of which look like the "mainland" or the surrounding countryside rather than a typical river island.
2. Or perhaps "The soil of this place is good but thin" is what is intended.
3. As with the other transcriptions and translations, I have preserved the orthography of the document, including this final capital E.
4. This sentence presents problems for translation because the Spanish appears to omit several words following "ground that can be cultivated," which is itself part of what appears to be a run-on sentence.
5. This could be read as "high valleys with much rock and boulders and sand bars . . . " if the literal order of the words is followed, but Bandera seems to have a habit of adding words after their logical place, by way of amplifying his thoughts. In this case, the "low" referring to "valleys" is added after the description of what the valleys contain.
6. The text apparently omits the end of this entry, probably because Bandera let his attention slip while copying the document we have.
7. I have translated *traje* thus because we know these women visited Pardo. *Traje* is probably the past perfect of *traer* used in the sense of entourage.
8. Again, the Spanish is cryptic, apparently with omissions.
9. The meaning of this reference to the sun is unclear but may refer to the Tropic of Cancer, the latitude at which the sun would be directly overhead at noon on the date of the summer solstice. If that is what is meant, the latitude is completely mistaken; the Tropic of Cancer is at 23.5 degrees north but these places are at approximately 30 degrees north.

The Pardo Relation

AGI, Patronato 19, R. 22 (document 1)

Introduction

Because of damage to the right margin of folios 1 and 2 that shows in the microfilm made for my use in 1988, I have checked my transcription against Ruidiaz's transcription in *La Florida*, II, 465–473, on the assumption that Ruidiaz's transcription, made ca. 1893, was based on a state of the document prior to when most of this damage occurred. I have incorporated Ruidiaz's readings of the broken parts of the text by using *italics* for those letters or words that are missing in the present state of the document. In one or two cases I have added notes to indicate where the correction is my own because Ruidiaz seems to have misread the text.

The reader should be aware that in checking my text against Ruidiaz's and the microfilm I found that he made a number of paleographic errors, consistently modernized spellings, and even made grammatical corrections, all of which I have rejected after careful restudy of the microfilm. I believe my text to be a more faithful copy of the original.

Transcription

Notado   1565   Florida

Yo parti del puerto de san Lucar del ano de sesenta y cinco en el tercio de sancho / de archiniega con diez y ocho nabios de alto bordo bispera de pascua de flores y en ellos / lleve mi conpania de docientos y cinquenta soldados y llegamos dentro de tres meses / a sant agustin los diez y siete navios vespera de san pedro porque el uno per/dio el viaje mas despues parescio y como d[ic]ho tengo desembarcamos en san/ta agustin toda la ynfanteria a donde no hallamos al adelantado pe[dro] /

melendez y vino el maeso[1] de canpo de sa[n] mateo y junto los capitanes / y el general sancho de archinega [sic] y dixo que conbenia que se prob*eyese* / luego sa[n] mateo de jente y ansi se probeyo y se enbio la conpania del / coronel y ansi mesmo conbenia que se enbiase a santa elena otra conpania y a/si me enbiaron a mi con dos naos la capitana y la nao de Çubieta donde / desde a pocos dias que ay[2] estavamos llego el adelantado pero melendez de abiles / y tomo muestra a mi conpania y hallo en ella dozientos y quarenta y ocho soldad/dos y hecho esto me mando que yo entrase el dia de santo Andres primero be/nydero la tierra dentro para dar a entender a los yndios como bibian Er*rado* / y que estubiese devajo de su santidad y de su mag[estad] y ansi venido el dia de san/to andres yo me parti con ciento y veynte y cinco soldados a donde en es[te] / Relacion no hago mincion de las quarenta leguas por ser toda tierra panteno/sa y aver pocos yndios y a ver benido parte dellos a santa Elena y abellos y / ya hablado de parte de su mag[estad] y de su santidad y ansi por mis jornadas [broken][3] /mas a donde ay un rrio grande y enbie a llamar los yndios porque los ca*ci*/ques alli estavan y les hize el parlamento de parte de dios y de S*[u] M[agestad]* / ansi como me hera mandado y ellos respondieron que [e]stauan pre*stos* / de obedezer a su santidad y a su mag[estad] com y de ay me parte la buelta *de* / canos a donde la primera jornada hize alta en canpana por no aver *pue*/blo y otro dia llegue al d[ic]ho canos donde halle mucha cantidad *de caci*/ques y yndios y les hize el parlamento acostumbrado de parte *de* / dios y de su mag[estad] y ellos quedaron muy contentos y obedien*tes* / al servicio de dios y de su mag[estad] tyene un rrio caudal y la tierra / es muy buena de ay[4] me parti a tagaya a donde hize jun*tar a* / los yndios y caziques y les hize el propio parlamento y an/si mesmo quedaron debajo del dominio de su santidad y de su / mag[estad] otro dia fui a tagaya el chico E yze[5] juntar animesmo[6] / los yndios y al cacique y les hize el proprio parlamento y que/daron devajo del dominio de su santidad y de su mag[estad] otro dia me pa*r*/ti y fui a un cazique que no me acuerde su nombre y hize el pro/pio parlamento y quedaron devaxo del dominio de *su san*/tidad y de su mag[estad] de aqui fui a ysa que [e]s un cazique grande don/de halle muchos caziques y gran cantidad de yndios a donde les hize el *par*/lamento acostunbrado y quedaron devaxo del dominio de su sant*idad* / y de su mag[estad] de ay[7] me parti otro dia y fuy a un casar [sic] del d[ic]ho ysa / hize juntar los yndios y les hize el propio parlamento y qu*e*/daron devajo del dominio de su santidad y de su mag[es-tad] toda es / muy buena tierra y tiene un rrio caudal otro dia me parti / fui a dormir a un despoblado porque no avia pueblo y o*tro* / dia me parti y fui a juada a donde halle mucha cantidad de y*ndios* FOLIO I VERSO y caciques y les hize el parlamento acostumbrado y todos que/daron del dominio de su santidad y su mag[estad] aqui estube / quinze dias porque me demandaron xpi[cristi]anos para que los / dotrinase y hize un fuerte a donde quedo boyano[8] my sargento y / ciertos soldados con sus municiones de polbora y querda y balas / y mayz para comer y pasados los quinze dias me parti la buel/ta del norte y hize noche en la canpana por no aver poblado junto / a un rrio caudal que passa por juada y toda esta tierra es muy / buena otro dia camine el rrio avajo y hize jornada ansimesmo / en de-

306   *The Juan Pardo Expeditions*

spoblado es tierra muy buena otro dia parti y fui a quinaha/qui y hize juntar el cazique y los yndios que tiene muchos y les hize / el parlamento acostumbrado y ellos quedaron devajo del dominio / de su santidad y su mag[estad] aqui estuve quatro dias tiene muy jentiles / Begas y passa el rrio caudal por el y a cabo deste t[iem]po me parti / y fui a otro cacique que no me aquerdo de su nombre y hize juntar / los yndios y al cazique y les hize el parlamento acostumbado y queda/ron devajo del dominio de su santidad y de su mag[estad] aqui estuve dos dias es tierra muy buena y pasa el rrio caudal por ella y otro dia / me parti y estube en un despoblado por no aver pueblo y otro / dia llegue a guatari a donde halle mas de treinta caziques y mucha / cantidad de yndios a donde les hize el parlamento acostumbrado y e/llos quedaron devaxo del dominio de su santidad y su mag[estad] aqui estube / quinze o diez y seis dias poco mas o menos a donde estos caziques / me demandaron que les dexase quien les dotrinase y asi yo les dexe / el clerigo de mi conpania y quatro soldados porque alli me /bino carta de esteban de las alas que diese la buelta de santa Elena / porque ansi cumplia al servi[ci]o de su mag[estad] porque avia nueba de franze/ses y otro dia me parti y fui a un despoblado d[ond]e estube aquella / noche y otro dia fui a guatariatique [sic] a donde hize el parlamento / acostunbrado y quedaron devajo del dominio de su santidad y de su mag[estad] / otro dia me parti y estuve en canpana por no aver poblado / toda esta tierra es muy buena otro dia llegue [a] aracuhilli [sic] a don/de halle cantidad de yndios y caçiques y les hize el parla/mento acostumbrado es tierra muy buena otro dia me / parti y fui a un cazique que no me aquerdo de su / nonbre y y hize el parlamento acostumbrado y quedaron / devaxo del dominio de su santidad y de su mag[estad] es tie/rra muy buena otro dia sali al propio camino que avia / llevado que [e]s a tagaya chiquito y les torne a hazer el par/lamento acostunbrado y tornaron se a confirmar en el y o/tro dia vine a tagaya y ansi mesmo se tornaron a confir/mar en lo primero otro dia bine a canos a donde estube / dos dias y les hize el parlamento acostumbrados y ellos / se tornaron a confirmar y de ay fui a guiomae En dos dias / a donde le torne a hazer el parlamento acostumbrado y se / confirmaron el lo primero de aqui ya tengo d[ic]ho que no hago FOLIO 2 caudal de las quarenta leguas por ser la tierra pantanosa / y estado devajo del dominio de santa Elena porque cada / dia nos beyamos[9] esto es lo de la primer jornada———

Llego el adelantado pero melendez de abiles el ano de sesenta / y seis[10] a la ciudad de santa Elena a donde me mando yo tornase / a proseguir la jornada y que me partiese el primer dia de / setienbre del dho año y que donde me demandasen algunos xpi[i.e., cristi]anos / para dotrinar los yndios los diese y asi yo me parti el prime/ro dia de setienbre ya tengo d[ic]ho que no hago mincion de las / quarenta leguas por ser la tierra como es y asi por mis jornadas / llegue a guiomae a donde halle muy bien recibim[ien]to y una ca/sa hecha para su mag[estad] que se la avia mandado hazer quando pase / aviendo estado ay dos dias me parti y llegue en otros dos [a] canos a don/de halle mucha cantidad de yndios y caziques y les torne a hablar / de parte de su santidad y de su mag[estad] y Respondieron que / Ellos estavan prestos y deuajo del dominio de su santidad / y de su mag[estad] de aqui me parti y fui a tagaya y hize el pro/pio parlamento y Respondieron que estavan prestos como / lo avian prometido la primera vez otro dia llegue a tagaya / chico y hize el propio parlamento y Respondieron que *es*/tavan[11] prestos como la primera bez de ay[12] fui a un cazique que no me aquerdo de su nonbre y hize el propio parla/mento y dixeron que estauan prestos como la primera / bez de ay fui [a] aracuchi y hize juntar los yndios y caziques / y les hize el propio parlamento y dixeron que esta*ban* / prestos como la primera bez otro dia me parti y fui a *un* / despoblado otro dia me parti y fui a guatari y ati*que* / a donde halle cantidad de yndios y caziques a donde les hize el / parlamento acostunbrado y dixeron que estava*n* / prestos como la primera bez de ay fui a un cazi*que* / que no me aquerdo de su nonbre y les hize el parlam*en*/to acostunbrado y Respondieron que estauan p*res*/tos como la primera vez de estar devajo del domin*io* / de su santidad y de su mag[estad] otro dia me parti y fui a qui[broken]/que[13] y junte los yndios y caziques y les hize el parlamento / acostunbrado y Respondieron que estauan como / primera bez devajo del dominio de su santidad y de su / mag[estad] otro dia fui a un despoblado y toda esta tierra / que tengo d[ic]ho es muy buena otro dia bino a otro despo/blado y es tierra muy buena otro dia llegue a Juada *don*/de halle que el sarjento boyano era hido del fuerte *donde* / yo le avia dexado y los soldados y que le tenian cercado *los* / yndios y con esta nueba yo hize el parlamento al d[ic]ho / juada y sus yndios acostunbrado y ellos Respondieron   FOLIO 2 VERSO que [e]stavan prestos de cunplir como la pri-mera bez devajo / del dominio de su santidad y de su mag[estad] y asi yo me par/ti luego y pase la sierra en quatro dias de despoblado / a donde llegue a tocae que [e]s un pueblo muy bueno y tiene / las casas de madera y alli avia gran cantidad de yn/dios y caçiques y les hize el parlamento de parte de / su santidad y de su mag[estad] y Respondieron que [e]llos / querian ser xpi[i.e., cristi]anos y ten por senor a su mag[estad] otro dia / me parti y dormi en despoblado otro dia me parti y / llegue a cauche a donde la tierra es muy buena y tiene / un rrio principal y tiene unas begas muy grandes y a/lli halle gran cantidad de yndios y caçiques y yo les / hize el parlamento de parte de su santidad y de su mag[estad] / acostunbrado y ellos Respondieron que querian ser xpi[i.e., cristi]a/nos y tener por señor a su mag[es-tad] aqui estube quatro dias / porque entendi que los yndios que se davan por Enemigos / heran ya amigos entendiendo como yo yba y otro dia me par/ti y fui a un despoblado y otro dia ansi mesmo otro dia / llegue a tanasqui a donde tiene un rrio caudal y el pueblo esta / çercado por una parte de muralla y sus torriones / y trabeses a donde hize juntar todos los ydios [sic] y cazi/ques y les hize el parla-mento acostunbrado y Res/pondieron que estauan prestos para hazer lo que / su santidad y su mag[estad] mandaua esta tierra es muy bue/na y creo que ay metales de oro y plata otro dia me / parti y llegue a chiaha que por otro nonbre se llama lameco [sic] / a donde halle al sarjento boyano y a los soldados a donde / me contaron de como los avian tenydo apertados los yndios y an/si yo hize juntar todos los yndios y caziques y les hize el / parlamento de parte de su santidad y de su mag[estad] / y ellos quedaron devaxo del dominio de su santidad / y de su mag[es-tad] como los demas aqui estube diez o doze dias para / que la jente descansarse a donde supe por los yndios ami/gos como me estauan aguardando en un paso seis o siete mill / yndios donde hera carrusa y chisca y costehe y cozca y con / todo esto yo determine de proseguir mi camino y me parti / la buelta de las zacatecas y minas de san martin / camine tres dias de despoblado a donde acabo de los tres / dias llegue a un pueblo que no me acuerdo del nonbre y jun/te los caziques y yndios y les hize el parlamento acostunbrado / y me Respondieron que estauan prestos de hazer lo que [inserted above the line: su santidad] su / [smudged: mag[estad]] mandauan y que querian ser xpi[i.e., cristi]anos esta tierra es muy buena / y creo ay metales en ella de oro y plata   FOLIO 3 otro dia me parti y llegue a satapo a donde halle mucha can/tidad de yndios y alli no fui bien Rescibido conforme y como has/ta alli me avian Rescibido porque el cazique se nego y asi yo llame / a la junta para dezilles lo que les cunplia de parte de Dios y de su / mag[estad] y se allegaron pocos aviendo muchos y no Respondieron / cosa ninguna sino antes se rreyan y avia muchos dellos que nos / entendian y asi aquella noche vinieron a mi las leguas [sic, he means lenguas] a de/zirme que no yrian comigo porque savia que avia gran canti/dad de yndios aguardandome para degollarme a mi y a los / mios y ansi mesmo vino un yndio del propio pueblo y me di/xo que le diese una hacha y que me diria una cosa que me inportuuo / mucho y asi yo se la di y el me conto de como los yndios de chisca / y carrosa y costehe y coza lo es nos estauan aguardando una / jornada de alli y que heran ciento y tantos caziques y tienen / conpetiencia parte dellos con los de la çacatecas y yo / iyendo [sic][14] esto junte mis oficiales y entramos en n[uest]ro consejo / y hallamos que ya que nosotros Ronpiesemos los Ene/migos no podiamos ganar nada por causa de las betuallas / que nos las davan Ellos propios y asi determinamos / de encomendallo a dios y dar la buelta a donde bolvimos E[broken][15] / En quatro dias a lameco [sic] que tiene por otro nombre chiaha y to/da esta tierra como d[ic]ho tengo es muy buena a donde en lameco [sic] / de parecer de todos y mio de hazer un fuerte para q[ue] si su / mag[estad] fuere servido de proseguir la jornada se hallase aquel [broken][16] / ganado y fue de parescer de todos como d[ic]ho tengo y ansi que/daron alli un cabo desquadra y treynta soldados con provi/sion y municion a cabo de quinze dias que esto fue hecho me / parti y llegue a cauchi a donde bine sienpre por despobla*do* / a donde el cazique demando xpi[i.e., cristi]anos para que los dotrinase / y de parecer de todos le quedaron doze soldados / y un cabo desquadra en una fuerza que se hizo En ocho dias / quedando le su polbora y municion de ay bolui a tocae en / dos dias a donde les torne a hablar a los d[ic]hos caziques E yndios / y todos estauan que obedezian a su santidad y su mag[estad] / aviendo estado aqui dos dias me parti para Juada y pa/se la sierra en quatro dias a donde halle mucha junta / de yndios y les hize el parlamento acostunbrado y di/xeron que [e]stauan prestos de hazer lo que avian pro/metido y alli dexe a mi alferez alberto escudero / con treynta soldados para que tubiese quenta con [smudged][17] / d[ic]ho fuerte que [e]staua hecho en el d[ic]ho pueblo para que / desde alli diese calor a los demas soldados e que quedauan de aquella parte de la sierra   FOLIO 3 VERSO

y aviendo estado diez dias en Juada como d[ic]ho tengo me parti / la buelta de

guatari y estube quatro dias en llegar a donde / halle los yndios y caziques juntos y les hize el parlam[en]to / acostunbrado y Respondieron que estavan prestos de / hazer lo que mandava su santidad y su mag[estad] y me deman/daron que les dexase xpi[i.e., cristi]anos y asi hize un fuerte a donde / dexe diez y siete soldados y un cabo de esquadra con ellos / a donde en este tienpo me detube en el d[ic]ho guatari diez y seis / o diez y siete dias poco mas o menos y biendo que / se concluia el termino que [me dio] el adelantado pero melen/dez de abiles me parti la buelta de Santa Elena por mis / jornadas esta tierra como d[ic]ho tengo guatari es una de las / buenas tierras que ay en el mundo y porque tengo hecha / Relacion en la primera jornada des[de] guatari hasta san/ta Elena q̶ ̶h̶u̶ no lo hago en esta por la prolegidad.
Ju[an] pardo

Notes

1. I.e., *maestro*; that is Bartolomé Menéndez, the Adelantado's brother, who served as his second in command.
2. I.e., *alli*.
3. Ruidiaz, *La Florida*, II, 466, also shows an ellipsis.
4. I.e. *alli*.
5. I.e., *hize*.
6. I.e., *asi mesmo* (*asi mismo*).
7. I.e., *alli*.
8. I.e., Hernando de Moyano.
9. I.e., *veiamos*.
10. An error by Pardo. Should be 1567.
11. I.e., *estaban*.
12. I.e., *alli*.
13. Probably *quinaha/que*; Ruidiaz, *La Florida*, II, 470, has *Quirotoque*. The "short" Bandera report published here (no. 2, above) indicates this is *Quinahaqui*.
14. I.e., *viendo*.
15. Ruidiaz, *La Florida*, II, 472, has *en* but omits the repetition found in the original.
16. Ibid., has nothing; apparently nothing is missing here.
17. Ruidiaz, *La Florida*, II, 472, has *el*.

Translation

Notado   1565   Florida

I left the port of San Lucar on the eve of Easter[1] in the year 65 [*sic*, 1566] in the *tercio*[2] of Sancho de Archiniega with 18 high-sided ships in which I carried my company of 250 soldiers. Inside of three months we arrived at St. Augustine on the eve of [the feast of] San Pedro³ with 17 ships because one was unable to make the trip but appeared later. As I have said, we disembarked all the infantry at St. Augustine where we did not find the Adelantado Pedro Menéndez. The Major General [Bartolome Menéndez] came from San Mateo and gathered the captains and General Sancho de Archiniega and said that it was suitable that San Mateo be provided with men. Thus it was done and the company of the Colonel⁴ was sent. Too, it was suitable that another company be sent to Santa Elena. Thus I was sent with two ships, the *Capitana* and the ship of Zubieta.⁵

Within a few days of our being there, the Adelantado Pedro Menéndez de Avilés arrived and mustered my company. He found 248 soldiers in it. When this was done, he ordered me to enter the land on the next day of San Andrés⁶ to give understanding to the Indians how they live in error and that they should be under [obedience to] His Holiness and His Majesty.

When the day of San Andrés came, I departed with 125 soldiers. In this report I will not mention the [first] forty leagues because it is all land and swamps and there are few Indians and part of them have come to Santa Elena and have been spoken to on behalf of His Majesty and His Holiness. Thus on my journeys [broken] more to where there is a great river and I sent to call the Indians because the Caciques were there. I made the speech to them on behalf of God and His Majesty as I was commanded. They responded that they were ready to obey His Holiness and His Majesty.

From there I departed on the way to Canos. As on the first trip [*sic*], I made a stop in the countryside because there was no town. The next day I arrived at Canos where I found a great number of caciques and Indians and I made them the customary speech on behalf of God and His Majesty. They were very content and obedient to the service of God and His Majesty. It has a high-volume river and the land is very good.

I departed from there for Tagaya where I gathered the Indians and caciques. I made the same speech and they remained under the dominion of His Holiness and of His Majesty.

The next day I went to Tagaya, the lesser, and also made the Indians and cacique gather. I make the same speech to them and they remained under the dominion of His Holiness and of His Majesty.

The next day I left and went to a cacique whose name I don't recall and made the same speech to him and they remained under the dominion of His Holiness and of His Majesty.

From there, I went to Ysa, who is a great chief. There I found many chiefs and a great number of Indians to whom I made the customary speech and they remained under the dominion of His Holiness and of His Majesty.

I left there the next day and went to a hamlet (*casar*) of the said Ysa. I had the Indians gather and made the very same speech to them and they remained under the dominion of His Holiness and of His Majesty. It is all a very good land and has a high-volume river.

The Pardo Relation

The next day I departed. I went to sleep in an uninhabited place because there was no town. The next day I departed [this camp site] and went to Juada, where I found a large number of Indians   FOLIO I VERSO and caciques. I made the customary speech to them and they all remained under the dominion of His Holiness and of His Majesty. I was there for 15 days because they demanded Christians from me to catechize them. I made a fort where Boyano [sic], my sergeant, and certain soldiers remained with their munitions of powder, matchcord, balls, and maize to eat.

When the fifteen days had passed, I departed on the route to the north and made [camp for] the night in the countryside, because there was no settlement, next to a high-volume river that passes by [or through] Juada. All this land is very good. The next day I went down [sic] the river and made a day's journey in uninhabited areas. It is a very good land.

The next day I left and went to Quinhanaqui. I made the cacique and Indians gather—he had many—and I made the customary speech to them. They remained under the dominion of His Holiness and of His Majesty. I was there four days. It has very fine bottomlands (*vegas*) and a high-volume river passes through it.

At the end of this time I left and went to another cacique whose name I don't remember. I made the Indians and cacique gather and I made the customary speech to them. They remained under the dominion of His Holiness and of His Majesty. I was there two days. The land is very good and a high-volume river passes through it.

The next day I departed and was in an uninhabited area, for there was no town. The next day I arrived at Guatari, where I found more than 30 caciques and many Indians, to whom I made the customary speech. They remained under the dominion of His Holiness and of His Majesty. I was there 15 or 16 days, more or less. These caciques demanded of me that I give them someone to catechize them. Thus I left them the cleric of my company[7] and four soldiers because there I received a letter from Esteban de las Alas that said that I should return to Santa Elena because that was suitable to His Majesty's service because there was news of Frenchmen.

The next day I left and went to an uninhabited place where I was that night. The next day I went to Guatari [and] Atique, where I made the customary speech. They remained under the dominion of His Holiness and of His Majesty.

The next day I departed and was in the countryside because there was no inhabited place. All this land is very good.

The next day I arrived at Aracuhilli [sic] where I found a number of Indians and caciques. I made the customary speech to them. It is a very good land.

The next day I left and went to a cacique whose name I don't remember. I made the customary speech; they remained under the dominion of His Holiness and of His Majesty. It is a very good land.

The next day I went out from there on the same road that I had taken, which is to Tagaya the Lesser. I again made the customary speech to them and they again confirmed it.

The next day I came to Tagaya and they confirmed [what they had said] on the first [occasion when I made the speech].[8]

The next day I came to Canos where I was for two days. I made the customary speech to them and they again confirmed [their obedience to His Holiness and His Majesty].

From there I went to Guiomae in two days' time. I again made the customary speech and they confirmed the first [pledge of obedience that they had made]. From here, as I have said, I do not make any account of the forty leagues because it is a swampy land and is under the dominance of Santa Elena [and] because every day we see it. This is the first trip.

In the year 66 [sic, 67] the Adelantado Pedro Menéndez de Avilés arrived at the city of Santa Elena where he ordered me to renew the prosecution of the trip and that I should leave on the first day of September of that year and that where they demanded some Christians of me to catechtize the Indians I should give them. Thus I left on the first day of September.

I have said that I will not mention the forty leagues because the land is as it is. Thus in my journeys I arrived at Guiomae. I found a very good reception and a house built for His Majesty, which I had ordered built when I passed [through there].

Having been there two days, I left and in another two days arrived at Canos where I found many Indians and caciques. I again spoke to them on behalf of His Holiness and His Majesty. They replied that they were ready and under the dominion of His Holiness and of His Majesty.

I departed from there and went to Tagaya. I made the same speech and they replied that they were ready, as they had promised the first time.

The next day I arrived at Tagaya the lesser. I made the same speech and they replied that they were ready, as the first time.

From there I went to a chief whose name I don't remember. I made the same speech and they said they were ready, as the first time.

From there I went to Aracuchi and made the Indians and caciques gather. I made the same speech to them and they said they were ready, like the first time.

The next day I left and went to an uninhabited place. The next day I left [there] and went to Guatari and Atique where I found a number of Indians and caciques. I made the customary speech to them. They said that they were ready, like the first time.

From there I went to a chief whose name I don't remember. I made the customary speech to them and they replied that they were ready, like the first time, to be under the dominion of His Holiness and of His Majesty.

The next day I left and went to Quinahaque and gathered the Indians and caciques. I made the customary speech to them. They replied that they were, like the first time, under the dominion of His Holiness and of His Majesty.

The next day I went to an uninhabited area. All this land I have mentioned is very good. The next day I came to another uninhabited place. It is a good land.

The next day I arrived at Juada where I found that the sergeant Boyano [sic] was gone from the fort [in which] I had left him and the soldiers and that the Indians had him under siege. With this news, I made the customary speech to the said Juada and his Indians. They replied   FOLIO 2 VERSO that they were ready to comply, like the first time, under the dominion of His Holiness and of His Majesty.

I then departed and passed the mountains in four days of uninhabited areas. I arrived at Tocae which is a very good town and has wooden houses. There there was a great number of Indians and caciques and I made the speech to them on behalf of His Holiness and of His Majesty. They replied that they wanted to be Christians and have His Majesty as their lord.

The next day I left and slept in an uninhabited place. The next day I left and arrived at Cauche, where the land is very good. It has a principal river and some very large bottomlands. There I found a large number of Indians and caciques. I made the speech to them on behalf of His Holiness and of His Majesty. They replied that they wanted to be Christians and have His Majesty as their lord. I was there four days because I understood that the Indians who had been enemies were now friends, understanding how I went [or "that I was leaving"].

The next day I left and went to an uninhabited place and the next day the same. The next day I arrived at Tanasqui where there is a high-volume river. The town is enclosed on one part by a wall and towers and traverses. I made all the Indians and caciques gather. I made the customary speech to them. They replied that they were ready to do what His Holiness and His Majesty commanded. This land is very good. I believe that there are metals of gold and silver.

The next day I departed. I arrived at Chiaha, which for another name is named Lameco [Olameco], where I found Sergeant Boyano [*sic*] and the soldiers. They told me how the Indians had had them hard pressed (*apertados*, that is, *apretados*). Thus I had all the Indians and caciques gather and I made the speech on behalf of His Holiness and His Majesty to them. They remained under the dominion of His Holiness and of His Majesty, like the rest. I was here ten or twelve days so that the men could rest. Here I learned from friendly Indians that six or seven thousand Indians were awaiting me at a pass, among whom were Carrusca, Chisca, Costehe, and Cozca. With all of that, I determined to continue in my road.

I departed on the way to Zacatecas and the mines of San Martín. I went for three days through an uninhabited area. At the end of the three days I arrived at a town whose name I don't remember. I gathered the caciques and Indians and I made the customary speech. They responded to me that they were ready to do what His Holiness and His Majesty commanded and that they wanted to be Christians. This land is very good. I believe there are metals of gold and silver in it.   FOLIO 3

The next day I departed. I arrived at Satapo where I found many Indians. There I was not as well received as I had been [earlier on the journey] because the cacique refused [to so receive me]. Thus I called [them] to a meeting for the purpose of saying to them what they should do for God and His Majesty. Few came, [although] there were many [in the town]. They did not reply anything but rather laughed. Many of them understood us.

That night the translators came to me to tell me that they would not go with me because it was known that there was a great number of Indians awaiting me, to decapitate me and mine. Also an Indian came from that very town and said to me that I should give him an axe and he would tell me something of great importance to me. So I gave it to him and he told me that the Indians of Chisca, Carrosa, Costehe, and Coza were awaiting us a day's journey from there and that they were one hundred and some caciques and they have an understanding (*competiencia*) between themselves and those of Zacatecas.

Seeing this, I gathered my officers and we entered our council. We found that even though we might break the enemies we would not gain anything because of the foodstuffs that they themselves gave to us. Thus we determined to commend it to God and return.

In four days we returned to Lameco [Olameco], which has as another name, Chiaha. As I have said, all of this land is very good. Everyone and I were agreed to build a fort at Lameco so that if His Majesty were served to continue the journey, that place would be found [already] gained. This was the opinion of everyone, as I have said.[9] Thus there remained there a corporal and thirty soldiers with provisions and munitions.

At the end of fifteen days when this [the fort] was finished, I departed. I arrived at Cauchi, to which I went always through uninhabited areas. There the cacique demanded Christians to catechize them. With the agreement of everyone, a corporal and twelves soldiers remained there in a fort that was built in eight days. Powder and munitions were left with him [the corporal].

From there I returned to Tocae in two days. I again spoke to the caciques and Indians and all were [ready] to obey His Holiness and His Majesty.

Having been there two days, I departed for Juada and crossed the mountains in four days. [At Joada] I found a great gathering of Indians and I made the customary speech to them. They said they were ready to do what they had promised. I left my Ensign, Alberto Escudero, and thirty soldiers there so that they could keep account of the fort that was built in the said town and so that from there they might give ardor to the other soldiers who remained in that part of the mountain range. FOLIO 3 VERSO

And having been ten days at Juada, as I have said, I left on the way to Guatari. I took four days to arrive there, where I found the Indians and caciques gathered. I made the customary speech to them. They replied that they were still ready to do what his Holiness and His Majesty commanded. They demanded of me that I leave them Christians. Thus I made a fort where I left a corporal and 17 soldiers. At this time I stopped for 16 or 17 days in the said Guatari.

Seeing that the term that the Adelantado Pedro Menéndez de Avilés [gave me] was ending, I departed on the way to Santa Elena for my days marches. This land, as I have said, [that is] Guatari, is one of the good lands that exists in the world. Because I have made a report of the first trip from Guatari to Santa Elena I do not make it in this one because of the prolixity [that it would entail].

Juan Pardo

The Pardo Relation

Notes

1. April 19, 1566.
2. The famous Spanish military formation consisting of about 1,000 men. Roughly equal to the later regiment.
3. June 29, 1566.
4. Colonel Hernando de Uruña.
5. These ships were *Tres Reyes* (Sancho Diaz, master) and the *San Salvador* (Sebastian de Zubieta, owner and master).
6. November 1, 1566.
7. Father Sebastian Montero. See Michael V. Gannon, "Sebastian Montero, Pioneer American Missionary, 1566–1572," *The Catholic Historical Review*, 51 (1965), 335–53.
8. Here and elsewhere Pardo seems to conflate his two visits to the interior—yet another example of his lack of care in preparing this document.
9. Pardo's insistence on the "opinion of everyone" as the basis for his leaving men at Chiaha and Cauchi suggests that this document was written after those posts had been overwhelmed by the Indians (probably in the spring of 1568). In essence, he is transferring responsiblity from himself to his officers. For the forts at Joada and Guatari, he assumes full responsibility. We know that some of the soldiers left in the interior returned safely to the coast after the rebellion, so this careful division of responsibility suggests that the garrisons of Joada and Guatari survived.

The Martinez Relation

AGI, Patronato 19, R. 22 (document 2)

Transcription

2º corr[egi]do
1566

Este es un traslado bien y fielmente saca/do de un traslado sinple que fue sacado de un libro / y memoria de la conquista y tierra de las provincias / de la florida que el Ill[ustr]e Senor garcia Osorio g[overnad]or y capitan / ge[ne]r[a]l de esa ysla por su mag[istad] dio a mi el escriv[an]o de yuso escripto / que fue sacado de un libro y memoria que ante / su m[erce]d hisio fran[cis]co martinez soldado de la conquista de la / dha florida que trata sobre la entrada y conquista / de la dha tierra y nueba / descubrimento della que su / tenor es le sigui[ent]e ———
 de la ciudad de santa elena salio el capitan juan pardo el primer / dia de nobienbre ano de mill e quinientos y sesenta y seis para entrar / la tierra dentro a descubrilla y conquistalla dende aqui hasta / mexico y ansi llego a un cacique que se llama Juada a donde hizo / un fuerte y dejo a su sargento con treynta soldados porque avia / tanta nieve en la sierra que no se pudo pasar adelante y el dho ca/pitan se bolvio con la demas gente a esta punta de santa Ele/na a donde agora al presente esta la tierra que hasta alli se a/via visto es buena en si p[ar]a pan y vino y todos los generos de / ganados que en ella se hecharen porque [e]s tierra llana y de muchos / rrios dulzes y buenas arboledas que son nogales y morales / y moreras y nyspolas y castanos liquidanbar y otros / muchos generos de arboledas y ansi mismo es tierra de mu/chas caças ansi de benados como de liebres y conejos y galli/nas y ossos y leones ——— A treynta dias de como llego / a esta punta de santa elena le vino una carta al capitan de / su sargento en que por ella le dezia que avia thenido gue/rra con un cacique que se llama Chisca que [e]s henemigo de los / espanoles y

que le avian muerto mas de myll yndios y que/mado cinq[uen]ta bohios y que [e]sto avia hecho con quinze soldados / y dellos no salieron mas de dos heridos y no heridas peligro/ssas y en la p[ro]pia carta dezia que si el senor estevan de las / alas y el senor capitan mandavan que pasarian adelante / y veria lo que abria y ansi el capitan rrespondio que de/xise diez soldados en la fuerte de Juada y una cabeça con e/llos y con los demas descubriese lo que pudiese y en el entre/tando que [e]sta carta llegava enbio un cacique de la sie/rra a amenazar al sargento diziendo que avia de venir / y comer se los a ellos y a un perro que [e]l dho sargento the/nia y visto esto acordo que hera mejor yr el a buscar los a / ellos que no hellos que no hellos[1] viniesen a buscar lo a el y an/si saliendo del fuerte de san Juan con viente soldados camino / q[ua]tro dias por la sierra y llego una mañana a los enemigos y / los hallo tan fortalecidos que se admiro porque estavan / cercados de una muralla de madera muy alta y con una pe/queña puerta con sus trabeses y biendo el sarjento que no avia / Rem[edi]o de entrar sino ser por la puerta hizo un pabesada / con que entraron con harto peligro porque hirieron al sarjen/to en la boca y a otros nueve soldados en diferentes p[ar]tes FOLIO I VERSO no fueron de peligro ninguna al fin ganando les el fuerte / se recoxeron los yndios a los buhios que tenian dentro del / estan debaxo de tierra dende donde salian a escaramucar / con los espanoles y matando les muchos yndios les ganaron / las puertas de los dhos buhios y les pegaron fuego y los / quemaron todos de manera que fueron los muertos y que/mados mill e quinientos yndios y alli llego la carta del ca/pitan al d[ic]ho sargento en que le mandava lo que arriva tengo / d[ic]ho que dexando diez soldados en el dho fuerte de san Juan fue /se con la demas gente a descubrir lo que mas pudiese y to/mando el camino de un gran cacique que esta de aquel cabo de / la sierra que se dize Chiaha llego a un pueblo suyo aviendo / caminado q[ua]tro dias donde lo hallo tanbien cercado de muralla / y con sus torreones en quadra muy fuertes de [e]stacada Esta/va este pueblo en m[edi]o de dos rrios caudalossos y mas de tres / myll yndios de guerra dentro porque no avia otra gente / ninguna de mugeres ni ninos donde los rrecivieron muy bien / y les dieron bien de comer otro dia se partieron la buelta / del cacique ya dicho y caminaron dos dias sienpre por / pueblos de [e]se cacique y dando les todo lo que avian menes/ter y yndios que los llevasen las cargas llegaron al pue/blo a donde estava el cacique principal el qual [l]es rre/zibio muy bien y les dio yndios p[ar]a que hiziese alli un fuer/te y aguardase al capitan porque este cacique dizia que/ria ser amigo del capitan y hazer lo que le mandase y an/si el d[ic]ho sargento hizo el fuerte donde aguardo al d[ic]ho ca/pitan que a de partir deste fuerte mediado agosto / a este fuerte de Santa Helena an venido a este fuerte muy / muchos caciques y yndios de la tierra dentro trayendo / cada uno lo mejor que tiene que [e]s gamuças y mandiles / y carne al capitan salian a rrezibir q[ua]tro y sies le/guas gran n[umer]o de yndios y el llevaban en una silla corriendo / hasta que llegavan al pueblo y alli le trayan todo el bas/tim[en]to que avian menester para su conpania de ma/hize venado y gallinas y pescados y el yndio que no lle/gava a la silla donde el capitan yba se tenia por afren/tado unos venian dancando otros vaylando muy pin/tados de muchas colores y la tierra es muy buena ansi esta / en que [e]stamos como la demas de adelante porque / hemos provado a sembrar trigo y cevada y se haze tan bu[en]o / como en Espana ansi de otras semillas de rrabanos / navos melones calabacas de una arrova y q[ua]lq[ui]er se/milla a prueva muy bien [slash] el fuerte que hizo el capitan / en Juada es de [e]sta punta de Santa Helena ciento y veinte / leguas y desde alli donde esta el sarjento ciento y qua/renta de manera que lo que [e]sta conquystdo dozien/tas y ses[en]ta leguas y todo esto que aqui esta escripto lo an visto / los t[estig]os que aqui van sus firmas y es verdad f[ec]ha en santa helena / a onze dias del mes de julio de 1567 anos al[ons]o gar[ci]a p[edr]o de / hermossa p[edr]o gu[tierre]z pacheco p[edr]o olivares ———

El qual dho traslado yo el dho scrivano de yu/so escripto hize sacar corregir y concertar   FOLIO 2 el dho traslado sinple por mandado del / dho senor governador alqual se lo doy entregue / en la v[ill]a de la havana desta ysla de cuva / en seys dias del mes de otu[br]e de mill e / q[uinien]tos y sesenta y siete anos el qual / le di firmado de mi n[onbr]e y signo siendo t[estig]os / alonso de Reyna y vernaldino de mata/

Por ende hize aqui mio signo a tal [sign made here] en testimonio de verdad / Bar[tolo]me de morales / es[criva]no de su mag[istad] pu[bli]co de n[umer]o.

Notes

1. This *que no hellos* is repeated in the document.

Translation

Second corrected
1566

This a good and faithful copy taken from a simple copy that was taken from a book and memorial of the conquest and land of the Provinces of La Florida that the Illustrious Gentleman Garcia Osorio, His Majesty's Governor and Captain General of this island, gave to me the undersigned notary, which [copy?] was taken from a book and memorial that Francisco Martinez, soldier of the conquest of the said Florida, made before his mercy, which treats of the entrance and conquest of the said land and new discovery of it. Its tenor is as follows:

The Captain Juan Pardo went out from the city of Santa Elena on November 1, 1566, to enter into the land to discover it and conquer it from here to Mexico. Thus he reached a cacique who is named Juada where he made a fort and left his sergeant with thirty soldiers because there was so much snow on the mountains that he could not go any further. The said captain returned with the rest of the men to this point of Santa Elena where he is at present.

The land that had been seen to there [Juada] is good in itself for bread and wine and all the sorts of herded animals that will be put into it because it is a level land and has many sweet rivers and good groves that are [of] walnut trees, mulberry trees, white mulberry trees, medlar trees, hazelnut trees, balsam trees and many other sorts of trees. It is also a land of much game, deer as well as hares and rabbits and hens and bears and lions.

Thirty days after he arrived at this point of Santa Elena, a letter came to the captain from his sergeant in which he told him that he had had a war with a cacique who is named Chisca, who is the enemy of the Spaniards. They [the Spaniards] had killed more than 1,000 [of his] Indians and burned 50 huts. This had been done with 15 soldiers of whom no more than two were wounded and [these were] not serious wounds. In that same letter he said that if Esteban de las Alas[1] and the captain ordered [it], that they would pass further on and see what was [to be seen]. The captain replied that he should leave ten soldiers in the fort at Juada and a coporal with them and with the rest [of the soldiers] he should discover what he might.

Before this letter arrived, a cacique of the mountains sent a threat to the sergeant saying that he would come and eat them and a dog that the sergeant had. In light of this he decided that it would be better to go to seek them than that they should come to seek him. Thus leaving the fort of San Juan with twenty soldiers he journeyed four days over the mountains and one morning arrived at the enemies. He found them so well fortified that it was admirable because they were enclosed by a very high wooden wall with a small door with traverses. The sergeant, seeing that there was no remedy for entering except by the door, made a covering of shields (*pavesada*) with which they entered with great danger because the sergeant was wounded in the mouth and another nine soldiers in different places. None [of the wounds] was dangerous. When their fort had been gained, the Indians took shelter in the huts that they had inside of it, which were under the ground, from which they came out to skirmish with the Spaniards. Killing many Indians, they [the Spaniards] gained the doors of the huts and set fire to them and burned them all. The dead and burned were 1,500 Indians.

There the captain's letter reached the sergeant. In it he ordered what I have said above, that leaving ten soldiers in the fort of San Juan he should go with the rest of the men to discover what more he could.

Taking the road of a great chief that was in that head of the mountain range, who is called Chiaha, he arrived at one of his towns, having traveled four days. He found it also surrounded by a wall with very strong square towers of pales. This town was between two heavily flowing rivers. [There were] more than 3,000 warriors inside it because there were no other persons, neither women nor children. They received them [the Spaniards] very well and gave them lots to eat.

The next day, they left on the route to that chief already mentioned. They journeyed two days, always in the towns of that chief and receiving all that they had need of and Indians to carry their burdens. They arrived at the town where the principal chief was. He received them very well and gave them Indians so that they might make a fort there and await the captain, because this chief said he wanted to be the friend of the captain and do what he might command. Thus the said sergeant made the fort, where he waited for the said captain. He [the captain] is to depart from this fort in the middle of August.

Very many chiefs and Indians have come to this fort from inland, each bringing the best that he has, that is deer skins, loincloths, and meat. A great number of Indians go out four and six leagues to receive the captain and carry him in a chair, running until they arrive in the town. There they bring him all the food that they need for his company, corn, venison, hens, and fishes. The Indian who does not reach the chair where the captain goes feels himself shamed. Some come whirling, others dancing, very painted up with many colors.

The land is very good both that in which we are as well as the rest further on because we have tested [it by] sowing wheat and barley, which were as good as in Spain, [and] also other seeds of radishes, turnips, mellons, and squashs of one arroba [in size]. Any seed whatsoever has proven very well.

The fort that the captain made at Juada is about 120 leagues from the Point of Santa Elena. From there to where the sergeant is [is] 140 [leagues] so that 260 leagues have been conquered. All that is written here the witnesses have seen, who here sign their names and it is true. Done at Santa Helena, 11 July 1567. Alonso Garcia, Pedro de Hermosa, Pedro Gutierrez Pacheco, Pedro Olivares.

The said copy I the undersigned notary caused to be copied, corrected, and concerted with the said simple copy, by order of the said lord governor, to whom I give it. Turned over at the Villa of Havana of this island of Cuba on 6 October 1567. Which I give signed with my name and sign, Alonso de Reyna, Vernaldino de Mata being witnesses.

Finally I made my sign here thus: [sign] in testimony of the truth. Bartolomé de Morales, Royal Public Notary.

Notes

1. Lieutenant Governor at Santa Elena.

Three New Documents from the Pardo Expeditions

AGI, Contratación 2929 No. 2, R. 7

Introduction

The three documents that follow in transcription and translation are heretofore unpublished and unused in studies of the expeditions of Captain Juan Pardo. In order, these documents are (1) a list of the persons to whom fiber sandals (*alpargatas*) and shoes were given at Chiaha/Olameco, October 8, 1567, (2) a charge (*cargo*)—discharge (*data*) account dated June 13, 1568, for supplies received by Pardo for his second expedition (September 1567 to March 1568), to which document number one forms an appendix because it lists the discharge of some of the fiber sandals and shoes mentioned in bulk in this document, and (3) a similar account of supplies received from the royal stores at Santa Elena for the first expedition (November 1, 1566, to March 7, 1567). The originals of these documents are preserved in the Archive of the Indies (Seville) in Contratación 2929, no. 2, R. 7.

These documents show how little in the way of food Pardo took with him on both expeditions and thus his need to live off the tribute and hospitality of the Indians whom he visited. The lists of munitions given to the garrisons left in 1567 and 1568 repeat materials found in the "long" Bandera report presented elsewhere in this volume and suggest just how precarious the positions of those garrisons were in the face of potentially hostile Indians. Despite the liberal distribution of gifts, which are not recorded in these three documents except for the shoes and fiber sandals given to the leading men at Chiaha/Olameco, the Spaniards left in the backcountry had little but the goodwill of their native hosts for defense, although the power of Spanish arms from behind defensive fortifications and in sudden rush attacks should not be underestimated.

The list of soldiers who received shoes and sandals is not a complete list of those who went on the second expedition. Forty-nine Spanish names are listed, whereas at least 110 and possibly as many as 150 Spaniards began the journey. Most of the men named were the garrison being left in the fort at Chiaha/Olameco, or possibly those who had served at Joara in Fort San Juan during the previous winter.

In sum, these three short documents cast new light on how important the Indians were to the survival of the men Captain Pardo took inland in 1566–68. Provided with scant supplies of food, munitions, and tools, Pardo's men were heavily dependent on the willIngness of the Indians to accept their presence and demands for tribute in corn, salt, and deer skins. The implicit extent of that dependence does not appear as strongly in any other contemporary record of the expedition.

Document 1

Transcription

El Cap[it]an Juan pardo   1568

En la muy noble y leal ciudad de Santa Helena q[ue] Es en las probincias de la florida veynte y uno dias / del mes de Abril año de mill E quinientos E sesenta y ocho años en precensia de mi Ju[an] de la / vandera escriv[an]o y de los test[ig]os di yuso esc[ri]ptos[1] parecio El muy magni[fic]o señor capitan Ju[an] pardo y dixo / se q[ue] ante mi El dho escrivano El ano proximo pasado de mill E quinientos y sesenta E siete anos / Es[tan]do con su m[erce]d la dha su compania en un lugar de yndios q[ue] le disze[2] olameco y por otro nombre chiaha q[ue] Esta ciento y / sesenta leguas desta ciudad la tierra adentro la parte de poniente dio cierto muni[ció]n de çapatos y alpargates / q[ue] abia recibido del thenedor de los bastimentos y municiones de su mag[estad] a algunos soldados de las / de su compania y a otros la data de los quales dhas çapatos y alpargatas le conbiene tenellos[3] en su poder p[ar]a lo / mostrar a quien como en quando se conbenga por tanto q[ue] me pide a mi El dho escrivano se lo de esc[ri]pto en limpio firmado y / signado en manera que haga fee lo qual yo El dho escrivano di q[ue] Es del thenor siguiente /

v. En el lugar de chiaha q[ue] por otro nombre se disze olameco q[ue] Es en las probincias de la Florida ciento y sesenta / leguas de la punta y ciudad de santa helena puerto del mar en estas dhas p[ro]bincias ocho dias del mes de / octubre del año del señor de mill E quinientos y sesenta y siete años yo Juan de la vandera escrivano digo / y doy fee como El dho senor Capitan Juan pardo en mi presen[çi]a saco de la dha ciudad de santa helena cierto / numero de alpargates y çapatos para dar y rrepartir al sargento y soldados de su compania que / Estaban Repartados en ciertos lugares de la florida de la tierra adentro los quales El dho señor / capitan ante mi el dho escri[van]o dio al dho sargento y a los soldados los nombre de los quales E / lo que a cada uno se les dio es como se sigue v.

Alpargates	Çapatos
v. a hernando moyano sargento quatro pares de çapatos.	iiii

324   *The Juan Pardo Expeditions*

	v. A Lucas de canizares cabo desquadra dos pares de çapatos	ii
I pares	v. A Baltasar de los reyes dos pares uno de çapatos y otro de alpargates	i
	v. A lope gonçales unos çapatos	i
	v. A pedro de bentura unos çapatos	i
i	v. a Juan de Castaneda un par çapatos y otro de alpargates	i
I	v. a fran[cis]co Enriques[4] de tenbleq[ue] otro tanto	i
I	v. a lucas hernandez otro tanto	i
I	v. A miguel Romano otro tanto	i
I	v. A Juan Carrasco otro tanto	i
I	v. A matheo Sanchez otro tanto	i
I	v. a Juan Lopez Espadero otro tanto	i
I	v. A Albaro de sevilla otro tanto	i
I	v. a pedro Alonso otro tanto	i
I	v. A luys de aguilar otro tanto	I
I	v. A Albaro de mendaña otro tanto	I
I	v. A Juan martín de badajoz otro tanto	I
I	v. a pascuale lopez otro tanto	I
I	v. a miguel sanchez calero otro tanto	I
I	v. a Juan perez de ponte de lima otro tanto	I
I	v. A Domingo martin otro tanto	I
I	v. A gabriel de paniagua otro tanto	I
I	v. A Juan gallego otro tanto	I
I	v. A Fran[cis]co perez otro tanto	I
I	v. a pedro martin de la çarça otro tanto	I
I	v. A Thome Albarez otro tanto	I
I	v. A Alonso bela otro tanto	I
I	v. A Antonio de Carrion otro tanto	I
I	v. A alonso martin otro tanto	I

FOLIO I VERSO

Alpargates		Çapatos
xxv pares		xxxiii pares
I	v. A miguel gonçalez otro tanto digo un par de çapatos y otro / de alpargates	I
I	v. A domingo miguelez otro tanto	I
I	v. A sebastian gomez otro tanto	I
I	v. a alonso de aguila otro tanto	I
I	v. A Juan Martin de olibares otro tanto	I
I	v. a Antonio de Vrito otro tanto	I
I	v. A pedro peralta otro tanto	I
I	v. A diego de morales otro tanto	I
I	v. A andres martin otro tanto	I

I	v. A anton muñoz otro tanto	I
I	v. A Juan grenon otro tanto	I
I	v. a barahona otro tanto	I
	v. a pedro H[ernan]d[e]z de ciudad Rodrigo unos çapatos	I
	v. Al cacique de chiaha unos çapatos	I
	v. a un mandador del mismo chiaha otros çapatos	I
	v. Al cacique de olameco otro tanto	I

{Ay chiaha por si y olamico / tiene dos nombres}

I	v. Al yniha q[ue] es como alguazil q[ue] manda el pueblo un par de çapatos / y otro de alpargates	I
I	v. a un mandador del mismo olameco otro tanto	I

v. A ciertos soldados biejos de la compania del s[eñ]or governador Estebano de / las Alas q[ue] son los q[ue] se siguen se les dio lo siguiente

I	v. a pedro de sierra unos çapatos y alpargates	I
I ojo	v. a pedro de barçena de miguel enrriquez otro tanto	I
I	v. a xptobal[5] de villalobos otro tanto	I
I	v. a Alonso de rrojas otro tanto	I
I	v. a fran[cis]co maldonado otro tanto	I
I	v. A melchior melendez otro tanto	I
I	v. A andres de Valdes unos çapatos	I

xlv pares   lviii pares

fueron testigos al ver dar y Recibir de todos los dhos çapatos y alpargates Re/partidos En la forma suso dho El señor alverto Escudero de villamar alferez her[nan]do moyano sargento y otros muchos sol/dados de la compania del dho señor capitan Juan pardo paso ante mi ju[an] de la vandera escrivano /

fueron presentes por testigos al ver pedir y dar corregir y conzertar desta dha escriptura con el original al[ons]o / g[arci]a allide[6] del fuerte de san felipe y Fran[cis]co hernandez de tembleq[ue] y fran[cis]co corrodor soldados de la dha compania / del dho señor capitan Juan pardo Estantes en la dha ciudad.

Juan de la vandera escrivano destas probincias doy fee de lo suso dho porq[ue] ante mi paso y / [a] todo fuy presente y soi testigo y en testimonio de verdad fize[7] mio sygno [sign here]
[end of document]

Document 2

FOLIO 2
Municiones que R[ecibi]o El cap[it]an Juan Pardo
1568
Armada

En la muy noble y leal ciudad de Santa helena q[ue] Es en las probincias de la florida treze dias / del mes de Junio año del nascimiento de nuestro salbador Jesu

Xpto[8] de mill E qui[nient]os / sesenta y ocho Años en presencia de mi Juan de la vandera escrivano y ante los tes[tig]os di yuso / esc[ri]ptos parescio El muy magni[fic]o señor Juan pardo capitan de ynfanteria Española / por su mag[estad] contino de su casa lugar theniente del governador destas probincias y dixo que / me pedia y rrequeria y pidio y rriquirio a mi el dho escrivano le diese por fee y testimonio de como ante mi El dho esc[ri]v[an]o / de su m[erce]d con los soldados de su compania y con otros de otras companias por horden del muy Ill[ustr]e señor pedro menen/dez de aviles Adelantado destas probincias En nombre de Su Mag[estad] El mes de septiembre proximo pasado del año / de mill e quinientos y sesenta y siete años salio desta ciudad a hazer como hiço cierta jornada y descubrimiento de/recho a la nueba España y para El dho hefeto[9] en otra Jornada que hiço antes desta y para Esta recivio del / thenedor de los bastimentos y municiones de Su mag[estad] q[ue] Esta en los fuertes desta ciudad cierto numero / de las dhas municiones y bastimento todo lo qual se a gastado y dystribuydo en Servi[ci]o de su mag[estad] ante my El dho / escrivano como persona que fuy en la dha jornada de que doy fee y le conbiene tenello en su poder para dar / quenta a su mag[estad] o a quien En su nombre se la tomare de que es lo que rrecivio y como lo distribuyo por tanto / q[ue] me pedia y pidio a mi El dho escrivano se lo de esc[ri]pto en limpio firmado y autorizado En manera que haga fee / en thenor de lo que el dho senor capitan rrecivio y como lo distribuyo disze como se sigue v.

R[ecib]o bizcocho	v. primeramente Recibio El dho señor capitan para hazer como hiço Esta sigunda jornada /demas de lo que rrecivio la primera Jornada ochenta y cinco libras de bizcocho	Lxxxv lbs
bino	v. mas rrecibio nobenta y seis quartillos de bino	xcvi q'es
plomo	v. mas recivio diez arrobas y cinco libras de plomo	x @ v lbs
cuerda	v. mas rrecibio seis arrobas y diez y ocho libras de cuerda de arcabuz	vi @ xviii lbs
polbora	v. mas rrecibio El dho señor capitan para hazer la dha jornada duze[10] arrobas y nuebe libres / de polbora de arcabuz	xii @ ix lbs
polbora	v. mas rrecivio El dho señor capitan treynta libras de polbora de canon	I @ v lbs
costales	v. mas rrecivio tres costales de mas de los de antes de agora rrecividos	3 costales
Alpargates	v. mas rrecibio El dho señor capitan para los soldados de su compania y de otras companias / y para dar a algunos caciques E yndios principales	cclxv pares de Alpargates

çapatos	dozientas y sesenta y cinco pa/res de Alpargates las ciento y sesenta y cinco pares de las de España y las cinq[uen]ta / de campeche	
	v. mas rrecibio el dho señor capitan Juan pardo ciento y setenta y ocho pares de çapatos	clxxviii pares de çapatos

Todo lo qual q[ue] esta dho y declarado q[ue] rrecibio El dho señor capitan Juan pardo de mas y allende de lo q[ue] antes de agora / abia rrecivido yo El dho escrivano doy fee q[ue] lo gasto y distribuyo En serbicio de su mag[estad] por la horden y forma sigui[ent]e

bizcocho [slash] bino	v. El bizcocho y bino dando lo y rrepartiendolo como lo dio y rrepartio a todos los soldados / y personas q[ue] con el dho señor capitan salimos desde Esta ciudad En su compania a hazer la / jornada que esta dho	
plomo	v. En quanto a lo que toca al plomo El dho señor capitan dexo catorze arrobas y nuebe libras / de plomo en balas en los fuertes que dexo hechos en la tierra adentro destas probincias / rrepartido en esta manera [slash] En el fuerte de chiaha [slash] tres arrobas y diez libras [slash] en el / fuerte de cauchi [slash] una arroba y onze libras [slash] en el fuerte de Joara [slash] quatro arrobas / En el fuerte de guatari [slash] dos arrobas y una libra [slash] En canos [slash] una arroba y treze / libras [slash] En orista diez y ocho libras y una arroba y seis libras q[ue] se perdieron / en un saco q[ue] traya un soldado que se le cayo pasando por un rrio grande	xiiii @ ix lbs
plomo	v. mas se repartieron y dieron a algunos soldados que abian perdido las balas / balas [sic] al pasar de ciertos rrios y pantanos grandes nuebe libras de balas	ix lbs

cuerdo	v. En quanto a lo que toca a la cuerda se dexaron en los fuertes arriba esc[ri]ptos / ocho arrobas y veynte libras rrepartido en esta manera [slash] En chiaha [slash] dos arrobas / En cauchi [slash] una arroba [slash] En joada dos arrobas y diez y ocho libras [slash] En guatari / una arroba y nuebe libras [slash] En canos una arroba [slash] En orista diez y ocho / libras	viii @ xx lbs
cuerda	v. Mas se gastaron quatro arrobas y media de cuerda desde primero dia del / mes de Septiembre hasta dos dias del mes de março del año siguiente que / fue quando El dho señor capitan bolbio con alguna parte de la gente que llebo / En la dha su compania a esta dha ciudad [slash] que llebaban los cabos de Esquadra FOLIO 2 VERSO de la dha compania enzendida de hordinario cada dia porq[ue] ansi conbino al serv[ici]o de su mag[estad]	iiii @ xiii lbs
cuerda	v. mas se gasto una arroba de cuerda que se rrepartio a los soldados de la dha compania / porque tubieron nescesidad della rrespeto de aver se les caydo la que tenian / pasando por pantanos trabaxosos donde se les caya y perdia todo quanto tenian /	I @
polbora	v. En quanto a lo que toca a la querda que el dho señor capitan rrecibio digo la polbora se dexaron rrepartidas en los dhos fuertes nuebe arrobas y beynte y dos libras / de polbora en esta manera [slash] en el fuerte de chiaha dos arrobas y diez libras [slash] en el de / cauchi [slash] una arroba [slash] En el de Joara tres arrobas y diez libras [slash] En el de guatari / una arroba y nuebe libras [slash] en canos una arroba [slash] en el de orista [slash] diez y ocho libras	ix @ xxii lbs

polbora	v. gastose y diese a los soldados que fueron en la dha compania en vezes quatro arrobas / de polbora rrespeto de que todo sienpre fuesen apercividos porq[ue] con las malas / noches q[ue] En todo El t[iem]po se llebaron en campaña y malos dias pasando rrios y pantanos / unas vezes se les caya de los frascos y otras vezes se les moxaba de suerte q[ue] no hera de / probecho de q[ue] yo El dho escrivano doy fee	iiii@
polbora	v. mas se gastaron dos arrobas de polbora en disparar El arcabuzeria algunas / vezes en lugares de yndios principales porque ansi conbino	ii @
costales	v. En quanto a lo que toca a ciertos costales que el dho señor capitan rrecivio fueron para / hazer como se hizieron muchas mochilejas y saquillos para llebar como se llebaron las / municiones que la dha compania llebo y el bastimento que de hordin[ari]o tubo con que / se rrompieron y acabaron	xxxvii costales
	v. mas Recibio dozientas y sesenta y cinco pares de Alpargates los quales gasto y distribu/yo En dar los como los dio a todos los soldados que fueron en hazer en la dha compania / la dha jornada y ansi mismo a los demas q[ue] quedaron en las fuertes desta ciudad y tan/bien En dar como dio a algunos yndios ciertos pares dellas como mas largamente / se dieze E declara en una memoria que el dho señor capitan tiene firmada del / nombre de mi el dho escrivano	cclxv pares
çapatos	v. mas rrecibio el dho señor capitan ciento setenta y ocho pares de çapatos los / quales dio y rrepartio a todos los soldados y personas q[ue] tubo en su compania / ansi della como de otras	clxxviii

companias [slash] como paresciera y
se declara en la memoria / arriba
declarada de q[ue] yo El dho
escrivano doy fee

v. Todo lo qual q[ue] dho es como esta dho y declarado yo El dho escrivano digo y doy fee q[ue] El dho señor capitan / lo rrecibio gasto dio y rrepartio En my presencia y de los tes[tig]os di yuso esc[ri]ptos [slash] T[estig]os q[ue] fueron presentes / a todo lo que Es dho ansi al rrescibir como albello[11] gastar y distribuyr y de todo Ello oyir y ver / pedir Esta fee a mi El dho escrivano pedro gutierrez pacheco cabo de Esquadra y Francisco hernan/des de tenbleq[ue] y Fran[cis]co corredor y Xptobal perez cavellos soldados de la dha compania y los de/mas soldados della Estantes en la dha ciudad/

Juan de la vandera escrivano doy fee de lo suso dho porq[ue] ante mi paso y soy T[estig]o [end of document]

Document 3

FOLIO 3

En la muy noble y leal ciudad de santa helena q[ue] Es en las provincias de la florida doze / dias del mes de junio año del nascimiento de nuestro salbador Jesu Xpto de mill E quinientos / y sesenta y ocho años en presencia de mi Ju[an] de la vandera escrivano y ante los t[estig]os di yuso esc[ri]ptos parescio El muy magn[ifi]co señor Juan pardo capitan de ynfanteria Española por su mag[estad] Contino de su casa / lugar teniente del gobernador destas probincias y dixo q[ue] para hazer como hiço El ano proximo pa/sado de mill y quinientos y sesenta y seys anos con los soldados de su compania con horden del muy Ill[ustr]e señor / p[edr]o menendes de aviles adelantado destas probincias en nombre de Su mag[estad] cierta jornada y entrada por la tierra de las/ dhas probincias adentro derecho a la nueba España [slash] su m[erce]d rrecibio cierto numero de municion y bastimento del tenedor / q[ue] esta en los fuertes desta ciudad por su mag[estad] y de la cantidad q[ue] rrecibio y como lo distribuyo le conbiene tener recaudo y / claridad para lo mostrar y presentar quando le sea pedido si acaso se le tomare quenta o se le hiziere cargo para dar / su descargo por tanto q[ue] me pedi y pidio a mi El dho escrivano le de una fee azerca de q[ue] cantidad de cosas Es la que rrecibio / y por el consiguiente de como lo gasto porq[ue] ansi le conbiene para su derecho y descargo [slash] El thenor de lo que El dho señor / capitan rrecibio y como lo distribuyo disze como se sigue /

R[ecib]os		
cuerda	v. primeramente rrecibio el dho señor capitan diez arrobas y diez libras de cuerda / de arcabuz	x @ x lbs

Three New Documents

plomo	v. mas rrecivio nuebe arrobas y beynte y dos libras de plomo	ix @ xxii lbs
polvora	v. mas rrecivio doze arrobas de polbora de arcabuz	xii @
hierro	v. mas rrecivio dos picos de hierro para las cosas q[ue] fuesen nescesarios	ii picos
hierro	v. mas rrecivio ziete palas de hierro	vii palas
hierro	v. mas rrecivio cinco azadones	v azadones
acero	v. mas rrecivio cinco vallestas para El dha biaje	v. ballestas
xaras	v. mas rrecivio veynte dozenas de xaras	xx dozenas de xaras
costales	v. mas rrecibio setenta y quatro costales de anxeo[12] para llebar bastimento Y otras cosas	Lxx [*sic*] costales de angeo
bizcocho	v. mas rrecivio El dho senor capitan Juan pardo al tiempo q[ue] le ubo de partir a hazer la / dha jornada para El sustento de su compania treynta y una arrobas y diez y ocho libras de bizcocho	xxxi @ xviii lbs
bino	v. mas rrecivio una arroba y diez y ocho quartillos de bino	I @ xviii q'es
quesos	v. mas rrecivio diez quesos	x quesos

De manera que lo que el dho senor capitan Recibio para El hazer de la dha jornada fue las cosas suso dhas y de/claradas parte de todo lo qual gasto y distribuyo por la horden y forma siguiente /

gasto		
pan binos queso	v. En quanto a lo que toca al pan y bino y queso en dar lo y rrepartir lo a los soldados de su compania / como se lo dio y rrepartio al tiempo q[ue] ubo mayor nescesidad quando se puso en efecto El hazer de la dha jornada v. como es cierto y muy notorio	
polbora	v. gastaron los soldados de la dha su compania en se ensenar A tirar algunos dias antes que / se partiese a hazer El dho Efecto tres arrobas de polbora como es not[or]io	3 @

The Juan Pardo Expeditions

cuerda	v. gastaron se dos arrobas de cuerda en llebar de hordinario cuerda enzendida	ii @
polbora	v. gastose En ciertas vezes que se disparo El arcabuzeria en algunos lugares de prin/cipales de la tierra adentro donde Estaban juntos muchos yndios por ponelles[13] temor / arroba y media de polbora	I @ xiii lbs

v. dexo El dho señor capitan En el fuerte de San Juan que se hiço al pie de la sierra En un / lugar que se disze Joara al sargento y treynta hombres que dexo En el de su compania / porq[ue] conbino asi para lo que toca al servicio de su mag[estad] las municiones y cosas siguientes /

polbora	v. seis arrobas de polbora	vi @
cuerda	v. cinco arrobas y ocho libras de cuerda	v @ viii lbs
plomo	v. cinco arrobas y diez libras de plomo	v @ x lbs
vallestas	v. quatro vallestas y veynte dozenas de xaras	
palas azadones picos costales	v. tres palas dos azadones y dos picos v. gastaronse y rrompiesen se En hazer muchilejas y saquillos para las municiones quarenta costales	

Todo lo qual que dho Es ansi rrescibo como gasto de la forma E manera q[ue] Esta dho y declarado yo El dho escrivano digo y doy fee / q[ue] El dho senor capitan lo rrescibio y gasto en servicio de Su mag[estad] a todo lo qual que dho Es ansi al rrecivo como al gasto / fueron presentes por t[estig]os pedro gutierrez pacheco cabo de Esquadra [slash] y francisco hernández de tenblique y / Fran[cis]co corredor soldados de la dha compania y todos los demas soldados della v.
Juan de la vandera escrivano doy fee de lo suso dho y soy testigo.

Notes

1. I.e., *de ayuso*, meaning "below" or "infrascript."
2. I.e., *dice*.
3. I.e., *tener ellos*.

4. An error. Should be Hernández.
5. I.e., *cristobal*.
6. I.e., *alcaide*, a warder or governor.
7. I.e., *hize*.
8. I.e., *Jesu Cristo*.
9. I.e., *hefecto*.
10. I.e., *doze*.
11. I.e., *haber ello*.
12. I.e., *angeo*, a type of linen cloth.
13. I.e., *poner ellos*.

Document 1

Translation

Captain Juan Pardo   1568

In the very noble and loyal city of Santa Elena which is in the provinces of Florida, April 21, 1568, the very magnificent gentleman, Captain Juan Pardo, appeared in the presence of me, Juan de la Bandera, notary, and of the witnesses noted below, and said that last year, 1567, in my presence, his company being with him in an Indian place called Olameco, or by another name Chiaha, which is 160 leagues inland from this city toward the west, he gave certain shoes and fiber sandals that he had received from the keeper of the King's stores and munitions to some soldiers of those of his company and to others. It suits him to have the discharge (*data*) of the said shoes and fiber sandals in his possession so that he can show it to whom as well as when it suits him. Therefore he asks me, the notary, to give it to him, written cleanly, signed, and sealed in a way which gives credence. I, the notary, gave it. The tenor of it is as follows:

In the place of Chiaha, which for another name is called Olameco, which is in the provinces of Florida, 160 leagues from the point and city of Santa Elena, seaport of these provinces, October 8, the year of the Lord 1567, I, Juan de la Bandera, notary, say and give certification how Captain Juan Pardo in my presence took from the city of Santa Elena a certain number of fiber sandals and shoes for the purpose of giving and distributing [them] to the sergeant and soldiers of his company who were divided out in certain places of inland Florida. These [shoes and sandals] the captain gave to the sergeant and soldiers before me, the notary. Their names and what he gave each is as follows:

Fiber Sandals		*Shoes*
	v. Hernando Moyano, Sergeant, four pair of shoes	iiii

	v. Lucas de Canizares, Squad Leader, two pair of shoes	ii
I pair	v. Baltasar de los Reyes, two pairs, one of shoes and another of fiber sandals	i
	v. Lope Goncales, some shoes	I
	v. Pedro de Ventura, some shoes	I
I	v. Juan de Castañeda, one pair of shoes and another of fiber sandals	I
I	v. Francisco Enriques [*sic*] de Tembleque, the same	I
I	v. Lucas Hernández, the same	I
I	v. Miguel Romero, the same	I
I	v. Juan Carrasco, the same	I
I	v. Matheo Sanchez, the same	I
I	v. Juan Lopez Espadero, the same	I
I	v. Alvaro de Sevilla, the same	I
I	v. Pedro Alonso, the same	I
I	v. Luís de Aguilar, the same	I
I	v. Alvaro de Mendana, the same	I
I	v. Juan Martín de Badajoz, the same	I
I	v. Pascuale Lopez, the same	I
I	v. Miguel Sanchez Calero, the same	I
I	v. Juan Perez de Ponte de Lima, the same	I
I	v. Domingo Martín, the same	I
I	v. Gabriel de Paniagua, the same	I
I	v. Juan Gallego, the same	I
I	v. Francisco Perez, the same	I
I	v. Pedro Martín de la Zarza, the same	I
I	v. Tomé Alvarez, the same	I
I	v. Alonso Vela, the same	I
I	v. Antonio de Carrión, the same	I
I	v. Alonso Martín, the same	I

FOLIO I, VERSO

Fiber Sandals   *Shoes*
25 Pair   33 pair

I	v. Miguel Gonzales, the same, I mean one pair of shoes and another of fiber sandals.	I
I	v. Domingo Miguelez, the same.	I
I	v. Sebastian Gomez, the same	I
I	v. Alonso de Aguila, the same	I
I	v. Juan Martín de Olivares, the same	I
I	v. Antonio de Brito, the same	I
I	v. Pedro Peralta, the same	I

1	v. Diego de Morales, the same	1
1	v. Andrés Martín, the same	1
1	v. Antón Múñoz, the same	1
1	v. Juan Greñon, the same	1
1	v. Barahona, the same	1
	v. Pedro Hernández, of Ciudad Real, some shoes	1

{*There is Chiaha itself and Olamico. It has two names.*}

	v. The Cacique of Chiaha, some shoes	1
	v. To a driver [*mandador*] of the same Chiaha, other shoes	1
	v. The Cacique of Olameco, the same	1
1	v. To the Iniha, who is like a sheriff, who commands the town, one pair of shoes, and another of fiber sandals	1
1	v. To a driver of the same Olameco, the same	1

To certain old soldiers of the company of the lord Governor, Estebano de las Alas, who are those who follow, the following were given:

1	v. Pedro de Sierra, some shoes, and fiber sandals	1
1 look	v. Pedro de Bazena, company of Miguel Enriquez, the same	1
1	v. Cristobal de Villalobos, the same	1
1	v. Alonso de Rojas, the same	1
1	v. Francisco Maldonado, the same	1
1	v. Melchior Meléndez, the same	1
	v. Andrés de Valdés, some shoes	1

45 Pair   58 Pair

Mr. Alberto Escudero de Villamar, Ensign, Hernándo de Moyano, Sergeant, and many other soldiers of the company of the captain, Juan Pardo, were witnesses who saw the giving and receiving of all the shoes and fiber sandals distributed in the way which is aforesaid. Done before me, Juan de la Bandera, notary.

Alonso Garcia, Warder of the Fort of San Felipe, Francisco Hernández de Tembleque, and Francisco Corrodor, soldiers of the company of Captain Juan Pardo, being in the said city, were present as witnesses who saw the asking and giving, correcting, and collation of this said writing with the original.

I, Juan de la Bandera, Notary of these provinces, give certification of the above said because it passed before me and I was present at all of it and I am a witnesses. In testimony of the truth I make my sign. [sign placed here.]
[End of Document]

Document 2

FOLIO 2
Munitions Which Captain Juan Pardo Received
1568
Fleet

In the very noble and loyal city of Santa Elena which is in the provinces of Florida, June 13 of the year of the birth of our savior Jesus Christ, 1568, in the presence of me, Juan de la Bandera, notary, and before the witnesses written below, there appeared the very magnificent gentleman Juan Pardo, royal captain of the Spanish infantry and *contino* of the King's household, lieutenant of the governor of these provinces, and said that he asked and required me, the notary, to give him certification and testimony of how in my presence, under orders of the very illustrious gentleman Pedro Menéndez de Avilés, adelantado of these provinces in the name of His Majesty, he left this city in the month of last September of the year 1567, with the soldiers of his company and others of other companies to make, as he did make, a certain journey and discovery straight toward New Spain. For that end, on another journey which he made before this and on this [one] he received a certain number of munitions and foodstuffs from the keeper of the King's foods and munitions in the fort of this city. All of this has been used and distributed in His Majesty's service before me, the notary, as a person who went on the trip, of which I give certification. It is convenient for him to have it [the certification] in his possession in order to give account, to His Majesty or to whomever in his name might take it, of what it was that he received and of how he distributed it. Therefore, he asked me, the notary, to give it to him written clearly, signed, and authorized so that it gives certification. The tenor of what the Captain received and how he distributed it says as follows:

Biscuit	First of all, the captain received for making, as he made, this second trip, besides what was received on the first trip, eighty-five pounds of biscuit	85 lbs
Wine	Further, he received 96 *quartillos* of wine. [48.4 liters].	96 q[*uartil*]los
Lead	Further, he received 10 arrobas and 5 pounds of lead [255 pounds].	10 @ 5 lbs
Matchcord	Further, he received 6 arrobas and 18 pounds [168 lb.] of matchlock cord.	6 @ 18 lbs
Powder	Further, the captain received 12 arrobas and 9 pounds [309 lb.] of matchlock powder for the trip.	12 @ 9 lbs
Powder	Further the captain received 30 pounds of cannon powder.	1 @ 5 lbs

Sacks	Further, he received 3 sacks besides those received before now.	3 sacks
Fiber Sandals	Further, the captain received 265 pair of fiber sandals for the soldiers of his company and of other companies and to give to some chiefs and leading Indians. One hundred and sixty-five pair were those from Spain, and 50 [pair] were from Campeche.	265 pair
Shoes	Further, Captain Juan Pardo received 178 pair of shoes.	178

I, the notary, certify that all that which it is said and declared that Captain Juan Pardo received over and above what was received before now was used up and distributed in the service of His Majesty in the following order and form.:

Biscuit, Wine	The biscuit and wine, giving and dividing it as he gave and divided it to all the soldiers and persons who went out from this city with the captain in his company to make the journey which is said.	
Lead	In regards the lead, the captain left 14 arrobas and 9 pounds [359 lb.] of lead in balls in the forts he left built inland in these provinces, divided in this way: in the fort of Chiaha, 85 lb; in the fort of Cauchi, 35 lb.; in the fort of Joara, 100 lb.; in the fort of Guatari, 51 lb.; in Canos, 38 lb.; in Orista, 18 lb.; 31 lb. were lost in a sack which a soldier carried that fell [into the river when] crossing a great river.	14 @ 9 lbs
Lead	Further, 9 lb. of balls were distributed and given to some soldiers who had lost them passing certain rivers and great swamps.	9 lbs
Cord	In regards the cord, 8 arrobas and 20 pounds [220 lb.] were left in the forts written above, divided in this manner: in Chiaha, 50 lb.: in Cauchi, 25 lb.; in Joada, 68 lb.; in Guatari, 34 lb.: in Canos, 25 lb.; in Orista 18 lb.	
Cord	Further 4.5 arrobas [112.5 lb.] of cord were used from September 1 until March 2 of the following year, which was when the captain returned to this city with some part of the men he took with him in his company. FOLIO 2 VERSO They were carried by the squad leaders of the company ordinarily burning every day because this was suitable for His Majesty's service.	4 @ 13 lbs

Cord	Further, an arroba [25 lb.] of cord was used that was distributed to the soldiers of the company because they had need of it due to that which they carried having fallen [upon] passing through laborious swamps where it fell and all they had was lost.	1 @
Powder	In regards the cord that the captain received, I mean the powder, 9 arrobas and 22 pounds [247 lb.] were left divided up in the forts, in this manner: in the fort of Chiaha, 60 lb.; in that of Cauchi, 25 lb. in that of Joara, 85 lb.; in that of Guatari, 34 lb.; in Canos, 25 lb.; in that of Orista, 18 lb.	9 @ 22 lbs
Powder	Four arrobas [100 lb.] were used and given at various times to the soldiers who were in the company so that all were always provided for because with the bad nights in all the time on campaign and the bad days passing rivers and swamps, sometimes the flasks fell and at others they got wet so that it [the powder in them] was not usable. I, the notary, certify this.	4 @
Powder	Further, 2 arrobas [50 lb.] of powder were used in shooting the matchlocks sometimes in the locales of leading Indians, because it was suitable to do so.	2 @
Sacks	In regards to certain sacks which the captain received: they were for making, as were made, many game bags and little sacks, for carrying as they carried, the munitions which the company carried and the food which it normally had, with which they were broken and used up.	37 Sacks
[Sandals]	Further, he received 265 pair of fiber sandals which he used and distributed in giving them, as he gave them, to all the solders in the company who went to make the journey and also to the rest who remained in the forts of this city and also in giving, as he gave, to some Indians pairs of them, as is said and declared at length in a memorandum which the captain has, signed with the name of me, the notary.	265 pair
Shoes	Further, the captain received 178 pair of shoes which he gave and distributed to all the soldiers and persons he had in his company, from it and from other companies, as may appear and is declared in the above declared memorandum, of which I, the notary, give certification.	178 pair

All of that which is said, as it is said and declared, I, the notary, say and certify that

Three New Documents

the captain received, used up, gave, and divided it in my presence and that of the witnesses written below. Pedro Gutierrez Pacheco, squad leader, Francisco Hernández de Tembleque, Francisco Corredor, and Cristobal Perez Cavellos, soldiers of the company and the rest of the soldiers of it present in this city were witnesses present at all that he said, and also at the receiving [of it] as well as in seeing it used up and distributed and at hearing and seeing the asking of this certification of all of that from me, the notary.

Juan de la Bandera, Notary. I give certification of the aforesaid because it passed before me and I am a witness.

[End of Document]

Document 3

FOLIO 3

In the very noble and loyal city of Santa Helena, which is in the provinces of Florida, June 12 [in] the year of the birth of our savior Jesus Christ, 1568, in the presence of me Juan de la Bandera, notary, and of the witnesses written below, the very magnificent gentleman Juan Pardo, royal captain of the Spanish infantry and *contino* of his [Majesty's] household, lieutenant governor of these provinces, appeared and said that in order to make a certain journey and expedition inland in these provinces on the way to New Spain with the soldiers of his company, on orders of the very illustrious gentleman, Pedro Menéndez de Avilés, adelantado of these provinces in the name of His Majesty, his mercy, received certain number of munitions and foodstuffs from the royal store keeper who is in the forts of this city. It suits him to have surety and clarity of the quantity he received as well as of how he distributed it so that [he may] show and present it in order to give his discharge when he might be asked, if by chance his account is taken or he is charged [with them]. Therefore he was asking me and asked me, the notary, to give him a certification about the quantity of items which he received and following it how he used them up because that is suitable for his right and discharge.

The tenor of that which the captain received and how he distributed it says as follows:

Received

Matchcord	First of all, the captain received 10 arrobas and 10 pounds [260 lb.] of matchlock cord.	10 @ 10 lbs
Lead	Further, he received 9 arrobas and 22 pounds [247 lb.] of lead.	9 @ 22 lbs
Powder	Further, he received 12 arrobas [300 lbs] of matchlock powder.	12 @
Iron	Further, he received 2 iron picks for things which might be necessary.	2 picks

Iron	Further, he received 7 iron shovels.	7 shovels
Iron	Further, he received 5 maddocks.	5 maddocks
Steel	Further, he received 5 crossbows for his trip.	5 crossbows
Bolts	Further, he received 20 dozen [240] bolts [for the crossbows].	20 dozen bolts
Sacks	Further, he received 74 linen sacks for carrying food and other things.	70 [sic] sacks
Biscuit	Further, the captain, Juan Pardo, received at the time that he was to leave to make the trip 31 arrobas and 18 pounds [793 lb.] of biscuit for the sustenance of his company.	31 @ 18 lbs
Wine	Further, he received 1 arroba and 18 *quartilios* [23.83 liters] of wine.	1 @ 18 q[uartill]os
Cheeses	Further, he received 10 cheeses.	10 cheeses

Thus what the captain received for making the journey were the things aforesaid and declared. Part of all of which he used up and distributed in the following order and form:

Expended

Bread, Wine, Cheese	In regards the bread, wine, and cheese, [it was used up] in giving it and dividing it to the soldiers of his company, as it was given and divided at times of greatest need when the journey was being made, as is certain and notorious.	
Powder	The soldiers of the company used 3 arrobas [75 lb.] in learning to shoot during some days prior to the departure for this purpose [i.e, for the trip], as is well known.	3 @
Cord	Two arrobas [50 lb.] of cord were used up in carrying lighted cord as an ordinary [matter].	2 @
Powder	One and a half arrobas [37.5 lb.] of powder were used at certain times when the matchlocks were fired in some places of principal [chiefs] in the interior, where many Indians were together, in order to make them fearful.	1 @ 12 lbs

The captain left the following munitions and things with the sergeant and 30 men he left in the fort of San Juan which was made at the foot of the mountain range in a place called Joara, because this was suitable for His Majesty's service.

Powder	6 arrobas [150 lb.] of powder.	6 @
Matchcord	5 arrobas and 8 pounds [133 lb.] of matchcord.	5 @ 8 lbs

Lead	5 arrobas and 10 pounds [135 lb.] of lead.	5 @ 10 lbs
Crossbows & Bolts	4 crossbows and 20 dozen [240] bolts.	
Bars, Shovels, Maddocks, Picks	3 shovels, 2 maddocks, and 2 picks	
Sacks	40 sacks were used up and broken in making game-bags and little sacks for the munitions.	

All of that which is said, what was received as well as what was used, in the form and manner which is said and declared, I, the notary, say and give certification that the captain received and used it up in the service of His Majesty. Pedro Gutierrez Pacheco, squad leader, Francisco Hernández de Tembleque, Francisco Corredor, soldiers of the company and all the rest of the soldiers of it were present as witnesses to all that which is said as well as to the receipt and using of it.

Juan de la Bandera, Notary. I give certification of the above said and I am a witness.
[End of Document]

Notes

1. I.e., *de ayuso*, meaning "below" or "infrascript."
2. I.e., *dice*.
3. I.e., *tener ellos*.
4. An error. Should be Hernández.
5. I.e., *cristobal*.
6. I.e., *alcaide*, a warder or governor.
7. I.e., *hize*.
8. I.e., *Jesu Cristo*.
9. I.e., *hefecto*.
10. I.e., *doze*.
11. I.e., *haber ello*.
12. I.e., *angeo*, a type of linen cloth.
13. I.e., *poner ellos*.

Afterword
Pardo, Joara, and
Fort San Juan Revisited

David G. Moore, Robin A. Beck Jr.,
and Christopher B. Rodning

> If the long time required to discover the site of Anhayca was discouraging, the recent discovery of what may be the site of Joara is encouraging. I, for one, had little hope that Joara would ever be found.
> —Charles Hudson[1]

During the mid-1980s, Charles Hudson and his colleagues from the University of Georgia published a series of articles that changed how archaeologists and historians view the routes of the Spaniards Hernando de Soto and Juan Pardo through the interior Southeast.[2] Although their research revived debate among scholars throughout the region traversed by these expeditions, in few areas did the reconstructions proposed by Hudson and his colleagues require as complete a re-drawing of earlier reconstruction efforts[3] as that of the south Appalachians and adjacent areas of the Piedmont. The differences between what have come to be referred to as the "Swanton Route" and the "Hudson Route" through this area hinged on whether De Soto and Pardo entered the Appalachians via the upper Savannah River (Swanton) or via the upper Catawba and its tributaries (Hudson).

As Hudson spearheaded the effort to reconstruct a more accurate route, archaeological research in the Savannah Valley[4] began to indicate an absence of well-populated, mid–sixteenth-century villages such as those described in the various accounts of the Spanish expeditions. Hence the Savannah Valley appeared to qualify as the *despoblado* between Ocute and Cofitachequi described in the De Soto documents. At this same time, the Catawba River Valley was virtually terra incognito, with archaeologists unable even to determine whether or not a Native American population occupied the valley in the mid-sixteenth century; identifying such sites as the native towns of Yssa, Guaquiri, and Joara seemed a remote possibility. It was toward the aim of testing Hudson's Catawba Valley route and developing an understanding of the valley's late prehistoric inhabitants that David Moore initiated excavations at the McDowell and Berry sites as part of his dissertation field research in 1986.[5]

Moore, at the time a doctoral candidate at the University of North Carolina–Chapel Hill, was spurred on in his research both by Charles Hudson himself and by the late Roy Dickens. Hudson and his associates had earlier suggested that the McDowell Site (31MC41), located near modern Marion, North Carolina, was the town of Joara (De Soto's Xuala); it was at Joara, in January 1567, that Juan Pardo founded the garrison of Fort San Juan, the earliest European settlement in the interior of what is now the United States.[6] Joara and Fort San Juan played leading roles in the story of the Pardo expeditions, as the present volume makes clear. Hudson well understood that the discovery of their locations by archaeologists could provide a crucial linchpin for his redrawn map of the sixteenth-century Southeast. Moore thus intended to focus his project at the McDowell site, but—due to unforeseen circumstances—he actually spent more of his time at the Berry site (31BK22), located about 40 kilometers east of Marion near modern Morganton. As is so often the case in archaeological fieldwork, this seemingly unfortunate turn of events would eventually prove to be a boon. Moore did, in the end, conduct extensive excavations at both sites, and has recently published a synthesis of his dissertation research.[7] Although his excavations were unable to confirm that either McDowell or Berry was the location of Joara, he did demonstrate that both were large villages with earthen mounds, and that the upper Catawba Valley had a substantial late prehistoric population consistent with that described in the Spanish accounts.[8]

In 1994, two independently unfolding events sealed the general course of Hudson's Catawba Valley route and strongly suggested that Berry, not McDowell, was the location of Joara and Pardo's Fort San Juan. First, John Worth[9] discovered and translated an account by Domingo de León, one of the interpreters who accompanied Pardo on his second expedition. León's account provides a detailed description of the route traversed by Pardo and his company, a description that only matches a course set along the Catawba-Wateree River. Second, David Moore and Robin Beck reported on the discovery of sixteenth-century Spanish ceramics and hardware at the Berry site.[10] During surface reconnaissance of Berry in early 1994, Beck had discovered several shards of Spanish Olive Jar and a wrought iron nail near the area of Moore's 1986 excavations. Subsequently, Beck and Moore re-examined the collections from the 1986 project, identifying additional fragments of olive jar, as well as molten lead sprue; further visits to the site yielded more olive jar and a single sherd of Caparra Blue majolica, the latter found in the New World at sites that date between 1492 and 1600. The assemblage of Spanish artifacts from the Berry site closely matches the list of supplies left for the soldiers stationed at Fort San Juan, as recorded by Pardo's scribe, Juan de la Bandera. The only other identified site in the interior Southeast with a similar assemblage of sixteenth-century Spanish artifacts is the Martin site in Tallahassee, Florida, location of De Soto's winter camp at Anhayca in 1539. Moore, Beck, and Worth identified Berry as the likely site of Joara and Fort San Juan, a location that fell within 40 kilometers of the site that Hudson and his colleagues had proposed in 1983.[11]

The identification of the Berry site as Joara led Beck to revisit the routes of De Soto and Pardo across the Appalachian Summit.[12] Specifically, if Berry was Joara, then the explorers likely crossed into the mountains by different routes, De Soto taking a northern course from Berry into the Nolichucky drainage and Pardo taking a western course into the French Broad drainage through the Swannanoa Gap. In 1996, as part of his Master's thesis at The University of Alabama, Beck conducted a systematic survey along Upper Creek-Warrior Fork, the tributary of the upper Catawba River on which the Berry site is located.[13] This survey identified five large villages or village pairs that appear to have been contemporaneous with Berry and that may represent the towns of the five *oratas,* or local headmen, known to have been under the authority of Joara Mico from 1567 to 1568, during the time that Pardo's men were at Joara.[14] Berry, which measures approximately five hectares (more than 12 acres), was twice the size of these other towns, suggesting that Joara Mico had forged a multicommunity polity—a chiefdom—by at least the mid-1500s. We feel that combining archaeological and documentary research in this manner is precisely what Hudson has demonstrated in tracing the routes of sixteenth-century Spanish expeditions across the Southeast. Hudson has considered route reconstructions as the beginning of the effort to map the social geography of the sixteenth-century Southeast, which can provide a cultural and historical baseline for looking backward into prehistory and forward into the seventeenth and eighteenth centuries.

In 1997, Beck, Moore, and the late Thomas Hargrove conducted a gradiometer survey of the Berry site that revealed evidence of at least four large, subsurface anomalies believed to represent the remains of burned buildings; subsequent auger testing confirmed the presence of burned materials in discrete areas corresponding to these anomalies.[15] In 2001, Moore, Beck, and Christopher Rodning (the latter a doctoral candidate at the University of North Carolina–Chapel Hill) formed the Upper Catawba Archaeology

Project and resumed excavations at the Berry site as part of an effort to investigate the effects of interactions between Spanish expeditions and the native chiefdoms in the Catawba Valley. Our work at the Berry site has focused on burned buildings located in a small area at the northern end of the site that coincides with the distribution of Spanish artifacts. These buildings are thought to represent the remnants of Fort San Juan.[16] Our excavations here, from 2001 to 2003, have uncovered the well-preserved remains of these burned buildings; each measures between 65 and 80 m^2 and together they form a distinct compound around a possible central plaza. Large pit features in the plaza area contain abundant quantities of native materials and several glass beads and brass lacing tips, or aglets, from Spanish clothing. A line of large posts near one of the buildings suggests that a wooden stockade may have protected the compound.[17]

Our excavations inside one of the burned buildings, Structure 1, have revealed an exceptional degree of architectural preservation, including intact features such as carbonized wooden posts that still remain upright and fallen roof timbers that still retain their bark. We have found burned sections of wooden wall benches made of split oak, with split cane matting still attached to the benches. Artifacts inside the building are lying in place where they fell or were left on the day that Fort San Juan was destroyed. We discovered fragments of chain mail armor on the floor of the structure, and some of its wooden timbers seem to be notched with metal tools in a European style of construction. We believe, in sum, that this was one of the buildings that quartered Pardo's soldiers stationed at Fort San Juan.[18] Our excavations have shown that Structure 1 was built in a style that was typical of native buildings, but with specific elements of its construction that exhibit non-native, European techniques and technologies. Also, the chain mail links found at the site suggest that Spanish soldiers spent time inside the building before its destruction.

These are the details we would expect for soldiers' houses built during the Pardo expeditions: Pardo's men brought only those personal items that they could carry on their backs, and therefore most of their material possessions at Joara were likely native in origin. Also, while the documents record that Pardo commanded the native people of Joara to build such houses, it is reasonable to expect that the soldiers of the expedition participated in their construction. We have yet to expose the contents of the other buildings in this area, but we have every reason to believe that they are just as remarkable in the preservation of their architectural details. That all four were burned serves as a chilling testament to how relations between these Spaniards and

Native Americans ended tumultuously in the spring of 1568, when the people of Joara destroyed Fort San Juan.

The Berry site, therefore, provides an archaeological context that is unique in several respects. First, it likely contains the first European settlement in the interior of what is now the United States, as well as the earliest site of a sustained interaction between Europeans and native peoples (18 months) in the interior of North America. Second, it may be one of the only places along the various routes of early Spanish exploration that can be linked to a particular point on the landscape, and to a specific archaeological site. Third, if Berry does indeed contain the remains of Fort San Juan, then its burned buildings offer a window onto a single historical event: the fiery destruction of the fort and the end of Spanish colonial ambitions in northern La Florida. Fourth, documentary sources indicate that Joara was still a major town during the early years of the seventeenth century, and the Berry site thus gives us an opportunity to uncover the history of this native town not only during the 27 years between the De Soto and Pardo expeditions, but also for at least four decades after the fall of Fort San Juan. Further investigations at Berry and other protohistoric sites in the upper Catawba Valley and surrounding areas will shed new light on the social landscape of the southern Appalachians, on the nature of early Spanish expeditions and settlements in this part of the Southeast, and on the nature of cultural interactions between these Europeans and Mississippian peoples. Charles Hudson has studied all of these topics over the course of his remarkable career, and we hope that our continued work will meet the standard that he has set. His research is at the very heart of our project—it is the catalyst that brought us all to the Berry site—and his tenacity and passion inspire us to help illuminate this long-lost corner of the forgotten century.

Notes

1. Charles M. Hudson, *Knights of Spain, Warriors of the Sun: Hernando de Soto and the South's Ancient Chiefdoms* (Athens: University of Georgia Press, 1997).
2. Chester D. DePratter, Hudson, and Marvin T. Smith, "The Route of Juan Pardo's Explorations in the Interior Southeast, 1566-1568," *Florida Historical Quarterly* 62(1983):125-58; Hudson, Smith, and DePratter, "The Hernando de Soto Expedition: From Apalachee to Chiaha," *Southeastern Archaeology* 3(1984):65-77; and Hudson, Smith, David Hally, Richard Polhemus, and DePratter, "Coosa: A Chiefdom in the Sixteenth Century Southeastern United States," *American Antiquity* 50(1985):723-37.
3. For example, John R. Swanton, Final Report of the United States De Soto Ex-

pedition Commission (1939; reprint, Washington, D.C.: Smithsonian Institution Press, 1985).
4. For a synthesis of the archaeological work in the Savannah Valley, see David G. Anderson, *The Savannah River Chiefdoms: Political Change in the Late Prehistoric Southeast* (Tuscaloosa: The University of Alabama Press, 1994).
5. David G. Moore, *Catawba Valley Mississippian: Ceramics, Chronology, and Catawba Indians* (Tuscaloosa: The University of Alabama Press, 2002).
6. See DePratter, Hudson, and Smith, "The Route of Juan Pardo's Explorations," *Florida Historical Quarterly*, 132; see also this volume.
7. Moore, *Catawba Valley*.
8. Janet E. Levy, J. Alan May, and Moore, "From Ysa to Joara: Cultural Diversity in the Catawba Valley from the Fourteenth to the Sixteenth Century," in *Columbian Consequences, Volume 2: Archaeological and Historical Perspectives on the Spanish Borderlands East*, edited by D. H. Thomas (Washington, D.C.: Smithsonian Institution Press, 1990), pp. 152-68.
9. John R. Worth, "Exploration and Trade in the Deep Frontier of Spanish Florida: Possible Sources for 16th-Century Spanish Artifacts in Western North Carolina" (Paper presented at the 51st Annual Meeting of the Southeastern Archaeological Conference, Lexington, Kentucky, 1994).
10. Moore and Robin A. Beck Jr., "New Evidence of Sixteenth-Century Spanish Artifacts in the Catawba River Valley, North Carolina" (Paper presented at the 51st Annual Meeting of the Southeastern Archaeological Conference).
11. Worth, "Exploration and Trade" and Moore and Beck, "New Evidence," both presented at the 51st Annual Meeting of the Southeastern Archaeological Conference.
12. Beck, "From Joara to Chiaha: Spanish Exploration of the Appalachian Summit Area, 1540-1568," *Southeastern Archaeology* 16(2)(1997b):162-68.
13. Beck, "The Burke Phase: Late Prehistoric Settlements in the Upper Catawba River Valley, North Carolina" (M.A. thesis, Department of Anthropology, University of Alabama, Tuscaloosa, 1997a).
14. Juan de la Bandera, "The Pardo Documents, transcribed and translated by Paul H. Hoffman," in Hudson, *The Juan Pardo Expeditions: Exploration of the Carolinas and Tennessee, 1566-1568* (Washington, D.C.: Smithsonian Institution Press, 1990; Tuscaloosa: The University of Alabama Press, 2005), pp. 205-321; and Beck and Moore, "The Burke Phase: A Mississippian Frontier in the North Carolina Foothills. *Southeastern Archaeology* 21(2)(2002):192-205.
15. Hargrove and Beck, "Magnetometer and Auger Testing at the Berry Site (31BK22), Burke County, North Carolina" (Paper presented at the 58th Annual Meeting of the Southeastern Archaeological Conference, Chattanooga, Tennessee, 2001).

16. Moore, Beck, and Christopher B. Rodning, "Joara and Fort San Juan: Culture Contact at the Edge of the World," *Antiquity* 78(299): March 2004; available at http://antiquity.ac.uk/ProjGall/moore/.
17. Megan S. Best and Rodning, "Mississippian Chiefdoms and the Spanish Frontier: An Overview of Recent Excavations at the Berry Site in Western North Carolina" (Paper presented at the 60th Annual Meeting of the Southeastern Archaeological Conference, Charlotte, North Carolina, 2003).
18. Beck and Caroline V. Ketron, "The Fall of Fort San Juan? Excavating a Burned Building at the Berry Site" (Paper presented at the 60th Annual Meeting of the Southeastern Archaeological Conference, Charlotte, North Carolina, 2003).

Index

Entries for the Preface to 2005 edition, Spanish transcriptions and translations in Part III, "The Pardo Documents," and the Afterword are not included in this index.

Aboyaca, 33, 44, 62, 68, 78-79; Orata, 62, 139
Achini, 76; Achini Orata, 74
Adamson site, 70, 73, 116n55, 117n63
Adini Orata, 90, 93
Ahoya, Indian village, 32, 44-46, 78, 128, 132, 139, 152, 176
Ahoyabe, Indian village, 32, 78; Ahoyabe Orata, 46
Aguilar, [Luís de?], Spanish soldier, 193, 201n78
Ajacán mission, 156, 170
Alas, Alonso de las, 180
Alas, Esteban de las, 16, 18, 23, 26, 35, 163-64
Almeydo, Portuguese pilot, 174, 191-92
Alonso, Pedro, 43
Altamaha region and peoples, 27, 180, 183-84
Alvarez de Pineda, Alonso, 6
Anduque Orata, 88-89, 96, 99
Ansohet. *See* Enxuete
Ansuhet Orata, 88-89
Anxuete Orata, 99
Apalachee region and people, 69, 95, 161, 195
Apalategui, Francisco de, 35, 153, 166n40
Aracuchi, 26, 34, 42, 76-78, 83, 140, 143-44, 151, 188
Aranbe, 82. *See also* Herape
Arande Orata, 100
Arasue Orata, 99-100
Archaic period (archaeology), 58-59
Archiniega, Sancho de, 18
Atache, Indian town/chiefdom, 13, 174
Athahache/Athahachi. *See* Atache
Atuqui Orata, 88-89, 96. *See also* Tucahe
Aubesan, 89; Aubesan Orata, 88
Auñón, Miguel de, 180
Avila, Francisco de, 180
Avilés, Pedro Menéndez de. *See* Menéndez
Ayllón, Lucas Vásquez de, 5-8, 20n12, 73, 81, 158, 169, 182
Aymay. *See* Guiomae
Ayo, 76; Ayo Orata, 74

Badajoz, Antonio de, 180
Baéz, Father Domingo Agustín, 108
Bandera, Juan de la, Pardo notary, xi, 4, 30-31, 33, 38, 40, 42, 47n2, 51-52, 61-63, 65-

66, 68, 75, 77-79, 82-84, 86, 88, 90-91, 93, 95-96, 98, 103, 107-109, 127-28, 131, 135-36, 140, 143, 149-51, 153, 162-64, 185, 190; as deputy governor, 155
Barcena, Pedro de, 43
Barnett (archaeological) phase, 101
Bela, Alonso, 133
Bell Farm site, 188
Berkeley, William, 184
Berry site, 94
Blanding site, 70, 73, 116n55, 160
Boykin site, 70
Buena Esperança, town. *See* Orista
building structures, 91, 127, 141-45; earthlodges, 93; barbacoas, 110, 117n69; Spanish forts, 146-52, 173, 175-77
Burguignon, Nicholas, 193; quoted, 138
burials. *See* mortuary practices
Burke culture, 109

Cabeza de Vaca, Alvar Núñez, 8, 192
Calderon, Andrés, 177
Calderon, Bishop, 82
Calusa Indians, 55, 58, 155
Canasahaqui Orata, 99-100
Caniçares, Cpl. Lucas de, 35, 128, 148, 151, 164
canoes, 43, 45, 132-34, 152
Canos, Indian town, 128, 143. *See also* Cofitachequi
Canos Orata, Cofitachequi Orata, 64, 74
Canosaca Orata, 74, 98
Canosaqui Orata, 98
Canosi. *See* Cofitachequi
Capachequi, chiefdom, 9
Caroline, Fort, 15, 17
Cartier, Jacques, 11-12
Casque Indians, 39, 106
Cataba, 83-84, 188
Catalan, Pedro, 175, 177
Catapa Orata, 74-75, 84, 88. *See also* Cataba
Catawba King, town, 188
Catawba Indians, ix-x, 52, 75, 114n22; language, 68, 74-77, 81, 83-84, 87-89, 109, 118n77, 184, 187
Catawba phase, 102
Catawba Trail, 87
Cauchi, Indian town, 36, 40, 66, 90, 96-103, 108, 110, 142-43, 191; Cauchi Orata, 97, 99-100; fort, 149
ceramics/pottery, 53, 71-72, 86, 88, 91-93, 101, 109, 160
Chalahume, Indian town, 38, 105, 131, 159, 191

Chantonne, Perrenot de, 13
Chara, 184-85; Chara Orata, 90, 93
Charlesfort, French settlement, 14-15
Cherokee Indians, 52, 86, 94, 97-102, 105, 120n135, 123n194, 189; language of, 67, 83, 87-89, 95-101, 103, 105, 109
Chiaha, Indian town and chiefdom, 4, 10, 28-29, 35-36, 39-40, 50n58, 62, 67, 87, 90, 98, 100-106, 127-28, 131, 148, 157, 159-60, 174, 190; Chiaha Orata, 103; language of, 108; rising against Spaniards, 177
Chichimeca Indians, 129-30, 172
Chicora, 81-82, 156; Legend of, 158-60
Chicora, Francisco de ("El Chicorano"), 6, 81-82, 156
chiefdoms, 55-56, 58, 60-65, 67, 71, 73, 79-83, 93, 109-110, 114n21-23, 170, 172-73, 189; levels of power, 61-66, 88, 103, 114n27, 115-16n47, 184-88; women as chiefs, 66-67; tribute to, 110-11, 119n113, 134
Chilhowee, Cherokee town, 105
Chisca Indians, 10, 27-29, 46, 87, 90-91, 100-101, 103, 106, 131, 148, 157, 159-61, 192
Cíbola myth, 192
Citico archaeological site, 39, 106, 196n11; town, 105. *See also* Satapo
Coçao, town, 43-44, 68, 78-79, 128, 131, 139, 185
Coçapoye, 78; Coçapoye Orata, 46, 79
Cofaqui, Ocute Orata, 67
Cofatache. *See* Cofitachequi
Cofitachequi, Indian town and chiefdom, 4, 9-10, 25, 34, 43-45, 61, 63-70, 73-75, 77-79, 82-84, 90-91, 98, 101-102, 104, 109, 111, 115n47, 116n47, 130-34, 140, 144, 151, 160-61, 170, 182, 185, 200n61; "Lady of," 63, 66-67, 73, 76, 90, 97, 144; sociogram of, 75, 117n67; Fort Santo Tomas, 44-45, 151-52; temple at, 157, 187; dissolution of, 187
Conch shells, 111, 141
Congarees, 187, 189
Coosa, Indian chiefdom, 10, 12-13, 27, 38-40, 65, 67-68, 78, 94, 97, 101-106, 108-110, 131, 142, 160, 170, 174-75, 182; great Orata, 62-63, 148; sociogram of, 104, 117n67
corn, 45, 52-53, 55, 58, 71, 76-77, 81, 84, 89, 110-11, 128, 132-34, 152, 155, 180; cornmeal, 79; corncribs, 110-11, 134, 143-44, 146
Corpa, Pedro de, 180
Cortés, Hernán, 5
Coste, town, 39, 104, 106-107, 174, 191
Creek Indians, 52, 65-66, 103, 109; language, 69-70, 74, 76-77, 79, 112; social organization, 100
Cuttagochi, Cherokee town, 99

dakwa, monstrous mythological fish, 95
Dallas culture, 60, 101-102, 104-105, 109
Datha, Duahe Orata, 82
de Soto, Hernando. *See* Soto
deerskins, as trade and tribute goods, 110-11, 140, 142

Dickens, Roy, 86, 101, 344
disease, 173, 181-82, 185-87
dogs, 97, 135
Drake, Francis, 138, 179, 193
Duahe. *See* Duhare
Dudca, Indian village, 41, 84
Duhare, 82, 156
Dyar archaeological phase, 60

economic system of native societies, 109-111, 132, 139, 156, 159, 187-89
Edista, Indian town. *See* Orista
Elasie. *See* Ylasi
Elvas, the gentleman of, 10, 37-38, 51, 84, 110
Emae Orata, Guiomae Orata, 62, 79
England, colonial activities of, 179, 181, 184, 195
Enjuete. *See* Enxuete
Enuque. *See* Anduque
Enxuete Orata, 88-89, 96, 99
Esaw people, 187-88. *See also* Yssa
Escalante Fontaneda, Hernando de, 201n73
Escamacu, village, 50n66, 82, 155
Escudero de Villamar, Ensign Albert, 41, 150, 163-64
Estate (Estatoe) Orata, 97

Fernandez de Chozas, Pedro, 183-84
Fernández de Ecija, Sgt. Francisco, 192
Ferry landing site, 74
Flores, Cpl. Pedro, 149-51
foodstuffs of Indians, 44, 58, 79-81, 110, 126, 156, 189; dogs as food, 97; native agriculture, 129, 134. *See also* corn
Fort Watson site, 79, 119n113, 134
fortifications by natives, 102-103
France, colonial activities of, 11-16, 26, 32, 125, 129-30, 158, 179, 181
Franciscans, 180-81
Francisco of Chicora. *See* Chicora, Francisco de
Fumica Orata, 104

gambling, 124n226, 155, 163, 166n46
gender roles among Indians, 98; transvestism, 122n176
gifts, 42, 46, 93-94, 100, 111, 128-29, 132, 134-40, 147
Gomez, Fernan, 43, 132
Gonzalez, Vicente, 179
Gran Copala, La, mythical city, 190-94
Great Indian Warpath trail, 40
Grenville, Richard, 179
Guale: region, 7, 81-82, 108, 125, 134, 180, 182-83, 195; Indians, 18, 119n113, 155,

177, 180; language, 78
Guanbuca Orata, 88, 99, 190
Guancamu, 100, 110
Guando Indians, 45, 80
Guanuguaca. *See* Guanbuca Orata
Guaquili. *See* Guaquiri
Guaquiri, Indian town, 35, 48n11, 84, 87-88, 143; Guaquiri Orata, 88-89
Guarero Orata, 100
Guaruruquet[e] Orata, 88, 99
Guasili. *See* Cauchi
Guatari, Indian town, 26, 35, 42, 46, 77, 91, 93-94, 109-110, 125, 143, 150, 153-54, 166n40, 182, 184, 187; Guatari Mico, Orata, 35, 62, 83, 90, 92-93, 153, 188; female chiefs of, 66-67, 90, 93-94; fort at, 151-52, 176; mission to, 177
Guatariyatiqui. *See* Otariyatiqui
Guaxule. *See* Cauchi
Gueça, 34, 74, 76-77, 83, 93, 111, 131, 134, 140, 143, 184; Gueça Orata, 76
Guenpuret, 89; Guenpuret Orata, 88
Guiomae, Indian town, 24, 33-34, 43-44, 62, 78-80, 82-83, 128, 131-35, 139, 141, 143-44, 152; Guiomae Orata, 144

Herape Orata, 74, 76. *See also* Ilapi
Hermosa, Sgt. Pedro de, 43, 45, 152, 164
Hitchiti language, 68-69
Hoffman, Paul, 47n2
Hollywood archaeological phase, 60
horses, 130-31, 183
Houstaqua people. *See* Yustega
Huaque, 82. *See also* Uca Orata
Huguenots, French, 3
Hymahi. *See* Guiomae

Ichisi, chiefdom, 9
Ilapi. *See* Ylasi
Irene archaeological phase, 60
Iron Horse archaeological phase, 60
Iroquois, 188; Iroquoian language, 83, 86-88, 98
Isaw. *See* Yssa

Jasire Orata, 104-105
Jesuits, 155-56, 170, 182
Jimenez, Luís, 43
Jimenez, Cpl. Marcos, 149-51
Joara, Indian town, 25, 26-27, 35, 40-41, 48n12, 83-84, 86-88, 93, 95, 97, 100-102, 107, 110, 142, 149, 176, 182, 185, 190; fort at, 28, 35, 46, 90, 126, 143, 146-48, 150, 159-61; Orata (Mico), 62, 66, 83, 88-90, 97, 100-101, 135, 153; sociogram of, 89, 117n67; Joara Chiquito Orata, 99-100, 102

Johnstone Farm site, 102
Jore, Cherokee town, 87
Jueca, 89; Jueca Orata, 88

Keyauwees, 188
Kiawah (Cayagua, Keyawah) Indians, 179, 185
King site, 102, 106, 175
Kituhwa, Cherokee town. *See* Quetua
Koasati language, 103, 108-109; people, 107
Kroeber, A. L., xii
Kussoo, town. *See* Coçao
Kymulga phase, 102

Lamar culture and phases thereof, 60, 68, 73, 85, 101-102, 109
languages, 52, 68-70, 74-84, 86-90, 95-96, 98
Lara, Juan de, 183
las Casas, Bartolomé de, 156
Laudonnière, René de, 15
Lawson, John, 121n153, 184-89, 200n58; quoted, 110
Lederer, John, 184-85, 198-200n52, 200n53
Le Moyne, Jacques, 136, 147, 158, 178
litters, as conveyances, 63, 65
Little Egypt site, 101
López, Baltasar, 181
López, Francisco, 193
López Avilés, Juan, 193
López de Gómara, Francisco, 158
Luna y Arellano, Tristán de, expedition of, 4, 12-13, 61-62, 102, 108-110, 169, 182

Mabila, battle of, 10, 13, 29, 49n28, 106-107, 174-75, 183
Marcos, Fray, 192
Martín, Teresa, 87, 91, 176, 197n20
Martín de Badajoz, Juan, 175-76, 197n20
Martínez, Francisco, 26, 47n2
Martínez, Jaime, 27, 114n23, 176
Martyr, Peter, 81-82, 156
Mathews, Maurice, 185, 187
Mati Yssa, *mandador* of Yssa, 66
McDowell site, 86, 88
Mendana, Alvaro de, 153-54
Méndez, Luisa, 87, 91, 190-91
Méndez Canço, Gov. Gonzalo, 87, 111, 174, 179-83, 190, 192, 194
Menéndez de Avilés, Pedro, 3-4, 15-17, 23, 26, 29, 31, 46, 94, 125, 128-29, 131, 134,
    147-49, 153, 169-70, 177-81
Menéndez Marqués, Juan, quoted, 161-62
Menéndez Marqués, Pedro, 35, 178-79, 193

metals, trade in, 91, 96, 156, 159, 190-92
Milfort, Louis Le Clerc, 112n3
minerals, search for, 38, 41-42, 46, 50n62, 128-30, 137, 156-64, 177, 180, 189-95
missions, 125, 134, 151, 153-56, 166n40, 170, 173, 177, 180-82
Mississippian transformation, 54-56, 58-59, 72-73, 85, 113n10
Montero, Sebastian, 26, 35, 48n9, 94, 151, 153-55, 166n40, 177
Moore, David, 94
Moore, James, 195
Morales, Diego de, 107
Morales, Pedro, 193
mortuary practices of Indians, 58, 64-65, 94, 117n69, 187
mounds, Indian, 53, 56-57, 64, 70-74, 79, 82, 85, 88, 93, 97, 116n55, 160, 189
Moundville culture, 191
Mouse Creek culture, 101
Moyano de Morales, Sgt. Hernando, 25, 26-29, 35-36, 38, 40-44, 46, 86, 88, 90, 98, 100, 103, 107-108, 126-28, 135, 148-50, 176, 192; as soldier, 154; prospecting of, 159-60, 163-64, 173, 177, 190, 192, 195
Mulberry phase/site, 60-61, 70, 73-74, 101, 116n55, 117n63, 160-61
Muñoz, Anton, 35, 153, 166n40
Muskogean languages, 68-70, 75-79, 82-84, 87, 90, 98-99, 101, 103-106, 109, 118n78; peoples, 98, 123n194

Nanipacana, Indian town, 13
Napochies, Indian people, 13, 102, 110
Narváez, Pánfilo de, 8, 169, 182
Natchez Indians, 187
Nelson, T. F., site, 94
Nequase (Nequasse) Orata, 97
Nuestra Señora, Fort, 45

Ochuse, 12, 13
Ocute, chiefdom, 9, 63, 66-68, 101, 112, 180-81, 183-84; wilderness of, 60-61, 84, 130-31
Ohebere Orata, 76-77
Olamico, town, 36, 38-40, 65-66, 103, 107-108, 148, fort at, 148-49, 173, 175-76, 190-91. *See also* Chiaha
Olamico, Chiaha Orata, 62, 103-104, 106, 128
Old Cherokee Path, 87
Olitifar, town, 109
Olmos, Alonso de, 156
Oluga, 39, 107
Orata Chiquini, Indian leader, 35, 93
Orista, Indian town, 15, 18, 44-46, 52, 78-81, 111, 118n102, 125, 127-128, 132, 139, 152, 155, 185; Orista Orata, 46, 80, 128, 132; fort at, 152
Orixa insyguanin, 82. *See also* Orista
Osuguen Orata, 88

Otape Orata, 104
Otari, 34-35, 90, 143, 153, 188; language, 48n14, 76, 84; Otari Orata, 74-75, 83
Otariyatiqui, Indian town, 26; Otariyatiqui Orata, 66

Pardo, Captain Juan, 18, 23-29, 31-32, 34-36, 38-46, 48n11, 50n58, 51-52, 59-62, 64-70, 72, 74, 76-80, 82-94, 96-111, 125-26, 129, 131-34, 139-40, 144, 146-53, 156, 159-61, 166n40, 169-70, 175, 180-82, 184-85, 188, 190-95; shares in prospecting, 163-64; stores and equipment for, 126-28, 131-32, 134-37, 147-52, 164n5
Pasque/Pasqui, 82-83, 134, Pasque Orata, 79, 134, 144
Pee Dee archaeological phase, 60
Perico, Indian guide, 63, 66, 160
Philip II, Spanish monarch, 12-16, 18, 63, 178
Pisgah culture, 60, 85-86, 88, 91, 97, 100-102, 109. *See also* Qualla culture
Pizarro, Francisco, 8
Plum Grove site, 28, 86, 91
Ponce de León, Juan, 6, 8, 182
Ponte de Lima, Juan Perez, 107-108
pottery. *See* ceramics/pottery
power, Spanish conceptions of, 62-63. *See also* chiefdoms
Prieto, Martín, 173
Pundahaque, 89; Pundahaque Oratas, 88

Qualla culture, 85-86, 88, 91, 100, 102
Quetua Orata, 97
Quinahaqui, Indian town, 25-26, 35, 42, 48n11, 66, 84, 87-89, 143, 150; Quinahaqui Orata, 84, 88
Qunaha, 89; Qunaha Orata, 88

Ranjel, Rodrigo, 10-11, 51, 69, 117n69
Rembert archaeological phase, 60
Ribas, Juan de, 27, 29, 106, 160, 162-63, 174, 190-91, 193, 196n12
Ribaut, Jean, 14, 17, 32, 111
roads, Spanish, 130. *See also* trails
Roberval, Jean François de la Roque, seigneur de, 12
Rodriguez, Blas de, 180
Rodriguez, Cpl. Gaspar, 45, 152
Rogel, Father Juan, 50n66, 80-81, 108, 111, 124n226, 134, 155-56, 176
Rojas, Hernando Manrique de, 15
Rouffi, Guillaume. *See* Rufin, Guillermo
Rufin, Guillermo, 14-15, 32, 39, 52, 78, 80, 82, 95, 106, 135
Rutherford's War Trace trail, 87

St. Augustine: foundation of, 3, 5, 125; Drake's attack upon, 179; state of, 180
St. Pa. *See* Sanapa
Salas, Gaspar de, 183

salt, 42, 75, 77, 87, 111-12, 195
San Juan, Fort, 25, 41
San Pablo, Fort, 40
San Pedro, Fort, 148-49, 173, 175-76, 190-91
Sanapa, town, 42, 185; Sanapa Orata, 74-75
Santa Elena, Spanish settlement, 3, 4-5, 12-14, 16-18, 23, 26, 29, 32, 35, 42, 44, 46, 125, 136-38, 152-53, 155, 158; sociogram of, 80, 117n67; Indians of, 177; efforts to resettle, 178-79
Santee Indians, 186-87. *See also* Sarati
Santiago, Fort, 42
Santo Tomás, Fort, 44, 45
Santos, Juan, 153-54
Sapona, town, 188
Sarati, 78, 118n86, 186; Sarati Orata, 76-77. *See* also Santee Indians
Saruti, 78; Saruti Orata, 77
Satapo, Indian town, 39-40, 46, 62, 66, 100, 104-109, 128, 131, 135, 148, 159, 173-75, 183, 191
Saturiba, Indian chief, 15
Saturiwa people, 95
scalping, practice of, 27
Sedeño, Antonio, 81, 134, 155, 182
Segura, Juan Baptista de, 156
Senneca, Cherokee town, 96
Sewee Indians, 185-86
Shinholser site, 183
Sierra, Pedro de la, 154
Simonds, Frederick W., 194
slavery, 16, 81, 100, 129, 174-76, 181, 185, 195
Sona, 82. *See also* Sanapa
Soto, Hernando de, 8; expedition of, x-xi, 4, 8-13, 18n2, 25, 28-29, 34-37, 39, 49n28, 51, 60-61, 63-69, 73-74, 78-79, 84, 87, 90-91, 95, 97-98, 101, 103-109, 111-12, 128, 131, 133, 135, 142, 156-58, 160-61, 169-70, 173-75, 180, 182-83
Square Ground phase, 116n49
Stewart, John, trader, 188
Suarez, Andrés, 38, 41, 159, 163
Subsaquibi, 84
Sugeree Indians, 77, 188. *See also* Suhere
Suhere, 184; Suhere Orata, 76-77
Suwali: trail, 95; people, 120n135
Suya Orata, 76-77

Tacoru Orata, 97
Tagaya, Indian town, 25, 34, 76, 140, 143; Tagaya Orata, 77, 140
Tagaya the Lesser (Tagaya Chico), Indian town, 25, 26, 34, 76
Talimeco, 63-64, 70, 74, 104, 117n63, 117n69
Talisi, chiefdom, 10, 13; Orata of, 175

Tama, region, 180-184, 192-193. *See also* Altamaha
Tanasqui, 36, 49n49, 102-103, 123n194, 190
Tascaluza, chiefdom, 10; Tascaluza Orata, 106, 119n111, 174
Tasqui, town, 107, 109
Tasquiqui, 109
Taucoe, Cherokee town, 96
temple, Indian, 73, 82; attendants, 82
Tequesta Indians, 155
Teran, Francisco de, 154
Timucua Indians, 15, 136, 158, 173; language, 69, 95
Toa, chiefdom, 9, 110
Tocae, Indian town, 35-36, 40, 65-66, 95-96, 98, 100, 110-111, 139, 142, 159; Tocae Orata, 88. *See also* Tucahe
Toteros, 188
Town Creek site, 57, 70, 78, 93, 145
trails, 40, 87, 95, 130-31
Tucahe (Tocae?) Orata, 99
Tugalo archaeological phase, 60
Tuscaroras, 194-95
Tuskegee, Creek town, 109

Uastique Orata, 95-96, 111, 139, 159
Uca Orata, 77
Uchi Indians, 39, 106
Uchiri, 184; Uchiri Orata, 84
Ulibahali, Indian town, 13, 102, 175
Unharca Orata, 76-77
Uniaca Orata, 76-77
Unuguqua Orata, 74, 76
Upper Saura Town, 91-93
Uraca Orata, 74, 76
Uscamacu, Indian town, 32, 44, 176, 177; Uscamacu Orata, 46, 80, 128; attack on Spaniards at, 177
Usi, 42
Ustehuque Orata, 88, 99
Utaca Orata, 97
Utahaque Orata, 88-89
Utina people, 95
Uwharrie complex, 71-72; culture, 85, 91-93, 109, 160

Váez, Br., 81
Valdés, Fernando, 180
Vastique. *See* Ustehuque Orata
Vastu, 89; Vastu Orata, 88
Vega, Garcilaso de la, 11, 51, 64, 66-67, 74, 90, 97
Vehidi, 76; Vehidi Orata, 74, 76

Velas, Alonso, 39, 107
Velasco, Luis de, 11-13, 156
Verascola, Francisco de, 180, 183-84
Verrazzano, Giovanni da, 11, 14
Villafañe, Angel de, 13-14
Villamanrique, Viceroy, 129
Villarreal, Francisco, 134, 155, 182
von Reck, Phlip G. F., 111, 142-43
Vora Orata, 74, 76

Warren Wilson site, 85
Wateree Chickanee, 187. *See also* Guatari
Watson, Fort. *See* Fort Watson site
Waxhaw people, 187, 200n6. *See also* Gueça
weapons: of the Spanish, 126-27, 147-51, 164n5
Westoes, Indians, 185
Woodland period (archaeology), 58-59
Woodward, Henry, 185

Xapira, 156
Xenaca Orata, 96
Xuala, Indian town, mountains. *See* Joara

Yamiscaron, 82
Yca, village, 43, 78
Yeardley, Francis, 194-95
Yamassees, 200n61. *See also* Guiomae. *See also* Emae
Yetencunbe Orata, 46, 80
Ylasi, 42-43, 72, 74, 77-78, 83, 128, 131, 133-34, 140, 143-44, 151, 185, 187, 200n61; Ylasi Orata, 74, 77
Yssa, Indian town, 25, 41-42, 66, 84-85, 89-90, 142-143, 150, 162-64, 184-85, 188; Yssa Orata, 62, 74-75, 83-84, 89
Yssa the Lesser, 42, 84; Yssa Chiquito Orata, 88
Yuchi Indians, 52, 143; language of, 83-84, 87-88, 101, 104, 109
Yustega people, 95-96, 158-59. *See also* Timucua Indians

Index 361

Errata

This second edition has been photographically reproduced from the first edition. The following are corrections for typographical errors in the latter.

Chap. 1

P. 7, lines 3, 7: "Allyón" should read "Ayllón"
P. 10, last line: "Rondrigo" should read "Rodrigo"
P. 13, line 22: "Athahachi" should read "Atahachi"

Chap. 2

P. 27, last paragraph: "Martinez" should read "Martínez"
P. 38, lines 10, 11 from bottom: "Andres" should read "Andrés"
P. 35, lines 5-7 from bottom: "Marques" should read "Marqués"
P. 41, line 4: "Andres" should read "Andrés"

Chap. 3

P. 65, lines 7, 9 from bottom: "Olameco" should read "Olemico"
P. 66, line 6: "Olameco" should read "Olemico"
P. 72, line 5 from bottom: "Llasi" should read "Ylasi"
P. 82, lines 4, 5, 13, 23: "Duahe" should read "Duhare"
P. 87, line 8: Add "de" after Méndez
P. 89, line 12: "Ansohet" should read "Ansuhet"
P. 92, illus.: "Upper Sauratown" should read "Upper Saura Town"
P. 100, line 6: "Canosahaqui" should read "Canasahaqui"
P. 104, line 2: "Talomeco" should read "Talimeco"
P. 106, line 2: "suprise" should read "surprise"
P. 108, line 22: "Agustin" should read "Agustín"

P. 111, last line: Add "de" after Governor
P. 114, n23: "Martinez" should read "Martínez"

Chap. 4

P. 128, line 2: "Canizares" should read "Caniçares"
P. 138, line 7: "Burgoigon" should read "Burguignon"
P. 139, line 11: "Cocao" should read "Coçao"
P. 150, line 11: "Jiminez" should read "Jimenez"
P. 151, line 11: "Jiminez" should read "Jimenez"
P. 163, line 12: "Bandera" should read "Moyano"

Chap. 5

P. 174, line 16: "Athahachi" should read "Atahachi"
P. 179, lines 1, 3 from bottom: "Méndez Canço" should read "Méndez de Canço"
P. 180, lines 2, 11, 19 from bottom: "Méndez Canço" should read "Méndez de Canço"
P. 181, line 11: "Méndez Canço" should read "Méndez de Canço"
P. 183, line 6: "Velascola" should read "Verascola"
P. 183, last line: "Solas" should read "Salas"
P. 184, line 1: "Velascola" should read "Verascola"
P. 189, line 8: "Congerees" should read "Congarees"
P. 193, line 20: "captain" should read "captive"
P. 200, n61: "Yemessee" should read "Yamassee"

About the Authors

David G. Moore (Ph.D., University of North Carolina–Chapel Hill 1999) has been actively involved in the archaeology of North Carolina's mountain and western Piedmont regions for more than 25 years. He served for 18 years with the Western Office of the North Carolina Office of State Archaeology and has been teaching full time at Warren Wilson College since 1999. His dissertation, *Catawba Valley Mississippian: Ceramics, Chronology, and Catawba Indians,* has been published by The University of Alabama Press (2002).

Robin A. Beck, Jr. (Ph.D., Northwestern University, 2004) is currently a Visiting Scholar at Southern Illinois University. In 1996, as part of his M.A. project at The University of Alabama, he directed a settlement survey of Upper Creek-Warrior Fork, the tributary of the upper Catawba River along which the Berry site is located. He co-directed a proton magnetometer survey at the Berry site in 1997, and it was during this survey that the burned structures were first identified. Since 1998, he has also worked in the Lake Titicaca Basin of Bolivia and Peru as part of his dissertation research at Northwestern.

Christopher B. Rodning (Ph.D., University of North Carolina–Chapel Hill, 2004) has been involved in the archaeology of western North Carolina and the Appalachian Summit area since 1994. Chris has published numerous articles on the late prehistoric and early historic native societies of the southern Appalachian region and in 2001 the University of Florida

Press published his edited volume *Archaeological Studies of Gender in the Southeastern United States* (co-edited with Jane Eastman). He joined the Department of Anthropology at the University of Oklahoma in January 2005.